Environmental Physiology
of Plants
SECOND EDITION

Environmental Physiology of Plants

SECOND EDITION

A. H. FITTER

Department of Biology, University of York, England

R. K. M. HAY

Plant Sciences Department, West of Scotland College, Ayr, Scotland

ACADEMIC PRESS

Harcourt Brace Jovanovich, Publishers

London   San Diego   New York   Berkeley
Boston   Sydney   Tokyo   Toronto

ACADEMIC PRESS LIMITED
24/28 Oval Road
London NW1 7DX

United States Edition published by
ACADEMIC PRESS INC.
San Diego, CA 92101

British Library Cataloguing in Publication Data
Fitter, A. H.
    Environmental physiology of plants.—
    2nd ed.
    1. Plant physiology   2. Botany—Ecology
    I. Title   II. Hay, R.K.M.
    581.1   QK711.2
    ISBN 0–12–257763–9
        0–12–257764–7 (Pbk)

Filmset by Latimer Trend & Company Ltd, Plymouth
Printed in Great Britain at the Alden Press, Oxford

Preface to the Second Edition

A second edition of a book gives authors the opportunity to correct and to update, but also to change the emphasis in certain areas. In this new edition some chapters have been extensively rewritten, while others retain most of their original structure. We have not, however, attempted to incorporate new material where we felt that the existing examples still made their point sufficiently strongly.

The subject matter of this book lies on the boundary between physiology and ecology, but we have chosen to write a physiology text from an ecological standpoint, hence the title. Since the physiology it discusses is that which affects the interaction of the plant with the environment, and since this interaction determines how well the plant functions in a given environment (and so ultimately its fitness), the book is also about adaptation, and associated concepts such as optimization are therefore central to its theme. Chapter 1 has been entirely rewritten to introduce these ideas explicitly and to place the rest of the book in this context. Elsewhere we have devoted more attention than formerly to photosynthetic pathways (C_3, C_4 and CAM), which are covered extensively in Chapters 2, 4 and 5, and to morphological as opposed to physiological responses to nutrient deprivation (Chapter 3). Other topics that have received particular attention in this edition are water movement in the soil–plant system (Chapter 4), the responses of plants to extreme temperatures (Chapter 5) and acid depositions (Chapter 7).

Environmental physiology is still a rapidly expanding field, and the extent of the literature is immense. We have consciously restricted our focus to studies in North America and Europe, both because of our personal experience in working in these areas (including Canada, France and Norway), and because we hope that this will give a greater coherence to the examples chosen. Although much excellent work is published in languages other than English, we have not relied heavily on this, since the book is intended primarily for students to whom such literature is often relatively inaccessible.

August 1987
A. H. FITTER
R. K. M. HAY

Acknowledgements

We are grateful for permission from the following authorities to use materials for the figures and tables listed. Academic Press Inc. (London and San Diego) and the appropriate authors—Figs. 4.7, 5.9 and 7.3; Agriculture and Food Research Council—Fig. 7.6; American Chemical Society—Fig. 7.7; Blackwell Scientific Publications, Oxford—Figs. 4.8, 4.9, 5.3, Table 4.6; Cambridge University Press and the appropriate authors—Tables 5.2, 7.2; Professor R. M. M. Crawford—Fig. 7.4; Professor M. C. Drew—Fig. 7.2; Professor T. C. Hsiao—Fig. 4.2; Macmillan Journals Ltd. and Dr J. N. B. Bell—Table 7.3; Macmillan Publishing Co. Inc., New York—Fig. 4.4; Professor T. A. Mansfield—Fig. 4.5, Table 7.2; Minister of Supply and Services, Canada—Fig. 5.4; Munksgaard International Publishers and Dr Y. Gauslaa—Fig. 5.5; Natural Environment Research Council—Fig. 7.8; Springer Verlag (Heidelberg and New York) and the appropriate authors—Figs. 5.1., 5.7, 5.8, Tables 4.8, 5.4; Stanford University Press—Fig. 4.12; University of California Press—Fig. 4.12; Professor I. F. Wardlaw—Fig. 5.2; Weizmann Science Press of Israel and Professor Y. Gutterman—Table 4.3; J. Wiley and Sons, New York—Fig. 5.10.

To
Rosalind and Dorothea

Contents

3. Mineral Nutrients

4. Water

PART III. RESPONSES TO ENVIRONMENTAL STRESS

5. Temperature

6. Ionic Toxicity

7. Gaseous Toxicity

8. Interactions Between Organisms

9. An Ecological Perspective

Contents

Part I

Introduction

1. Introduction

A. Plant Growth

This book is about the way in which the physiology of plants responds to their environment. Plants obtain resources (energy, CO_2, water and minerals) from their environment, and in Chapters 2 to 4 we discuss the responses they make to situations where resources become limiting. Other environmental factors affect plants directly, by either chemical or physical alterations to metabolism; these factors are the subjects of the later chapters. Whether the constraint is the lack of a resource, the presence of a toxin or an extreme temperature, the plant usually responds by an alteration in the rate or pattern of growth, since growth is the outcome of the affected metabolic processes. In this sense then, environmental physiology is the study of plant growth.

Plants grow in a very distinctive way. A seed has two localized areas of cell division—the tips of the young shoot and root, the meristems. Whereas in most animals, growth involves many sites of division leading to the differentiation of the various organs, in a plant virtually all cell production takes place in these meristems. Even in very short-lived annual plants, however, new meristems are formed as growth proceeds. A root system, for example, may initially consist of a single main axis, but in time laterals will be formed, each with its own meristem; these may give rise to other branches and so on. Such branching structures are common in nature and not just in plants— think of your own lungs, blood vessels and neurones, or of many colonial invertebrates, or even in the inanimate world of river networks. In each case the daughters are copies of the parent branches that gave rise to them.

Such a mode of construction is often called modular (Harper, 1986), and it has important consequences. First it means that growth is indeterminate. The number of modules is not fixed and there is no necessary endpoint to a branching system: whereas a cow always has four legs and two ears, a pine tree may have almost any number of needles or root tips. This in turn means that plant growth is very flexible; since growth does not proceed to a fixed endpoint, it is much more open to environmental influence. As we shall see, this is of fundamental importance in environmental plant physiology.

Despite this, of course, plant growth does follow some rules. It is simple to distinguish an oak tree from a poplar, and other contrasts are far greater. Though a weed such as groundsel *Senecio vulgaris* may vary in size from a stunted stem a few centimetres high with a single flowerhead, to a luxuriant branching plant half a metre high with 200 heads, it will never look even remotely like an oak tree, a cactus or a rose. These differences in growth form reflect different rules of growth and have evolved in response to distinct selection pressures.

Some environments contain abundant resources and offer few physical or chemical constraints on growth. In such favourable environments, free from major disturbance, the species which can grow largest and, by means of over-shadowing leaf canopies and widely ramifying root systems, obtain the largest share of the environmental cake, will dominate—in simpler terms, trees. Over large areas of the earth trees are the main growth form, but their life-cycle is long and they are at a disadvantage in areas of intense human activity or other forms of disturbance. In such circumstances herbaceous vegetation appears, characterized by fast growth rather than large size. In between these two extremes are other environments and habitats, each favouring to a greater or lesser extent the various growth forms.

Environmental favourability does not only affect size. It is also true that the fastest growing plants are found in productive habitats, whereas unfavourable and toxic sites support slower growing species (Table 1.1). The measure of growth used in Table 1.1 is the relative growth rate, a concept introduced to describe the exponential phase of growth of annual crop plants (Blackman, 1919). It assumes that new growth is simply related to the existing biomass and is therefore exponential (the bigger the plant, the greater will be the growth increment). The relationship between plant weight and time (i.e. growth rate) can therefore be described as:

$$\mathbf{R} = \frac{\mathrm{d}\ln W}{\mathrm{d}t} = \frac{1}{W} \cdot \frac{\mathrm{d}W}{\mathrm{d}t}$$

so that **R** represents, at an instant in time, the rate of increase in plant weight per unit of existing weight per unit time. If growth were truly exponential **R** would be constant and a plant characteristic, but in reality this is only the case for short periods. What is normally

TABLE 1.1.   Mean frequency (%) of species of different maximum relative growth rates (R_{max}) in different habitats (from Grime and Hunt, 1975).

	R_{max} (week^{-1})			
Habitat	< 1	1–1·24	1·25–1·44	> 1·44
Acidic pastures	23	17	10	3
Limestone pastures	22	23	19	17
Rocks	6	12	8	7
Cliff	5	8	6	4
Soil heaps	6	10	21	17
Arable	2	11	19	20
Manure heaps	0	13	18	23

calculated is the mean value of R over a period of time, which can be calculated (Radford, 1967) as:

$$\bar{R} = (\ln W_2 - \ln W_1)/(t_2 - t_1)$$

This equation is useful when comparing the growth of plants of different size, but since plant growth is usually only exponential in the very early stages, the value of R is continually changing, usually declining.

This approach assumes that the growth rate of a plant is in some way related to its mass, as is generally true of annuals, and is dramatically illustrated by the duckweed *Lemna minor*, growing in uncrowded culture (Fig. 1.1). If perennials, particularly long-lived woody perennials, are considered, however, it is at once apparent that the trunk of an oak tree contributes to the tree's welfare by supporting the leaf canopy in a dominant position and does not directly augment the growth rate . If relative growth rate were calculated for a tree in the way outlined above, ludicrously small values would result. Alternative models have been proposed to take account of such points (for example Dudney, 1973), but it is still important to bear in mind the ecological limitations of the concept of relative growth rate. Plants use the carbohydrate produced by photosynthesis for a range of functions, such as support, resistance to predators, reproduction, and so on, which reduce growth rate from its potential maximum—indeed that would be attained by a plant consisting solely of leaves. It is no accident that the fastest growth rate measured in an extensive survey by Grime and Hunt (1975) was for *Lemna minor*, a plant comprising one leaf and a single root a few

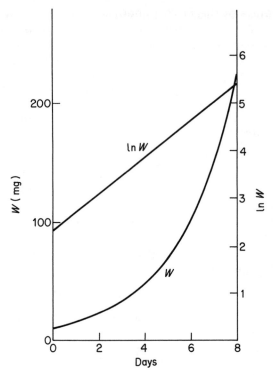

FIG. 1.1   Growth in dry weight (W) of duckweed *Lemna minor* in uncrowded culture. Hypothetical data based on maximum growth rate of 0.39 mg mg^{-1}d^{-1}.

millimetres long. Logically, then, one would expect (and indeed finds) that unicellular algae, the closest approximations to free-living chloroplasts, are the fastest-growing of all green plants.

Relative growth rate is therefore a useful indicator of the extent to which a species is using its photosynthate for growth and further photosynthesis—the production and functioning of more chloroplasts—as opposed to secondary functions, such as defence, support, reproduction, and nutrient and water gathering. In many habitats, usually unfavourable or toxic ones, such growth is disadvantageous and acquisition of water and nutrients or protection from grazing or disease is the first priority. These are the characteristic features of plants from very infertile (Chapter 3), very dry (Chapter 4), cold (Chapter 5) or toxic (Chapters 6 and 7) environments, and may be termed conservative.

B. The Influence of the Environment

When plants are grown in culture the aim is often to achieve the best possible growth, by removing all environmental constraints. In other words, we provide optimal environmental conditions where the plant's inherent growth rate can be expressed. Most physiological work is carried out in conditions that are at least an attempt to create such an optimum. Even so, various plant species differ by nearly an order of magnitude in the maximum rate that they can achieve, largely because some re-invest the carbon they fix in more photosynthetic machinery, while others use it for other purposes. Many umbellifers, such as pignut *Conopodium majus* for example, scarcely develop above ground beyond the production of the cotyledons in their first year of growth; the rest of the photosynthate goes to produce an underground storage organ. In the next season these stored resources enable it rapidly to produce leaves and flowers in early spring. In contrast annuals tend to direct virtually all their photosynthate first into leaf and then into flower and fruit production.

In optimum conditions, therefore, plant species vary in their use of resources and their growth patterns. In natural environments such conditions are rarely, if ever, reproduced, and in particular the supply of the various resources is typically unbalanced. For example, many environments are well supplied with radiant energy, at least for the topmost leaves in a leaf canopy, since the radiant flux density of full sunlight, even in temperate regions, is well above the saturation point for photosynthesis. However, to utilize this energy the plant must be able to obtain a supply of CO_2 and the concentration of this in the atmosphere at around 300 mg l^{-1} is so low that it cannot diffuse into the intercellular spaces in the leaves fast enough (Fig. 1.2). Even if it could, this would mean that the stomata would have to remain open and this could lead to damaging water loss if the supply of water in the soil was inadequate (cf. Chapter 4).

Under most environmental conditions, therefore, there is likely to be a specific environmental factor that is limiting growth. Alleviation of the effect of this factor should bring about an increase in growth up to the point where some other factor becomes limiting, although in many cases several factors will simultaneously limit growth, and only when both or all are supplied will there be a response (Fig. 1.3). Such interactions make the adaptive responses that plants must make to their environment extremely complex.

The situation is worse for the plant where multiple limitations exist which require conflicting responses, as in the example already given,

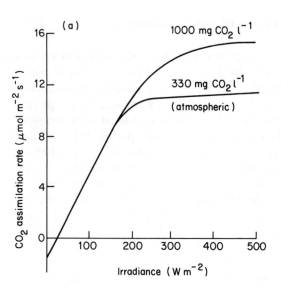

(a)

1000 mg CO_2 l^{-1}

330 mg CO_2 l^{-1}
(atmospheric)

CO$_2$ assimilation rate (μmol m^{-2} s^{-1})

Irradiance (W m^{-2})

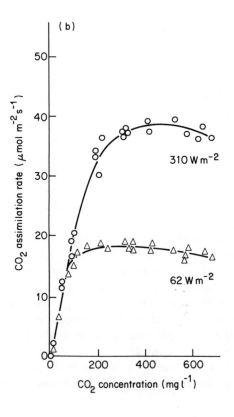

(b)

310 W m^{-2}

62 W m^{-2}

CO$_2$ assimilation rate (μmol m^{-2} s^{-1})

CO$_2$ concentration (mg l^{-1})

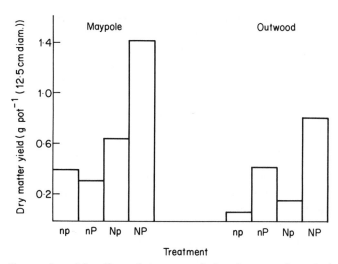

Fɪɢ. 1.3   Interaction of the effects of nitrogen and phosphorus on the growth of *Lolium perenne* in two extremely nutrient-deficient colliery spoils in a pot experiment. N and P represent applications equivalent to 25 kg ha^{-1}, n and p to 2·5 kg ha^{-1}. Note the failure to respond to high N applications in the absence of high P (data from Fitter, A. H. and Bradshaw, A. D. (1974). *J. appl. Ecol.* **11**, 597–608).

where the maintenance of an adequate CO_2 supply necessitates open stomata which may lead to excessive water loss (Chapter 4, p. 140) It is further complicated by the variation of environmental factors with time. In the course of a single day a plant may experience in effect the full gamut of possible environments. In a tropical montane environment, for example (see Chapter 5, p. 204), night temperatures can be extremely low, so that the ability to resist frost damage may be essential. After sunrise, irradiance levels will rise rapidly along with the temperature and rapid photosynthesis will be possible, limited possibly by CO_2 or by mineral deficiencies. At mid-day, under very high radiant energy flux, the plant may experience a water deficit and stomata may close; alternatively low cloud could produce conditions where photosynthesis was limited by irradiance.

Fɪɢ. 1.2.   (a) Effect of CO_2 concentration in air on photosynthesis in wheat leaves. At low irradiance, CO_2 is not limiting and has no effect on the slope of the response (quantum efficiency), see p. 53). (b) Effect of *PAR* on photosynthetic rate in attached leaves of *Sorghum sudanense*, as a function of CO_2 concentration in air. This is the alternative view of the process in (a): at low CO_2 concentrations, *PAR* does not affect photosynthetic rate, but as soon as CO_2 becomes non-limiting, there is a large response to PAR.

Such variability demands enormous flexibility of the physiological systems of plants, and it occurs on all time-scales from the almost instantaneous upwards. Within the life-time of an individual there will certainly be significant environmental fluctuation. Where this is sufficiently predictable, it may be catered for by rhythmic behaviour (for many diurnal environmental fluctuations) or by predetermined onto-genetic changes, such as the progressive increase in leaf dissection that occurs in many seedlings with successive leaves. This can be interpreted as a change from a shade-adapted entire leaf at the base of the canopy, to the more dissected leaf form adopted by most sun plants (cf. Chapter 2). The timing of such ontogenetic changes and the duration of the life-cycle may be highly plastic and an important part of adaptation to temporal fluctuation. Thus the autumn-shedding of leaves by deciduous trees is wholly under environmental control, and continuous leafiness can be maintained under long days.

1. Damage and Response

The effects of environment on plant growth may be divided into enforced damage effects, caused by environment, and adaptive res-ponses, controlled by the plant. Damage, which may be manifested as death of all or part of the plant, or merely as reduced growth rate due to physiological malfunction, is a common phenomenon and the agents are various: wind, ions, temperature, grazing and many others.

Clearly, however, the occurrence of damage implies a lack of resistance on the part of the plant, and plants differ greatly in their resistance to damage. Resistance may be conferred by molecular, anatomical or morphological structure, or by phenology (the timing of growth and development), and is a fundamental component of a plant's physiology and ecology, being responsible for all major differ-ences in plant distribution. The critical feature is that such resistance is constitutive: a particular enzyme will be capable of operating over a certain temperature range and beyond that range damage will occur (cf. Table 5.4). Resistance can in a sense be viewed as a form of homeostasis, permitting the plant to maintain function in the face of an environmental stimulus, without apparent physiological or morpho-logical change.

Adaptive responses, however, are the fine control on this constitutive resistance to damage. They involve a shift of the range over which resistance occurs, which may be reversible (and usually, therefore, physiological) or irreversible (usually morphological). Levitt (1980) has used the terminology of physics to distinguish elastic strains, which are reversible, from irreversible plastic strains, in response to environmental stress. Thus the same stress, for example shade, may induce a reversible physiological response in photosynthetic activity in a woodland herb, but an irreversible morphological response in a weed or crop plant (cf. Chapter 2). Both responses are adaptive within the context of the plant's normal environment, and both require phenotypic flexibility.

Phenotypic plasticity of morphology is a universal feature of plants (Bradshaw, 1965, 1973) and conspicuous examples, such as the heterophylly of water buttercups (*Ranunculus aquatilis*), are well known. The ubiquity of reversible, adaptive, phenotypic changes is less clearly appreciated, but good examples are changes in amounts of enzymes, particularly inducible enzymes such as nitrate reductase (cf. Chapter 3), and behavioural responses, such as the opening and closing of flowers (daisy, *Bellis perennis*, or *Mesembryanthemum*) and compound leaves (white clover, *Trifolium repens*), and sun-tracking by flowers (buttonweed, *Cotula coronopifolia*, or sunflower, *Helianthus annuus*; cf. Chapters 2, p. 24, and 5, p. 206). These may be more or less direct responses to environment, or may in some cases have developed an endogenous rhythmicity that permits them to continue without the environmental cue.

Each individual plant is thus capable of a range of response to environmental fluctuation. Clearly all must have some molecular basis but it is possible to classify them according to whether the molecules directly provide the adaptation or act by creating structures or behaviour patterns that are adaptive. Resistance to injury can be classified simply on this basis: either the molecules themselves are resistant to stress, as with the enzymes of thermophilic bacteria, or the molecules are protected from damage by other molecules, special structures, or behaviour patterns. Plastic responses are, however, more complex and a simple classification is suggested in Table 1.2. Which of these response types are utilized by a plant depends upon the way in which the environmental stimulus is presented.

C. Population Responses

Where an environmental fluctuation occurs over a time-scale similar to or less than the life-span of an individual, plasticity permits each individual to adopt an appropriate phenotype. On a longer time-scale, however, gradual change in an environmental factor will tend to bring about genetic, and therefore evolutionary, change in the population. Indeed the whole evolutionary history of life on earth has been a

TABLE 1.2.   Classification of response types.

I   Adaptations explicable in molecular terms
(a) Changes in molecular structure—Qualitative
    (i) to resist denaturation if molecule would not survive stress—e.g. reduction in number of SH groups in proteins of cold-hardened plants (Chapter 5).
    (ii) to improve operation—i.e molecule would operate sub-optimally in changed conditions—e.g. changes in affinity of uptake systems of nutrient-deficient plants (Chapter 3).
(b) Changes in amounts of molecules—Quantitative
    (i) to adjust capacity if insufficient or excess molecules would otherwise be present—e.g. differences in ribulose bis-phosphate carboxylase activity in shaded and unshaded leaves (Chapter 2) or synthesis of phytoalexins in response to fungal attack (Chapter 8).
    (ii) to protect metabolic system from injury—e.g. osmotic balance by amino acid synthesis in halophytes or toxic ion chelation (Chapter 6).
(c) Changes in functional types of molecules
    (i) to operate alternative pathway—i.e. one set of molecules not appropriate for changed conditions, but no synthesis required—e.g. switch of anaerobic respiratory pathways in flood-tolerant plants (Chapter 7).

II   Adaptations requiring explanation in morphological or behavioural terms
(a) Adjustments to temporal changes in the environment
    (i) immediate reactions—i.e. behavioural responses such as leaf and stomatal movements (Chapter 4).
    (ii) long-term reactions: seasonal differences in growth form, including dormancy, juvenile and mature leaves, etc. (Chapters 2, 4, 5).
(b) Adjustments to spatial changes in the environment
    (i) qualitative responses: different structures produced, such as sun and shade leaves, aquatic heterophylly (Chapter 2); aerenchyma in flooded plants (Chapter 7).
    (ii) Quantitative responses: changes in resource allocation, such as morphological reactions to density and shading (Chapter 2) or to nutrient stress (Chapter 3).

---

sequence of such shifts in gene frequencies, largely in response to long-term changes in environmental conditions.

Similar adaptation by change in gene frequency is possible where spatial variation occurs, but as with temporal fluctuations, the effect is related to the scale. Clearly where the change occurs over a distance less than the size of the individual (or its relevant organ system), a plastic response is likely. Such distances may be remarkably large: trees such as *Eucalyptus, Pseudotsuga* (Douglas fir) and *Sequoia* typically grow to 60–80 m in height at maturity, while root systems may reach to depths as great as 40 m (alfalfa; Meinzer, 1927, quoted by Daubenmire, 1947). *Prosopis velutina* forms a small tree on flood plains with tap root descending to 12 m or more, and a shrub on desert soils, with a small tap-root but a lateral spread of more than 15 m (Cannon, 1911; see also Chapter 4, p. 156). Over such distances one must expect variation. Indeed all plants which produce leaf area indices greater than 1 are likely to experience some self-shading, and for many trees the contrasts in light levels are sufficient to elicit the production of morphologically and physiologically distinct "sun" and "shade" leaves on each individual (cf. Chapter 2). Other classic examples of such variation are the heterophylly of *Ranunculus*, already mentioned (cf. Fig. 2.9), and of the aerial and appressed branches of ivy, *Hedera helix* (Fig. 1.4).

The definition of an individual is notoriously difficult in higher plants, many of which reproduce vegetatively, and nowhere more so than in the grasses, which grow by producing new shoots (tillers), which are effectively new individuals with their own root system and virtual physiological independence. Harberd (1961) has shown that single clones of the common grass *Festuca rubra* can be spread over a distance as great as 240 m; in this instance 51% of the tillers from a 100 square yard (*c.* 80 m^2) quadrat were of a single genotype, and 82%

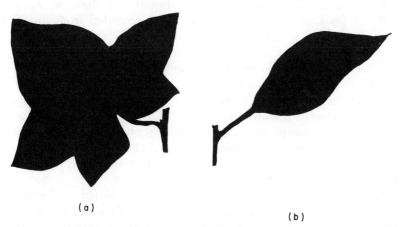

(a)

(b)

FIG. 1.4.   Silhouettes of ivy leaves, showing the difference between the palmate leaves from appressed, vegetative stems (a) and the oval leaves of flowering stems (b).

were accounted for by the six commonest genotypes. Any spatial variation within this quadrat must therefore be coped with by each genotype by plasticity, and the same is true for many water plants. For example, Canadian pondweed *Elodea canadensis* was introduced into Great Britain in the nineteenth century, but for some reason male plants are almost never found, though females are very common. Most populations must therefore be huge clones derived vegetatively from a single parent, and the same is true for the duckweeds (*Lemna* spp.), which are free floating. In both cases, the spatial scale of the environment must be vast.

For sexually reproducing species and where the spatial scale of variation is greater than that of the individual or clone, the situation is more complex. Certainly here differentiation into genetically distinct, locally adapted populations (ecotypes) by adaptive changes in gene frequency is possible. Such ecotypes will develop if the strength of the selection is sufficient to overcome gene flow between such contrastingly selected individuals. This gene flow will usually result from pollen flow, but can also occur as a result of seed dispersal, if selection pressure is insufficient to eliminate the "foreign" genotype prior to reproduction. There is good evidence that if selection coefficients are sufficiently high, local pockets of distinct genotypes can be maintained as little as 10 cm apart, even in outbreeding species (Snaydon, 1970; Davies and Snaydon, 1974). In an inbreeding species, where gene flow is negligible, this alone can account for adaptation to the environmental mosaic, as can readily be observed in the tiny crucifer, whitlow grass *Erophila verna*, where genetically distinct inbred lines may co-exist. By contrast *Festuca rubra* is predominantly an outbreeder, but in the study by Harberd (1961) at least 17 distinct genotypes were recorded from an 80 m^2 quadrat and two of the most abundant of these showed contrasting responses that correlated with their distribution over a soil pH range of 5·0 to 6·6. In this case local adaptation was maintained by vegetative reproduction. In other words all plant species, except those relatively rare obligate outbreeders lacking means of vegetative reproduction (e.g. the mayweeds *Anthemis arvensis* and *A. cotula*, Kay (1971)), can by these means maintain sharp genotypic discontinuities where corresponding environmental boundaries occur. The exceptions are generally weeds of transient habitats, where such spatial adaptation is not necessary.

The importance of the breeding system is underlined by the demonstration that discontinuities such as these may be sharp to the windward, but blurred to the leeward of an environmental boundary in anemophilous grasses (McNeilly and Bradshaw, 1968). In fact so strongly leptokurtic is pollen dispersal in most plants (Fig. 1.5), and so powerful are the selection pressures acting on such stationary organisms, that even in many outbreeding plants the population size in terms of gene flow can be measured in metres or even centimetres (Bradshaw, 1972). Of course many species have breeding systems which permit or encourage inbreeding, and the extent of this may be related to environment.

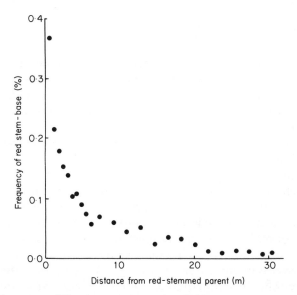

F IG. 1.5. Frequency (%) of a genetic marker (red stem-base) in an experimental population of *Lolium perenne*, as a function of distance of the white-stemmed parent from a block of red-stemmed plants. The distribution implies a strongly leptokurtic pollen distribution. (Data of Griffiths (1950), as plotted by Gleaves, T. J. (1973). *Heredity* **31**, 355–366).

There is, of course, an important interaction between temporal and spatial variation, since any environmental factor must show variation on some scale in both respects simultaneously. Nevertheless, these arguments perhaps lead us to imagine that environmental variation on a vertical axis, and for clonal species horizontally as well, will be responded to by plasticity, but that other horizontal variations may be more conducive to genetic differentiation, so that both will be essential components of the armoury of all plant species. Again, as with variation in time, predictability must be considered as determined by seed dispersal. Many species, such as groundsel *Senecio vulgaris*, are fugitive (r-selected in the terminology of population dynamics), relying on wide dispersal and fast growth rate to exploit ephemeral habitats. For these species the predictability of spatial variation is minimal and a high degree of plasticity may be necessary to permit offspring to colonize a wide range of habitats, which implies that species of unstable habitats should display more plasticity than those of mature communities.

D. Adaptability and Adaptedness

Plants that survive in their habitats are clearly adapted. To that extent the word is effectively meaningless. This does not, however, mean that all aspects of their biology are adaptive. Selection acts on organisms by differential survival and not on organs or responses, and so various parts, functions or activities of an organism may be well or poorly adapted in some mechanistic sense, as long as the sum of their effects is sufficiently suited to the habitat.

The extent of phenotypic plasticity for any character in an organism is another character which during evolution will have contributed to the fitness of particular individuals. By extending the environmental range over which an individual can survive it may increase fitness by increasing adaptability. In many, perhaps most environments, fitness will be maximized not by a character which is most highly suited to a particular environmental state, but by one which allows the organism to track environmental fluctuations. This is particularly true of plants because they are sessile and so liable to experience greater temporal variation than more mobile animals.

By its nature environmental physiology or physiological ecology is concerned with such adaptation and adaptability. It would be perverse to study the interaction between environment and physiological process without considering the evolutionary framework in which the changes in the process might have come about. The subject is essentially a comparative one: it is essential to examine a wide range of species growing in a variety of habitats in order to see the diversity of physiological response which has evolved. This approach has its dangers, however: it is easy to ascribe every difference to adaptation, without defining precisely the criteria by which one can recognize it.

In reaction to this, several prominent biologists have criticized this "adaptationist" programme and have even coined new words to indicate that not all features of an organism may be evolutionarily adaptive. So Gould and Vrba (1982) speak of "exaptation", by which they mean to draw attention to the fact that organisms have to make do with what their ancestors handed down to them: they are not designed from scratch each generation. If, as in one of Gould's most vivid examples, a panda is to have a thumb, it may have to get it by devious means, by putting to new use a quite different bone from the one used by apes. Harper (1982) prefers the term "abaptation" to make an essentially similar point, namely that organisms are not adapted to their present environment (latin ad = to) but from their ancestral ones (ab = from). Selection acting on living organisms can only alter their descendants.

It is doubtful that these neologisms will catch on, but they serve a useful purpose in emphasizing the way in which selection works and the nature of adaptation. If we are to talk about adaptation and adaptive responses, it is necessary to erect criteria by which we can judge adaptedness, for we must be able to tell which of two or more possible structures or activities is the most adapted or, in effect, the best. The problem becomes one of optimization, and therefore of benefits and costs.

1. Cost–benefit Analyses and Optimization
One solution that ecologists have adopted is to borrow the clothes of economists. It is no accident that the two words are so similar, for they both refer to the manipulation of resources. A plant can be viewed as an economic system, with resource inputs and outputs, and an efficiency or a profit margin (output–input) can be calculated. A leaf, for example, has a fixed construction cost and a continuing maintenance cost (Fig. 1.6). These can be measured using carbon as a currency. Against these can be set its output—the amount of carbon it fixes and eventually exports. At the end of its life a certain amount of carbon may be recovered during senescence, thus defining the net cost. Several variables can alter the profit achieved by the leaf, including changes in leaf longevity (which affects maintenance costs, the total amount of carbon fixed, and the probability of the leaf being eaten and so lost: see Chapter 3); photosynthetic efficiency (which may alter either or both of the construction and maintenance costs: see Chapter 2); and the quantities of defensive compounds (which affect both costs and longevity: see Chapter 8).

The problem with this economic analogy, however, lies in the definition of the currency. Economics is in effect to do with the movements of money, an independent and artificial variable, which is not consumed or utilized as it is transacted. Plant resources are not strictly comparable. Carbon seems to be an obvious choice as a

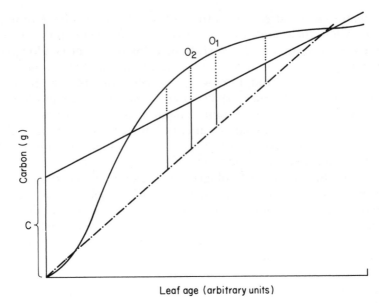

Leaf age (arbitrary units)

Fig. 1.6. Hypothetical scheme to illustrate cost–benefit analysis for a leaf. The straight solid line represents the cumulative maintenance costs of the leaf and its intercept (C) represents the construction costs. The broken line represents the true cost, allowing for the proportion of the construction cost that can be retrieved during senescence—this declines as the leaf ages. The curved line represents the cumulative photosynthetic carbon gain. The optimum leaf life-span is given by O_1 if retrieval is not allowed for, and by O_2 if it is taken into account.

currency, but in practice plants are often not limited in their growth by carbon, since photosynthesis is frequently sink-limited (i.e. its rate is determined not by the supply of energy but by the rate of utilization of fixed carbon by the plant; see Chapter 2). As a result various minerals, such as N, P or K, might be thought of as alternative currencies, but the allocation of these and of carbon to different plant functions (leaves, roots, fruits, etc.) may be quite different (Abrahamson and Caswell, 1982).

As an alternative to such global economic models, which seek to explain various plant functions in terms of some general currency, several workers have developed local models. These aim to explain the operation of a particular plant function simply in terms of its own consequences. We have already pointed out the conflict experienced by plants in relation to carbon dioxide acquisition and the control of water loss. Since the rates of movement of CO_2 and water vapour through the stomata are not equivalent, the relationship between carbon assimilation and water loss is non-linear, CO_2 diffusing in air at only about two-thirds the rate of water vapour (cf. Chapter 4). Cowan (1977) and Cowan and Farquhar (1977) have defined a parameter λ which is in effect the slope of this relationship, and have developed a theory of optimal stomatal function based on the premise that assimilation should be maximized with respect to water loss (or that water loss should be minimized with respect to assimilation). Cowan (1982) makes the analogy between λ and the economic variable, unit marginal cost. In the optimization model, this cost determines particular diurnal patterns of transpiration and assimilation (Fig. 1.7) and these seem to correspond to those observed in actual plants, though as Cowan (1982) admits, the improvement in water use efficiency achieved by optimally functioning stomata may be small.

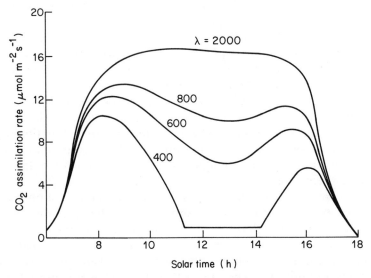

FIG. 1.7. Optimal diurnal time-courses of carbon fixation rate for different values of the parameter λ, the slope of the relationship between transpiration (E) and assimilation (A), i.e. $\delta E/\delta A$. The results are appropriate to a hot, dry, clear day in an arid sub-tropical environment, and indicate that stomata close (and gas exchange therefore ceases) during the middle of the day where λ is small and unit marginal cost is small. From Cowan (1982).

In this later paper, Cowan has extended the model to cover long-term water use, considering whole plants growing in environments of unpredictable rainfall and even situations where plants are growing with their root systems partially overlapping those of others. In each case the predictions are intuitively appropriate to what is known of the water use economy of plants growing in such conditions (see Chapter 4), and are based on immediate or prospective costs implied in losing water and the parallel benefits from fixing carbon. There is, however, a real problem in such an approach, for it is at best difficult to test the predictions against data. Cowan (1982) explicitly states in his opening paragraph that "we may *presume* that selective adaptation has tended to optimize this relationship" (italics added). Clearly if this is a presumption initially, it cannot be a conclusion eventually.

It is of course true that the assumption that selection is responsible for the patterns of the living world underlies virtually all the worth-while biology of the last 100 years, and environmental physiology or physiological ecology is no exception. It is concerned specifically with the interface between phenotype and environment where selection may be presumed to be acting, and in this it differs from other areas of biology, even those such as population genetics which are explicitly evolutionary, but deal with genotypes and are not concerned with adaptation. Environmental physiology deals with the functioning of physiological processes in ecological conditions, the way in which the one is fitted to the other, or in other words with adaptation in a neutral, non-evolutionary sense. There is danger in confusing this interpre-tation of the word with the evolutionary sense in which selection has acted on organisms in such a way as to improve their functioning, to adapt them to some particular environment. This evolutionary mean-ing implies optimization, though not of course that an optimum is necessarily achieved.

In the chapters that follow we discuss the adaptations of physiologi-cal processes to environmental conditions. Underlying all these discus-sions is the assumption that they are the result of natural selection. One should never assume, however, that selection acts on the processes: it acts on organisms, on entire phenotypes. A plant with the most exquisitely optimized phenotype with respect to water use efficiency may not survive to reproduction if, either as a consequence of this or for some unrelated genetic reason, it has an inefficient defence against grazing animals. Alas, the entire genotype will be lost, whereas another apparently inferior (in terms of adaptation) phenotype may in practice have greater fitness. Equally, it would be a mistake to assume that selection acts on a single function: the hairs on a leaf may improve plant fitness by altering its energy balance (Chapters 2, 4 and 5) or by deterring insect herbivores (Chapter 8), or (and most probably) for both reasons. The net result in all cases, however, is an improvement in the matching of physiology to environment.

Part II

The Acquisition of Resources

2. Energy and Carbon

Photosynthesis is fundamental to plant metabolism, and so the acquisition of radiant energy and CO_2 is critical to the ecological success of a
plant. Radiation that is photosynthetically active (PAR) roughly
corresponds to visible light, but both represent only a small part (*c.*
400–700 nm) of the full solar radiation spectrum (Fig. 2.1) and plants
are also sensitive to other wavelengths: for example, the role of far-red
radiation (far-red "light" is a convenient misnomer) of wavelength *c.*
700–800 nm in morphogenesis is well documented elsewhere (Morgan
and Smith, 1981). Radiation affects organisms by virtue of its energy
content and is only active if absorbed. Thus, ultra-violet "light" is
strongly absorbed by proteins and can cause damage; blue light is

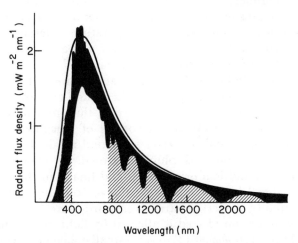

Fig. 2.1. Solar radiation flux. The outer solid line represents the ideal output for a
"black body" at 6000 K (the solar surface temperature); the upper rim of the black
area is the actual solar flux outside the earth's atmosphere; and the inner open and
cross-hatched area the flux at the earth's surface. Only the open part is photosynthetically active radiation (*PAR*).

absorbed by carotenoid pigments and chlorophyll, red light by chlorophyll, and both red and far-red by phytochrome. The existence of pigments, therefore, is basic to any response and most plants appear green simply because most plant pigments absorb green light weakly.

At longer wavelengths one can no longer think in terms of pigments (which of course strictly refer to only the visible range), since long-wave radiation is absorbed by all plant tissues with consequent heating. The energy budgets of plant organs are discussed in Chapter 5; they are of great importance in regulating the temperature of plants, particularly in extreme climates. In many situations there is a conflict between the need to intercept light for photosynthesis and the resulting increases in leaf temperature. Energy loss, by convection and evaporation, then becomes paramount; consequently there may be benefits from both changes in leaf morphology which increase convective loss, and changes in transpiration rate which increase evaporative loss of energy, despite their often deleterious effect on the absorption and utilization of radiant energy for photosynthesis.

Because of this dual effect of solar radiation—in supplying the energy for metabolism and in controlling the temperature of plants—responses to sunlight may have no photosynthetic or morphogenetic basis. For example, flowers in Arctic regions, such as *Dryas integrifolia* and *Papaver radicatum*, are saucer-shaped and follow the sun, acting rather in the manner of a radio telescope, so concentrating heat on the reproductive organs in the centre of the flower and attracting pollinating insects to these "hot spots". A temperature differential of 7°C or more is frequently attained between flower and air, with a maximum temperature of 25°C being recorded (Kevan, 1975).

Physiologically, light has both direct and indirect effects. It affects metabolism directly through photosynthesis, and growth and development indirectly, both as a consequence of the immediate metabolic responses, and more subtly by its control of morphogenesis. Light-controlled developmental processes are found at all stages of growth from seed germination and plumule growth to tropic and nastic responses of stem and leaf orientation, and finally in the induction of flowering (Table 2.1). There may even be remote effects acting on the next generation by maternal carry-over; Shropshire (1971) has shown that dark germination of seed is affected by light quality incident on the flower-head in *Arabidopsis thaliana*, a small ephemeral crucifer. Dark germination of seed was much greater when the parents had been grown in fluorescent light than in incandescent light, which contains more far-red. This effect has considerable ecological significance (see below, p. 36).

These responses are mediated by three main receptor systems:

TABLE 2.1   Some light-controlled developmental processes.

Process	Control
Germination	Light-requiring seeds are inhibited by short exposure to far-red (FR) light. Seeds capable of dark germination may be inhibited by FR irradiation.
Stem extension	Most plants show etiolation in dark. Red (R) light stops this but brief FR irradiation counteracts R. Prolonged FR irradiation can have similar effects to R.
Hypocotyl hook unfolding	Occurs with R or long-term exposure to FR.
Leaf expansion	Require prolonged illumination for full expansion.
Chlorophyll synthesis	Short-term FR inhibitory, long-term may or may not be inhibitory.
Stem movements	Blue light most effective.
Leaf movements	Blue and red light active. R/FR reversible.
Flower induction	In short-day plants, R can break dark period. FR reverses effect.
Bud dormancy	Usually imposed by short-days. Behaves as for flowering.

chlorophyll with several absorption peaks at 650, 670, 680 and 695 nm (and others in the blue region) for photosynthesis, phytochrome absorbing in two interchangeable forms at 660 and 730 nm for many photomorphogenetic responses, and (probably) flavins absorbing at around 450 nm for tropisms and high-energy photomorphogenesis. All plants contain a wider variety of compounds capable of absorbing radiation, and no function is known for many; it is likely that in some the absorption is chemically fortuitous. In algae, however, the accessory pigments are known to play an important auxiliary role in photosynthesis.

I. THE RADIATION ENVIRONMENT

A. Radiation

Radiant energy is measured in joules (J) and its rate therefore in $J\ s^{-1}$ or watts (W). The rate at which surfaces intercept energy is therefore expressed in W m^{-2}. For most purposes in physiological ecology, however, it is only photosynthetically active radiation (PAR, i.e. 400–700 nm) that is of importance and so a measurement that takes this into account is appropriate. This can be achieved in practice by using filters to measure the irradiance within this band.

According to duality theory, radiation can be described either as waves or streams of particles, but for radiometric purposes it is most conveniently treated as if particulate and discretely packaged in photons, whose energy content (quantum) depends upon wavelength. The quantum energy (in J) of a photon is $h\nu$, where h is Planck's constant ($6 \cdot 63 \times 10^{-34}$ J s) and ν (which is the Greek letter n, pronounced nu or 'new') is the frequency of the radiation. Since:

$$\text{quantum energy} = h\nu \text{ and} \qquad (2.1)$$

$$\nu = c/\lambda \qquad (2.2)$$

where c is the speed of light (radiation) (3×10^8 m s^{-1}) and λ is wavelength (in nm), then:

$$\text{quantum energy} \simeq \frac{2 \times 10^{-16}}{\lambda} \text{ J} \qquad (2.3)$$

It is common now to distinguish photosynthetic irradiance, which is the total energy falling on a leaf in the waveband 400–700 nm, and measured in W m^{-2}, and the photosynthetic photon flux density, which is the number of photons in the same waveband. The latter can be more usefully related to physiological processes in photosynthesis. The relationship between the two is given by using molar terminology: a mole (or $6 \cdot 022 \times 10^{23}$, which is familiar as Avogadro's number) of photons is now referred to as an Einstein (E). From equation 2.3, therefore, 1 E of a given wavelength carries $1 \cdot 2 \times 10^8 / \lambda$ J. If the wavelength distribution of radiation is known, conversion from W m^{-2} to E s^{-1} m^{-2} is therefore possible: for daylight the appropriate factor is 1 W m$^{-2} = 4 \cdot 6$ μE s^{-1} m^{-2} (McCree, 1972). Workers in environmental physiology use both Einstein and mole terminology about equally: the two are of course equivalent.

B. Irradiance

Radiant energy input reaches a maximum on cloudless days with minimum particulate matter or water vapour in the atmosphere, and when the sun is at its zenith. The differences in irradiance between this situation and that on a cloudy winter day, and between that and bright

TABLE 2.2 Variation of radiant flux density in the natural environment and of plant response to it (adapted from Salisbury, F.B. (1963), "The Flowering Process").

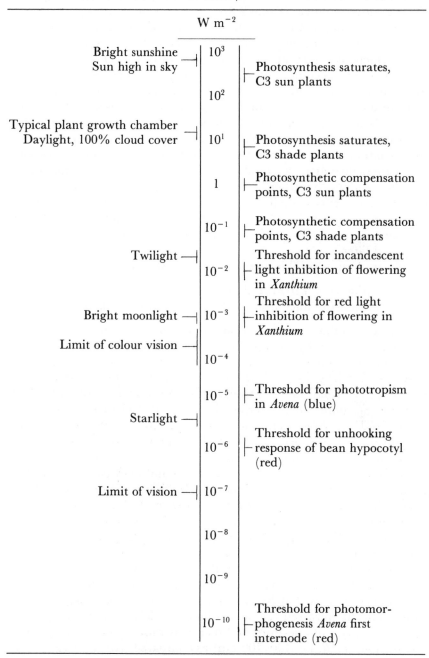

W m^{-2}

Bright sunshine Sun high in sky	10^3	
		Photosynthesis saturates, C3 sun plants
	10^2	
Typical plant growth chamber Daylight, 100% cloud cover	10^1	Photosynthesis saturates, C3 shade plants
	1	Photosynthetic compensation points, C3 sun plants
	10^{-1}	Photosynthetic compensation points, C3 shade plants
Twilight		Threshold for incandescent light inhibition of flowering in *Xanthium*
	10^{-2}	
Bright moonlight	10^{-3}	Threshold for red light inhibition of flowering in *Xanthium*
Limit of colour vision		
	10^{-4}	
	10^{-5}	Threshold for phototropism in *Avena* (blue)
Starlight		
	10^{-6}	Threshold for unhooking response of bean hypocotyl (red)
Limit of vision	10^{-7}	
	10^{-8}	
	10^{-9}	
	10^{-10}	Threshold for photomorphogenesis *Avena* first internode (red)

moonlight, encompass several orders of magnitude. Plant responses cover a parallel range (Table 2.2).

The main effects of changes in irradiance occur on the process that uses radiation as an energy source—photosynthesis—rather than on those which use it as an environmental indicator. For most plants photosynthesis becomes saturated at flux densities well below the maximum they occasionally experience, largely due to the problems of CO_2 supply, but in shaded conditions photosynthesis is often limited by the level of radiant energy. Variation in irradiance is a universal feature of habitats colonizable by plants and the complex nature of this variation is well shown in forests where any point under the canopy will experience first, seasonal variation; secondly, a diurnal cycle; thirdly, random "weather" effects due to cloud cover; and fourthly, canopy shade effects such as sunflecks. In addition to this temporal variation, immediately adjacent points may differ radically in the last two factors (Anderson, 1964). Leaf canopy effects on radiation are discussed later.

Solar radiation reaching vegetation has two components:

  (i) irradiance of direct sunlight (I), and
  (ii) diffuse irradiance from both clouds and clear sky (D).

Diffuse irradiance increases in importance as the solar beam is attenuated, either by actual obstruction (clouds, leaves, etc.) or by scattering due to particles and molecules in the atmosphere. Scattering is affected by the density of these particles, and also by the path-length of the direct solar beam through the atmosphere, both of which increase the chances of scattering occurring. Particles such as dust and smoke, and molecules such as water vapour, cause scattering in inverse proportion to the wavelength, following a power law relationship; the power function depends upon particle size, but the net effect is to reduce the blue content of direct radiation and increase that of diffuse radiation since this is entirely composed of scattered radiation. Thus, although the sunset is red, as a consequence of blue-scattering along the long path-length of the beam at such a low solar angle, the overall radiation load is blue-shifted at that time, since diffuse radiation predominates.

In effect the reductions in irradiance caused by occlusion of the direct solar beam are partially offset by the enhanced blue component of diffuse radiation and by the fact that water vapour in particular absorbs in the infra-red region, radiation which is not photosynthetically active. About one-third of direct solar radiation is photosynthetically active (PAR, i.e. 400–700 nm) as compared with over two-thirds of diffuse radiation. Under most meteorological conditions, therefore, *PAR* as a fraction of total solar radiation remains virtually constant at 0.5 ± 0.02 (Szeicz, 1974), so that it is convenient to disregard wavelength differences between direct and diffuse irradiance. Direct sunlight will, of course, be more intense than diffuse light, but at least in temperate climates, diffuse radiation is a major part of the total. Theoretical calculations by Avaste show that even under cloudless skies diffuse radiation (D) may account for between one-third and three-quarters of the total (T), and in a series of measurements in Cambridge the ratio D/T was always greater than 0.5 (Szeicz, 1974).

The maximum flux density of bright sunlight (for most purposes) depends on the solar constant, the radiant flux density at the outer margin of the earth's atmosphere, confirmed by satellite measurements to be about 1400 W m^{-2}. Typical instantaneous values for *PAR* at vegetation surfaces are 2500–4500 µE s^{-1} m^{-2} (*c.* 500–1000 W m^{-2}) for sunlight and 250–1000 µE s^{-1} m^{-2} (*c.* 50–200 W m^{-2}) for overcast skies (Table 2.2).

Irradiance is reduced not only by cloud, dust, water vapour, and other atmospheric obstructions that increase the D/T ratio, but also through shading by terrestrial objects. Some shadows are caused by selective filters, which let a part of the spectrum through, such as leaves (considered below), water (cutting out long wavelengths), soil, and so on; other objects are opaque or act as neutral filters, such as tree trunks and rocks. The two are clearly not exclusive; in sites to the north of vegetation a mixed shade occurs, having both the enhanced far-red component of leaf-transmitted shade and the enhanced short-wave content (mainly blue) of diffuse radiation. It has been termed "open shade" (Stoutjesdijk, 1974).

C. Temporal Variation

Temporal variation in irradiance is a universal feature of ecosystems: all places experience day/night fluctuations and the relative duration of these varies systematically with latitude. This is the basis of photoperiodism. Outside the tropics daylength is the most reliable indicator for predicting, and hence avoiding or resisting, unfavourable conditions. Virtually all temperate zone plants exhibit photoperiodic responses, many for flower initiation, but also for seed germination, bud-break, stem elongation, leaf-fall and many other phenomena. Although there is abundant evidence for the photoperiodic preferences of many individual species or cultivars, and although it is known that phytochrome is almost certainly the photoreceptor normally involved, no clear picture has emerged of the mechanisms involved. There is even doubt as to the significance of the "redness" of end-of-day irradiance, which earlier seemed an obvious signal, given the involvement of the red/far-red reversible pigment phytochrome (Vince-Prue, 1983).

D. Leaf Canopies

Many substances absorb radiation preferentially at some wavelengths, for example water and soil components, both of which reduce long wavelengths, increasing the red/far-red (R/FR) ratio. Leaves, on the other hand, absorbing in blue and red (Fig. 2.2), greatly reduce the ratio. Leaf canopies, therefore, produce a radiation climate varying in both irradiance and spectral distribution in a most complex manner.

Leaf canopies are not solid sheets, but loosely stacked. Radiation

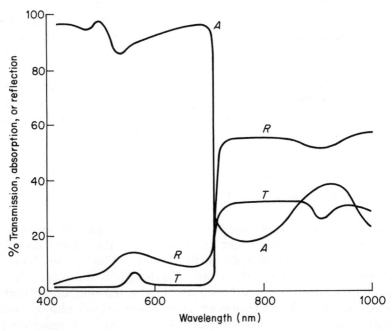

FIG. 2.2. Generalized spectral characteristics of plant leaves between 400 and 1000 nm (from various sources). Abbreviations: A, absorption; R, reflection; T, transmission.

therefore penetrates canopies in four ways, through actual holes as well as by reflection and transmission.

(i) Unintercepted direct irradiation may penetrate either as diffuse or direct radiation, the latter appearing as sunflecks, which largely have the characteristics of direct irradiation, although gaps in the canopy may act as diffusing lenses, spreading the beam (Anderson and Miller, 1974) and reflected radiation (see below) may be focused into the sunfleck, adding some of the high FR component (Holmes and Smith, 1977). In densely shaded environments, direct sunlight is likely to be of less value to most sub-canopy species than might be expected, since their photosynthetic response curves tend to saturate at lower irradiances (see p. 53), though induction of the system may occur. Nevertheless, over half of the total daily radiation flux may be accounted for by sunflecks (Pearcy, 1983). Where spreading occurs, however, the sunfleck will occupy a larger area at lower flux density and will be used in photosynthesis more efficiently. Sunflecks are by nature transitory, and Anderson (1970) found that a photocell of 1 mm diameter placed below a crop canopy can give readings fluctuating by as much as 80% several times a second, whereas a larger cell gave an averaged, more or less uniform reading. Flashing light can in fact be as effective for photosynthesis as a continuous source (Emerson and Arnold, 1932).

(ii) Unintercepted diffuse radiation is the diffuse skylight counterpart of the sunfleck. It makes a very large contribution to the total irradiance beneath canopies.

(iii) Transmission—the degree of shade clearly depends upon the amount of light absorbed and reflected by the leaves. The radiation flux passing through a leaf is not simply reduced in density, but is also radically altered in terms of spectral quality, due to the action of the various leaf pigments. Typically leaves transmit a small portion of incident *PAR* (*c.* 1–20%) in the green band at around 550 nm, but are otherwise effectively opaque in the visible range. There is almost invariably a dramatic change from opacity to near transparency above 700 nm, so that transmitted light has a very high FR/R ratio. This is shown for various canopies in Fig. 2.3. Detailed analyses of spectral energy distribution under leaf canopies are given by Holmes and Smith (1977).

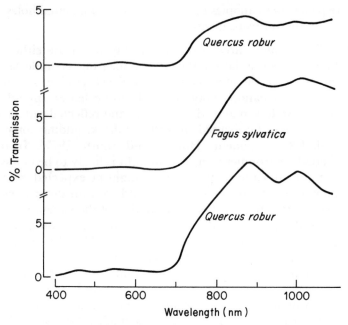

Fig. 2.3.   Transmission spectra of three woodland canopies. Note the sharp transition at about 700 nm (redrawn from Stoutjesdijk, 1972).

(iv) Reflection—leaves do not simply transmit light, but in common with all other biological surfaces reflect a certain proportion. The amount reflected will depend, for example, upon leaf shape and the thickness and shininess (chemically determined) of the cuticle. Reflected light is altered spectrally in much the same way as transmitted light (Fig. 2.2).

The relative contribution of these four components, and hence the depth and nature of the shade, depends upon the number, thickness, distribution and type of leaves in the canopy. The number is usually expressed as the leaf area index (LAI or $\mathbf{L}$), a dimensionless parameter representing the area of leaf surface over unit area of ground. It is possible to relate the rate of attenuation of solar radiation through a canopy to $\mathbf{L}$; differences between canopies in such a relationship reflect differences in leaf geometry, arrangement, and thickness (Fig. 2.4). This relationship is approximately exponential and can be simply described by the formula:

$$I_{\mathbf{L}} = I_0\, e^{-k\mathbf{L}} \qquad (2.4)$$

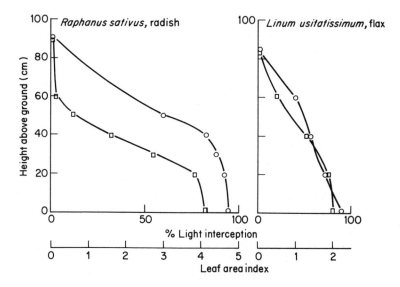

Fig. 2.4. Correlation between interception of light with depth in stand (O) and cumulative leaf area index (□) for two contrasting crops (from Newton and Blackman, 1970).

where I_0 and I_L are the irradiance respectively above the canopy and at a point above which are **L** layers of leaf, and k is the extinction coefficient, varying between canopies as just described. A plot of ln I_L/I_0 against **L** therefore should reveal a straight line of slope k.

A wide variety of more sophisticated models have been advanced to describe this relationship, taking into account such important additional factors as the angle of the leaves (which may itself alter with the angle of the sun in heliotropic species), the solar angle, anisotropy of the radiation field, and the distribution of leaves (Ross, 1970, 1981). These approaches average out a most complex situation, since closely adjacent points within a canopy will vary widely in radiation characteristics, depending on the balance between direct diffuse, transmitted and reflected radiation reaching them. This balance in turn will vary in time as both the sun and the leaves move.

Most work concerned with characterizing the radiation climate within plant canopies has been on agricultural, mono-specific stands, with the aim of maximizing photosynthetic production. In a natural situation, however, we must also consider the environment of subordinate layers in a complex community. For instance, a temperate oak wood (*Quercus robur* on mull soil) may have a closed canopy of oak leaves with a leaf area index (**L**, the area of leaf per unit ground area) of between 2 and 4, a lower, usually more broken canopy of shrubs such as *Corylus avellana* (hazel), a herb layer whose density will be dependent partly on the upper layers, but which, in the case of *Mercurialis perennis* forming mono-specific stands, may itself have $L = 3.5$, and beneath all these a layer of mosses such as *Atrichum* and *Mnium* species and possibly soil algae and cyanobacteria as well, which must still achieve net photosynthetic production.

Radiant flux density beneath the tree and shrub canopies can be as low as 1% of levels in the open; beneath the herb canopy it will be even less. In deciduous woods, however, and in most herbaceous plant communities (reedbeds, tall grasslands, etc.) this massive reduction is a seasonal phenomenon. Salisbury (1916) pointed out the importance to the ground flora in oak–hornbeam (*Quercus–Carpinus*) woods in Hertfordshire, England, of the switch from light to shade conditions that occurred at the time of leaf expansion, and Blackman and Rutter (1946) showed that the distribution of the bluebell (*Hyacinthoides nonscriptus*) was profoundly influenced by it. Coombe (1966) gives seasonal irradiance data for a shaded site in Madingley Wood, Cambridge, showing a clear peak of irradiance from March to May. The effects of this peak on the distribution and phenology of woodland herbs is immense; no analogous situation occurs in evergreen or continuously moist tropical forests.

Under leaf canopies, therefore, all three aspects of the radiation environment, irradiance, spectral distribution and periodicity, are modified to produce a distinct environment to which photosynthetic and morphogenetic adaptation has been necessary.

II. EFFECTS OF SPECTRAL DISTRIBUTION OF RADIATION ON PLANTS

A. Ultra-violet Radiation

Ultra-violet radiation (UV) is a significant component of extra-terrestrial solar radiation, but all of the UV below 280 nm (UV-C) and a large part of the 280–320 nm band (UV-B) are absorbed in the atmosphere by ozone. This is important because nucleic acids, in particular, absorb strongly in the UV-C band, and because of the high quantum energy of photons of such high frequency radiation (see p. 26), major damage to DNA can occur. This is why UV lamps ($\lambda = 254$ nm) are used to eliminate bacteria in sterile culture conditions. A serious concern at the use of chlorofluoromethanes, particularly as spray propellants in aerosol cans, is their possible effect in catalysing the destruction of ozone in the stratosphere by chlorine released in their decomposition by UV-C. Reductions of 15–30% in stratospheric ozone have been predicted (National Academy of Sciences, 1979), and this could result in damaging levels of UV-A reaching the earth's surface (Caldwell, 1981).

Longer wavelength UV (above about 300 nm, i.e. UV-A and some UV-B) is attenuated through the atmosphere by normal processes of scattering and absorption (see p. 28). The result of the action of a variety of phenomena including atmospheric ozone production, distribution and accumulation, solar angle and cloud cover, means that UV-B penetration is greatest at low latitudes and high altitudes. There is some evidence that plants at high altitudes in the tropics have less UV-transparent epidermal layers, largely through the accumulation of phenolic compounds such as flavonoids and flavones (del Moral, 1972; Caldwell, 1981), or surface waxes (Mulroy, 1979). As a result estimated UV penetration to mesophyll layers differs little in tropical alpine regions compared to those where UV-A and UV-B flux is lower. In general, these wavebands have rather slight effects on plant physiology and ecology (Caldwell, 1968, 1981), and the most significant concern in this field is of a reduction in stratospheric ozone leading to a damaging increase in UV-C.

B. Germination

The seeds of many species will not germinate in the dark—they require a light stimulus. This phenomenon is found in two distinct ecological situations: disturbed ground and under canopies.

Weeds, such as many *Chenopodium* species (goosefoots: Cumming, 1963), which are characteristic of ground which is frequently disturbed, typically have small seeds, which can easily become buried, and favourable conditions for germination occur when they are returned by subsequent disturbance to the surface. It is a commonplace that ploughed land quickly comes to support a crop of such plants as chickweed *Stellaria media* and the speedwell *Veronica persica*. All these species have a light requirement for germination. Similarly, for those species which produce seeds under canopies (either their own or those of other species), germination is most likely to be successful when the canopy is either temporarily or permanently removed. This is true of such species as tufted hair-grass *Deschampsia cespitosa* and foxglove

Digitalis purpurea; the latter is a species that can become extremely abundant in clear-cut woodland, though soon disappearing when the vegetation closes. In many of these cases seed is initially insensitive to light, but exposure to FR under a canopy induces a light requirement. Light stimulation of germination is, however, a red/far-red reversible phenomenon, implying the involvement of phytochrome. Typically light-sensitive seed will germinate when exposed to red light, but this stimulation can be erased by subsequent treatment with far-red; in this case it is the final exposure which determines germination.

Several workers have shown that light-sensitive seed will not germinate under leaf canopies (Black, 1969; Stoutjesdijk, 1972; Gorski, 1975). King (1975) showed that several annuals found specifically on the relatively bare ground of anthills in otherwise closed chalk grassland (*Arenaria serpyllifolia*, *Cerastium holosteoides* and *Veronica arvensis*) were completely inhibited from germination by leaf-filtered light.

There is, however, a problem in interpreting such experiments because of maternal effects on light control of germination (Shropshire, 1971; see also p. 24). For example, of three dock species, *Rumex crispus*, *R. obtusifolius*, and *R. sanguineus*, only the latter, a woodland species, shows appreciable germination under FR, but seed from *R. obtusifolius* growing in shade germinated better in both darkness and FR than that from the same species grown in the open (Table 2.3); such differences could clearly be due either to ecotypic differentiation or to direct environmental effects on the seed during maturation of the parent plant.

It seems, therefore, that light controlled germination is an adaptation of plants which are intolerant of shade. If the seed of such plants

TABLE 2.3   Germination of seed of three dock species, *Rumex crispus*, *R. obtusifolius*, and *R. sanguineus* from open and shaded habitats, under different light conditions. Only *R. sanguineus* is normally a woodland plant. Fitter (unpublished).

Species	R. crispus		R. obtusifolius		R. sanguineus	
Habitat	Open	Shade	Open	Shade	Open	Shade
Treatment						
Dark	89	95	74	94	96	99
Far-red light	12	14	7	26	55	49

Figures are percentages of germination in full light, which was between 90 and 97% in all cases.

is exposed to high FR levels, in nature implying that it is under a leaf canopy, or if it is in the dark, which would normally mean burial, it does not germinate. Clearly such a mechanism could be used to adjust the timing of germination to take advantage of seasonal "windows" in the canopy. Seed of shade-resistant plants such as *Rumex sanguineus* on the other hand, will be at less of a disadvantage if it germinates under a closed canopy, as this is the normal habitat for the adult plant. In addition floral structures can alter the germination characteristics of the enclosed seeds. Cresswell and Grime (1981) have shown that a light requirement may be induced not just by exposure to leaf-filtered light after shedding, but by light filtered through the green bracts or sepals surrounding the seeds in the parent plant. Even the seed capsule may act in this way. Seeds of the grass *Arrhenatherum elatius*, which has no chlorophyll in the surrounding structures, germinate fully in the dark, but those of the composite *Tragopogon pratensis*, surrounded by bracts still green when the seeds are mature, have an absolute light requirement. The greenness of fruiting heads may therefore be an important focus for selection in relation to germination behaviour.

C. Morphogenesis

Raised FR:R ratios may markedly alter plant development, particularly affecting stem extension rates and apical dominance (Mohr *et al.*, 1974). Most such physiological work has been carried out using briefly raised levels of FR, presumably to simulate sunrise and sunset effects. For example, tomatoes given 5 min FR at the end of each day produce no side shoots (Tucker, 1975). Kasperbauer (1971) has suggested that such short-term changes are responsible for the fact that tobacco plants in the centre of a stand have longer internodes and paler, thinner leaves than those at the edge. As the sun passes through very low angles at dawn and dusk, light received by central plants would be filtered through the leaves of the marginal ones and so have enhanced FR levels. He was able to mimic all the responses of mid-row plants by growing them in controlled conditions with a short pulse of FR at the end of each light period. Kasperbauer and Peaslee (1973) subsequently demonstrated that FR-treated plants (i.e. analogous to mid-row plants) had longer, narrower, lighter leaves with fewer stomata and less chlorophyll per unit area. Carbon dioxide assimilation was the same on an area basis, but greater on a leaf weight basis, showing that FR-treated plants had maintained photosynthetic assimilation at a lower flux density by increasing leaf area.

The mechanism and ecological significance of the elongation response are presumably similar to that of etiolation, allowing plants shaded by other leaves to grow into better illuminated regions. The situation under a plant canopy, however, has two important characteristics:

(i) the alteration in R/FR ratio is permanent (at least in the "leafy" season) and not a 10 min dawn or dusk effect;

(ii) the distance from the ground to the unfiltered light may be as much as 100 m, though in most forests 15–25 m is more likely; no etiolation responses will be of value to a herbaceous plant in these conditions, although sustained extension growth might be important for tree seedlings.

Under natural conditions plants detect shading by changes in phytochrome state. Holmes and Smith (1975) measured the photostationary state of phytochrome under leaf canopies, finding it to be similar to that under an FR filter. They showed that incandescent lights (producing much FR) stimulated greater stem extension growth in two arable weeds than did fluorescent tubes (which produce almost no FR), and were able to relate this to measurements showing a shift of phytochrome to the P_R (red-absorbing) form in etiolated seedlings. In plants characteristic of shaded conditions, however, the situation may be different. Fitter and Ashmore (1974) used two species of speedwell, *Veronica montana* and *V. persica*, the latter an arable weed, the former growing typically in deep shade. Under neutral shade both species showed reductions in various growth parameters, but under continuous FR at the same flux density, *V. persica* exhibited a further decline in leaf area, whereas *V. montana* was not affected differently by the FR and neutral shade treatments (Fig. 2.5). In very deep shade cast by a tobacco leaf canopy both species grew very poorly, but whereas *V. persica* became etiolated, *V. montana* showed no alteration to its development pattern. Stem extension rates, leaf weight ratio and petiole length are all characteristically more labile in plants adapted to open conditions than in those found growing in shade: Morgan and Smith (1979) measured the stem extension rates of four woodland and four arable weed species as a function of FR:R ratio, and hence of the phytochrome photoequilibrium ($P_{FR}:P_R$). Generally the woodland species showed little response while the weeds were very plastic. *Mercurialis perennis*, a woodland species that stays leafy throughout the summer in deciduous woods, was almost completely unresponsive to FR.

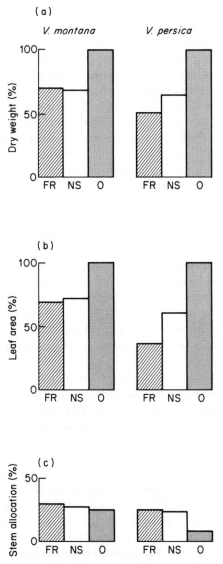

Fɪɢ. 2.5. (a) Dry weight and (b) leaf area of *Veronica montana* (a shade plant) and *V. persica* (a sun plant) in far-red supplemented shade (▨), neutral shade (□) and open (▤) conditions. Both variables are expressed as a percentage of the value in open conditions. (c) Percentage of biomass allocated to stems in the same experiment. From Fitter and Ashmore (1974).

D. Photoperiodism

Athough strictly a different aspect of the light factor from quality, periodic effects are mediated *via* the red/far-red reversible pigment phytochrome. Most temperate region plants are photoperiodic; in equatorial regions daylength shows little seasonal march and so photoperiodism would be of less value. Since the beginning and end of the day are marked and measured by changes in the R/FR ratio, it may well be that effects such as those shown by Kasperbauer for tobacco (1971; and see p. 37), could have an important influence on processes that are under photoperiodic control; apparently the question has still not been raised experimentally.

Keeping a photoperiodic plant under a constant daylength will usually maintain it in one developmental stage, although if the photoperiod promotes flowering, it will normally be stopped eventually by other processes. Thus *Epilobium hirsutum* and *Lythrum salicaria* flower if given 16 h days, but will remain vegetative indefinitely under a 9 h photoperiod (Whitehead, 1971). Not all temperate zone plants are photoperiodic, however: many weedy species, such as groundsel (*Senecio vulgaris*) will flower in all months of the year if the weather is favourable, and *Impatiens parviflora* is an example of a large group of plants which flower at a particular stage of development, irrespective of photoperiod.

The photoperiodic response enables the plant to time vegetative and floral growth to fit seasonal changes in environment. If a plant is moved to a different latitude it will then be out of phase and may die from, for example, attempting vegetative growth in the winter or too late in spring. This was what led to the first recognition of the phenomenon (Garner and Allard, 1920) and its importance was quickly seen by silviculturalists, who discovered that a wide-ranging tree species may have well-marked photoperiodic races or ecotypes as for example in birches (*Betula*) (Vaartaja, 1954; Håbjørg, 1978a). The same is true of grasses (Olmsted, 1944; Håbjørg, 1978b) and of other herbaceous plants such as *Oxyria digyna*, the mountain sorrel, a circum-polar arctic–alpine species that descends as far south as California in the North American Rockies (Fig. 5.4). Mooney and Billings (1961), in a comprehensive study of the different arctic and alpine races of this species, found that the arctic populations required a much longer day to induce flowering; arctic plants could be kept vegetative in growth room photoperiods that stimulated flowering in alpine populations. Despite these important investigations, the ecological role of photoperiodism within communities, for example in determining flowering time phenologies (see Chapter 8) is completely unknown.

III. EFFECTS OF IRRADIANCE ON PLANTS

A. Patterns of Response

Irradiance exerts its primary effect on photosynthesis, acting secondarily on morphogenesis; most morphogenetic responses are to low flux densities, but some require greater energy (Mancinelli and Rabino, 1978). The ecology of a plant with regard to radiant flux density is governed by two considerations:

(i) the position of the leaves in the canopy. The upper leaves will experience unintercepted radiation, but in a complex community most of the leaves will be shaded, and will therefore receive reduced flux densities; and

(ii) the ability of the leaves to utilize intercepted radiation, especially where this is below the light-saturation point for normal photosynthesis, so as to remain in net positive carbon balance (photosynthetic CO_2 fixation greater than the sum of respiration and carbohydrate export). A leaf in negative C-balance will need to import sugars from the rest of the plant and will reduce overall fitness.

Patterns of leaf placement are more complex than a first glance suggests. Horn (1971) has suggested two basic architectures for forest trees—monolayer and multilayer. The monolayer is defined as a complete, uniform layer of leaves that lets through little photosynthetically active radiation (*PAR*), but it will have low productivity since it can have fewer layers of leaves (low **L**). The multilayer species has a more dispersed canopy but relies on the facts that:

(i) an individual leaf only casts a shadow for a certain distance (Horn uses 70 diameters), and

(ii) light-saturation for most species occurs at flux densities well below full sunlight

so permitting several layers of leaves (high **L**) all operating at high photosynthetic rates. The multilayer can therefore grow faster, but since it lets through more *PAR*, it is more open to invasion. Even though Horn's theory does not take account of variation in photosynthetic response to irradiance (see p. 52), it has important implications.

Within this basic dichotomy control of both leaf production and placement are important. The morphogenetic responses involved are closely tied to the phytochrome system. Typical responses include stem elongation and leaf orientation.

1. Etiolation

Bringing leaves up into more brightly illuminated regions is only of value to plants of low-growing communities, or to those that typically form the canopy, except in the case of hypocotyl elongation in hypogeal seeds. Significantly, etiolation is normally demonstrated with crop plants (beans, mustard, etc.) and is absent from species adapted to and resistant to shade such as *Veronica montana* (Fitter and Ashmore, 1974) or *Mercurialis perennis* (Morgan and Smith, 1979). Grime (1966) has suggested that resistance to etiolation is found in two ecologically constrasted groups of plants—those adapted to dense shade, and those that normally grow in open habitats. In his experiments *Arenaria serpyllifolia* and *Hieracium pilosella*, species of open communities, showed no elongation growth in shade, apparently owing to exhaustion of carbohydrate reserves, whereas *Betonica officinalis*, a grassland species, became very etiolated. In a low but closed canopy this would enable *Betonica* to reach the light, where, since it is a relatively large plant, with leaves capable of overshadowing competitors, it is likely to be successful; once it has penetrated the canopy the consequent increase in irradiance would inhibit further extension growth. Trenbath and Harper (1973) showed that the elongation response of oats (*Avena sativa*), a species selected for growth in herbaceous stands ecologically similar to tall grassland, resulted in about 20% extra weight in each seed when grown in mixed culture with *A. ludoviciana*, as compared to the estimate for a shaded plant. Similar responses are shown by aquatic plants: internode lengths of *Hippuris vulgaris* can be as much as 30 cm at depths of 1–2 m, but only 3–6 cm at depths less than 0·4 m (Spence, 1981).

2. Phototropism and Orientation

Tropic responses remain a physiological enigma. Much of the groundwork was laid by Darwin and an apparently adequate explanation was propounded by Went and his co-workers in the 1920s, based on a theory of auxin movement. It is now known that adequate auxin movement does not occur, but as yet no general model of phototropism can be accepted (Firn, 1986). What is certain is that the action spectrum is complex, with as many as four peaks between 370 and 470 nm, suggesting the involvement of flavins and carotenes, and that the response *is* affected by IAA. Whatever the physiology, the adaptive significance is clear, and is enhanced by the partial reversibility of the process, since although it involves differential growth, this can be reversed by complementary growth opposite the original site.

Generally, however, phototropism will be a response to a more less fixed spatial pattern in the light environment.

Phototropic responses are adaptive; differences between plants adapted to contrasting habitats should therefore occur. Genotypic differences in the patterns of leaf arrangement occur in forage grasses, and the same phenomenon was recognized early for uncultivated plants (Turesson, 1922). Plants with prostrate leaf arrangements have very much higher extinction coefficients for light (k, see equation 2.4, p. 32) within the canopy than those with erect leaves (Cooper and Breeze, 1971), a situation analogous (for a very different growth form) to Horn's (1971) distinction between mono- and multilayer tree strategies. As a result, under agricultural conditions, long-leaved, erect plants of low k can grow faster and are more productive (Table 2.4). Indeed in terms of plant breeding it seems that much more can be obtained by alteration in canopy architecture than in photosynthetic rates, which Sheehy and Peacock (1975) found to be unrelated to growth rate.

TABLE 2.4   Crop growth rates of six common forage grasses in relation to the extinction coefficient of light in the canopy and the leaf area index (Sheehy and Cooper, 1973). Note the correlation between growth rate and k.

Species and cultivar		Crop growth rate $(\mathrm{g\ m^{-2}\ d^{-1}})$	Extinction coefficient (k)	Leaf area index
Festuca arundinacea	S170	43·6	0·34	11·2
Dactylis glomerata	S 37	40·5	0·23	13·7
	S345	25·0	0·91	14·9
Phleum pratense	S 50	36·4	0·30	15·5
	S 48	28·9	0·39	10·4
	S352	21·9	0·55	14·5

In natural conditions, however, the prostrate form (high k) will often be favoured, since the advantage of the erect types increases with longer intervals between cutting, and under very frequent cutting regimes (intervals of less than 10–14 days), the short-leaved genotypes are favoured (Rhodes, 1969). Under these conditions, which could be produced in the wild by grazing animals, the erect types never produced an adequate canopy. Very severe environments will also enforce the same result, as shown by Callaghan and Lewis (1971) for the grass *Phleum alpinum* in Antarctica: plants from sheltered habitats allocated more biomass to leaf growth than those from exposed sites, but the latter compensated for this by apparently adaptive changes in photosynthetic physiology.

Where changes in the spatial pattern of light are more short-lived, however, more rapid (nastic) responses will be required. Leaf and petiole movements, operated by turgor changes, occur more or less continuously in controlled conditions, as can be shown by time-lapse photography. The function of these rather slight movements in the field is unclear, as they will often be swamped by air turbulence, though they can be regarded as a mechanism for sensing the environment, particularly in climbing plants. Nevertheless, nastic movements as dramatic as those of the leaves of the sensitive plant, *Mimosa pudica*, whose leaves collapse on physical contact, or the compass plant, *Lactuca serriola*, whose leaves tend to orient themselves north–south, can readily be observed in the field. For plants growing in low flux densities these movements typically follow the sun and ensure maximum illumination; for plants in strong light they are normally avoidance reactions which reduce the heat load on the leaf and allow subordinate leaves in the canopy to receive more *PAR*. When the sun is far from the zenith such movement can also markedly influence the effective leaf area index. In *Lactuca serriola* both effects are achieved—greater photosynthesis in the morning and lower radiation loads at mid-day (Werk and Ehleringer, 1984).

B. Responses to Low Irradiance

1. Temporary Stress

As shown above, there is a distinct peak of irradiance in temperate deciduous woods in early spring, and a well-marked group of plants takes advantage of this. Deciduousness in trees in response to seasonal temperature changes (as opposed to variation in water supply), is, however, an adaptive response to avoid frost damage to leaves capable of high productivity, so that the herbs which photosynthesize during this brief seasonal window need to be frost resistant (see Chapter 5). To take full advantage of the radiation peak they must also have fully expanded leaves by April at the latest, so that leaf growth must take place at very low temperatures in February and March, when photosynthetic activity will be low. This growth therefore requires stored reserves, and almost all these species are perennials with underground storage organs, whether bulbs (*Hyacinthoides non-scripta, Allium ursinum*), corms (*Cyclamen*), tubers (*Ranunculus ficaria*), or rhizomes (*Anemone nemorosa*). These storage organs are re-charged during the radiation peak.

Occupation of this particular niche, therefore, requires modification to all other parts of the life-cycle. Whereas some species, such as *Hyacinthoides*, complete their annual growth during the light phase and remain dormant for the rest of the year, others such as *Oxalis acetosella*, remain active for a large part of the shade phase. This activity occurs in very dim light and so requires a change in physiology, which is brought about plastically in two distinct ways. *Oxalis* adjusts both physiologically, by altering photosynthetic affinity in surviving leaves (Daxer, 1934) and morphologically by producing new leaves (Packham and Willis, 1977), and *Aegopodium podagraria*, a garden weed in northern Europe, but a woodland herb in central Europe, also has two leaf types produced by a single plant: thin broad summer leaves, and thicker spring leaves, adapted to higher light levels. These two mechanisms are of general significance in the adaptation of plants to low irradiance and the physiological background to the ecological switch will be examined below.

Of course, any changes in physiology which maintain carbon balance involve changes in respiration rate too. This typically means a lowering of the compensation point, where respiration exactly balances photosynthesis (Fig. 2.6).

2. Long-term Stress

If a plant grows in a shaded environment and does not restrict its activity to periods of high illumination, selection will act more on the photosynthetic process itself. This is inescapable for the lower leaves of a plant forming a multilayer canopy, and it follows that both plastic (within a genotype) and genetic (between genotypes) differences must exist in the photosynthetic system.

The problem faced by a shaded leaf is to maintain a positive carbon balance and the flux density at which this is reached is the compensation point (Fig 2.6). Three mechanisms are available to balance this equation:

  (i)   reduced respiratory rate, which lowers the compensation point;
  (ii) increased leaf area, which provides a greater surface for *PAR* absorption; and
  (iii) increased photosynthetic rate per unit radiant energy and leaf area.

All these three courses are adopted by shaded plants, but they impose particular restraints.

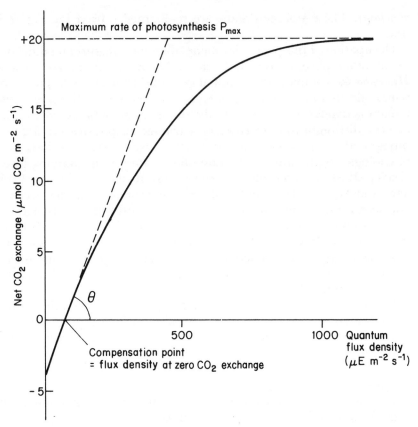

FIG. 2.6. Representative photosynthetic response curve showing the compensation point, the flux density at which net CO_2 exchange is zero, as the point at which gross photosynthesis equals respiration.

(a) Reduced respiratory rate. At the compensation point photosynthetic carbon fixation equals respiratory loss. A reduction in respiration will therefore lower the compensation point, but a reduction in respiration rate is likely to reduce growth rate, which could lower the competitive ability of the plant in relation to faster growing species. As a response to low irradiance it is therefore only likely to be advantageous in severe shade where growth rates are sufficiently reduced to minimize competitive interactions. It is certainly true that plants normally found in deep shade have low relative growth rates (Grime and Hunt, 1975): Mahmoud and Grime (1974) found that at extremely low illumination (down to 0.07 W m^{-2}) *Deschampsia flexuosa* showed zero growth rate, whereas *Festuca ovina* and *Agrostis capillaris (tenuis)* had large negative **R**

below their compensation points at about 0.7 W m^{-2} (Fig. 2.7). *Deschampsia flexuosa* survived four weeks in deep shade; the two latter species senesced. Although respiratory rates were not measured, the clear implication is that *D. flexuosa* owed its resistance to a virtual cessation of respiratory activity. Certainly McCree and Troughton (1966) found that white clover, a species unable to survive in deep shade, when grown at 65 W m^{-2} and transferred to a range of intensities from 88 to 3.7 W m^{-2}, had respiration rates varying from 7.0 down to 0.7 mg CO_2 dm^{-2} h^{-1}.

A small reduction in respiratory rate is then a fairly general response to reduced irradiance, but as a major adaptive response it is only of value to severely shaded plants, and particularly as a survival mechanism for long-lived plants to persist during periods of temporary stress (cf. above). This is clearly the value of the near-dormant behaviour of *D. flexuosa* and has previously been recognized, for example by Chippindale (1932), who observed that seedlings of *Festuca pratensis* were able to survive long periods without growing when under severe

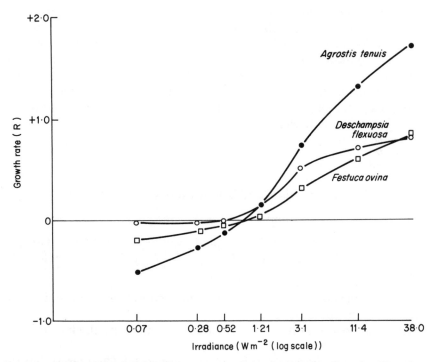

Fig. 2.7. Effect of very low irradiance on growth rate in three grass species. Note the very small negative **R** of *Deschampsia flexuosa* (from Mahmoud and Grime, 1974).

competitive stress from older plants of *F. pratensis* and *Lolium italicum*, but could resume normal growth when the stress was removed. He termed the phenomenon inanition.

A similar case has been reported by Cross (1975) for *Rhododendron ponticum* seedlings, and Hutchinson (1967) showed that woodland plants, such as *Digitalis purpurea* and *Bromus ramosus* could survive for 5–6 months in complete darkness, and very strikingly that *Deschampsia flexuosa* could persist for as long as 227 days. He was able to demonstrate that small-seeded pioneer plants, such as *Betula pubescens* and *Erophila verna*, were generally least resistant to this stress, and that plants grown on nutrient-deficient soils, which reduced growth and probably also respiration rates, could survive for longer in complete darkness than those grown on fertile soil.

(b) Increased leaf area. Plant growth can be described by the classic growth equation (see p. 4):

$$\mathbf{R} = \mathbf{E} \times \mathbf{F} \qquad (2.5)$$

The immediate, enforced response of a plant removed to a lower flux density will be a reduction in $\mathbf{R}$ caused by a lowering in $\mathbf{E}$ (net assimilation rate or *NAR*), reflecting the effect of *PAR* on photosynthesis. To maintain $\mathbf{R}$ therefore, assuming no change in the light-dependence of photosynthesis, requires an increase in the leaf area ratio ($\mathbf{F}$ or *LAR*). $\mathbf{F}$, as the ratio of leaf area to plant weight, is a complex function without any obvious biological interpretation. It can, however, be thought of as:

$$\mathbf{F} = LWR \times SLA$$

$$A_{\mathrm{L}}/W = W_{\mathrm{L}}/W \times A_{\mathrm{L}}/W_{\mathrm{L}} \qquad (2.6)$$

Leaf area ratio = leaf weight ratio × specific leaf area

$$(\mathrm{cm}^2\ \mathrm{g}^{-1} = \mathrm{g}\ \mathrm{g}^{-1} \times \mathrm{cm}^2\ \mathrm{g}^{-1})$$

LWR is the ratio of leaf weight to plant weight, *SLA* that of leaf area to leaf weight. Leaf area described by *LAR* or $\mathbf{F}$ can therefore be discussed in terms of two components:

(i) the proportion of plant weight devoted to leaf material; i.e. how much leaf is there?
(ii) the area:weight ratio of the leaf itself, i.e. how thick is it?

Changes in leaf dry weight. It is well established (cf. Chapter 3) that the ratio of root weight to shoot weight is very plastic—for example, plants grown in infertile soils tend to have very high root:shoot ratios. Since for most herbaceous plants the leaves are a large proportion of the shoot weight, and since the stem acts to place leaves in an appropriate radiation environment, one would expect the leaf weight ratio (LWR) to be equally variable. Evidence on this point is surprisingly unclear. Evans (1972) suggests that LWR is susceptible only to changes in temperature, daylength, and soil factors, and not to flux density, the daily total of irradiance, or the spectral composition of radiation. This is largely based on experiments on one plant *Impatiens parviflora*. Fitter and Ashmore (1974) found LWR to be unaltered by severe shade stress in *Veronica montana*, a shade-resistant species, but to be reduced by shade in *V. persica*, an arable weed. In contrast, LWR increases in response to shading in several semi-shade species, such as *Chamaenerion angustifolium* (Myerscough and Whitehead, 1966), *Filipendula ulmaria* (Whitehead, 1973) and several trees (Loach, 1970). Most probably LWR is a reflection of the plant's ability to maintain its normal developmental pattern, and it will be found to be constant over the range of flux densities to which a plant is adapted. Non-adapted plants in shade, however, exhibit etiolation and LWR is then reduced.

Specific leaf area. SLA is a much more variable parameter than LWR; in other words, leaf area is more plastic than leaf weight. A striking illustration is that, immediately subsequent to germination, there is a marked rise in SLA caused by expansion of the first leaves, whose dry weight, of course, changes only slightly during expansion. SLA also responds to environmental changes. Newton (1963) showed that the leaf area of cucumber plants (*Cucumis sativa*), a light-demanding species, was proportional to the total radiation, with a maximum at about 100 cal cm^{-2} d^{-1} (equivalent to 4·2 MJ m^{-2} d^{-1} or about 350 μE s^{-1} m^{-2} for a 16 h day) and that changes in both daylength and irradiance (as components of total daily radiation) had no effect if the daily total was the same. The reduction in leaf area at the highest irradiances was due to a reduction in cell size, whereas cell number increased to a plateau (Milthorpe and Newton, 1963).

Apparent effects of irradiance may therefore be responses to total radiation load; this is probably the case for *Impatiens parviflora* which shows an almost three-fold increase in SLA when grown in 7% of full daylight (Evans and Hughes, 1961). When plants are grown under field conditions, responses are even more dramatic, SLA changing from 4·9 to 12·7 dm^2 g^{-1} for plants growing under 48 as opposed to 13 cal cm^{-2} d^{-1} (Hughes, 1959; see conversion above). The plasticity of the character is emphasized by the rapidity with which SLA adjusts when plants are transferred from one light regime to another.

These differences are for whole plants. Similar effects can be found between leaves of a single plant exposed to heterogeneous light, as in a forest canopy. Such differences are the basis of the sun and shade leaves, first recognized by Haberlandt. In beech (*Fagus sylvatica*) the phenomenon is very clear-cut (see Table 2.5), but as with *LWR* the degree of plasticity appears to be related to the ecological niche of the species. Some shade-resistant species, such as *Veronica montana* (Fitter and Ashmore, 1974) and *Rhododendron ponticum* (Cross, 1975) show much less striking shade-induced changes in *SLA* than species from open habitats.

TABLE 2.5   Characteristics of sun and shade leaves of mature and young beech trees *Fagus sylvatica*, from Nordhausen, quoted by MacGregor Skene (1924) in "The Biology of Flowering Plants".

	Mature		Young	
	Sun leaves	*Shade leaves*	*Sun leaves*	*Shade leaves*
Leaf thickness (μm)	210	108	117	90
Number of palisade layers	2	1–2	1–2	1
Thickness of upper palisade layer (μm)	60	28	39	24

Leaf morphology. The relative constancy of *LWR* and plasticity of *SLA* imply that the plant has an optimum developmental pattern in terms of dry weight distribution, achieving adaptation to irradiance by changes in leaf morphology. The increased ratio of leaf area to weight (*SLA*) must imply important anatomical changes in the mesophyll and palisade layers and this is clearly shown in *Mimulus* (Hiesey *et al.*, 1971) and in a wide range of deciduous trees (Jackson, 1967; see also Table 2.5). In all cases the palisade layer is reduced from 2–3 cells to 1 cell in the shaded or shade-resistant leaves. Such thin leaves produce high values of *SLA*.

These changes have been shown to affect CO_2 diffusion, but there is little evidence that CO_2 diffusion inside the leaf limits photosynthesis except at very high flux densities (Björkman, 1981). The main effect of a change in leaf thickness is the concomitant change in the quantity of photosynthetic apparatus per unit leaf area, with internal geometry being much less important.

The environmental trigger for most changes in leaf morphology is low irradiance, and a more satisfactory explanation rests on a

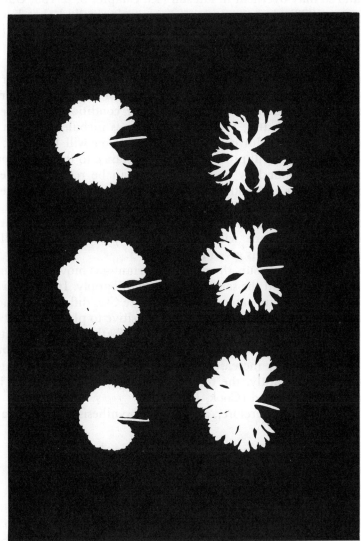

Fig. 2.8. Silhouettes of successive leaves from base to top of a stem of musk mallow *Malva moschata*, showing the progressive increase in leaf dissection.

consideration of the energy balances of different leaf morphologies. Shade leaves tend to be larger and less dissected; as a result they have greater boundary layer resistance, are less susceptible to convective cooling and so may overheat in full sun (see Chapter 5, p. 195; Gates, 1968; Grace, 1983). It is common, for example, to find that lower leaves of a plant have these "shade" characteristics (Fig. 2.8) and it is notable that cotyledons are rarely dissected.

Leaf cooling is brought about both by evaporation of water as vapour through the stomata and by convective heat loss. The former is controlled by factors such as size and frequency of stomatal apertures and both are affected by the boundary layer conditions of the leaf surface atmosphere (Gates, 1968). If this layer is turbulent (which is likely except in very calm conditions) evaporation will be faster; a stable boundary layer will provide a large resistance to the movement of water vapour. Boundary layer resistance can be reduced by leaf dissection, as long as each individual leaf segment maintains a discrete boundary layer. Where the distance between segments is too small, a common boundary layer will form over the whole complex and boundary layer resistance will be considerably increased. A fuller discussion of leaf cooling is given in Chapter 5.

Apart from water vapour, the other important gas moving through the stomata is CO_2, and the same constraints will apply. It becomes of particular importance in water, however, where CO_2 diffusion is much slower and where its concentration is very sensitive to pH as a result of bicarbonate ion formation; the characteristic feathery leaf of water plants (*Myriophyllum, Ranunculus* subgenus *Batrachium, Hottonia palustris*, see Fig. 2.9) effectively increases the surface area and allows greater CO_2 uptake. Significantly, the response is usually light-triggered, by perception of photoperiod (Cook, 1972).

Leaf morphology, therefore, affects photosynthesis in four main ways:

  (i)   interception of *PAR*;
  (ii) temperature regulation;
  (iii) water balance;
  (iv) CO_2 diffusion.

(ii) and (iii) are probably the most important. In all cases the environmental stimulus to which the plant responds is radiation, whether as irradiance, duration, or their product.

(c) Increased photosynthetic activity. Since shading tends to cause an increase in *SLA* (and hence, at constant *LWR* in **F** also), relative

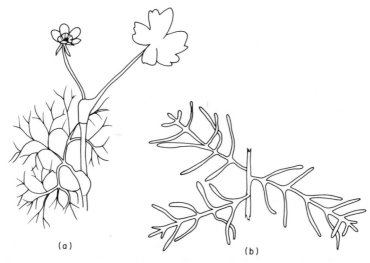

FIG. 2.9.　Leaves of (a) *Ranunculus trichophyllus* and (b) *Hottonia palustris*, showing dissection of underwater leaves.

growth rate (**R**) can theoretically be maintained, even with falling **E**. It is generally true that *SLA* and **F** are inversely correlated with irradiance, while **E** is directly proportional to it (Blackman and Wilson, 1951; Newton, 1963; Coombe, 1966), since **E** is a manifestation of the efficiency of the photosynthetic system.

One possible mechanism of shade resistance, therefore, is an increased responsiveness of the photosynthetic system to irradiance. Sun and shade plants do indeed have very different light response curves (Fig. 2.10). Shade plants are characterized by lower compensation points and light saturation levels, and by steeper angles of response. Light response curves of this type can be analysed in terms of maximum photosynthetic rate (P_{max}), the initial slope of the response curve or quantum efficiency, and the saturating irradiance above which no further increase in photosynthesis occurs (see Fig. 2.6). The last of these and also P_{max} are therefore both measures of light saturation, and the quantum efficiency is an indication of efficiency in low irradiance.

Generally a sun physiology has high P_{max} and saturating irradiance and low quantum efficiency; a shade physiology is the converse. *Cordyline rubra*, an extreme shade plant found in Australian rainforests has P_{max} of 2 μmol CO_2 m^{-2} s^{-1} and saturates at around 80 μE m^{-2} s^{-1}. Corresponding values for the sun species *Atriplex triangularis*, are over 30

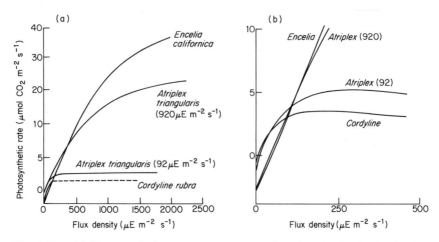

FIG. 2.10.   (a) Photosynthetic response curves to radiant flux density for the C_4 plant *Encelia californica* grown at 40 E m^{-2} d^{-1}, the C_3 sun plant *Atriplex triangularis* grown at high (920 μE m^{-2} s^{-1}) and low (92 μE m^{-2} s^{-1}) irradiance regimes, and of the extreme shade plant *Cordyline rubra* grown under a rainforest canopy at 0·3 E m^{-2} d^{-1}. (b) An expansion of the bottom left-hand corner of (a) to show the greater quantum efficiency of the shade-grown plants. Redrawn from Björkman (1981).

and 1000 respectively. In this case there is a clear distinction between sun and shade species on physiological grounds, and similar data are given for species of beechwood ground flora by Schulze (1972).

    A similar physiological distinction can be found between individual leaves on a single plant. These sun and shade leaves are morphologically distinct, as shown above (cf. Table 2.5), and show parallel differences in response to irradiation (Boysen-Jensen and Muller, 1929). This plasticity can also be manifested as a temporal adaptation and as such appears to be widespread. Woledge (1971) found that leaves of plants of *Festuca arundinacea* grown previously at over 40 W m^{-2} (i.e. around 180 μE m^{-2} s^{-1}) had considerably faster rates of photosynthesis at high illumination (higher P_{max}) than those from plants previously grown in dim light. The same is true for *A. triangularis* in Fig. 2.10; when grown at 92 as compared to 920 μE m^{-2} s^{-1}, P_{max} fell to 5 μmol CO_2 m^{-2} s^{-1} and the saturating irradiance to around 400 μE m^{-2} s^{-1}.

    Shade resistant plants do not possess this plasticity and may in fact suffer damage which leads to reduced CO_2 fixation rates if exposed to high irradiance. Even where damage does not occur there may be a lag period of several minutes to a few hours before a previously shaded leaf can adjust its maximum photosynthetic rate to the new high radiant flux—this is known as induction. The induction appears to be related to the activation of ribulose bisphosphate carboxylase. Two species native to deeply shaded rainforests in Australia, *Alocasia macrorrhiza* and *Toona australis*, illustrate this well. *Alocasia* is an understorey herb, which never grows in full sun, while *Toona* is a tree that develops in gaps caused by tree-fall. Their maximum photosynthetic rate declines by around 80% if they are maintained in low irradiance conditions ($10 \, \mu E \, m^{-2} \, s^{-1}$), but exposure to 500 (*Alocasia*) or 1200 (*Toona*) $\mu E \, m^{-2} \, s^{-1}$ for 30–40 minutes restores full activity (Chazdon and Pearcy, 1986a). Brief periods of high irradiance, simulating sunflecks and lasting 30–60 s, are just as effective (Fig. 2.11); the more shade-tolerant *Alocasia* benefits more from very brief illumination—the first 30 s fleck results in twice as much carbon fixation as in *Toona*—but a series of short flecks or as few as two longer (60 s) ones benefit *Toona* more (Chazdon and Pearcy, 1986b). For such plants to utilize sunflecks effectively, they must either be of long duration, or occur in rapid succession. In fact, flecks often do come in groups and may be of major importance in the carbon economy of shaded leaves.

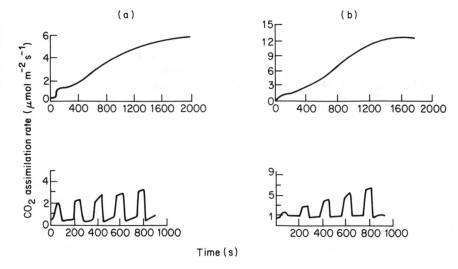

FIG. 2.11. A comparison of the photosynthetic induction of (a) *Alocasia macrorrhiza* and (b) *Toona australis* in response to continuous illumination or a series of 60 s lightflecks. In each case the upper figure represents the time-course of induction when irradiance was raised from 10 to 400 (*Alocasia*) or from 20 to 1200 $\mu E \, m^{-2} \, s^{-1}$ (*Toona*) permanently; the lower figure represents the induction under 60 s flecks. From Chazdon and Pearcy (1986a).

3. Mechanisms

Populations of plants from habitats differently illuminated, and individuals or parts of individuals similarly distinguished have differing photosynthetic responses to irradiance. Since these differences are expressed on a leaf area basis, they could have three causes:

 (i) Improved access to substrate; i.e. higher CO_2 diffusion rates.
 (ii) Increased capture of photons; i.e. more chloropyll per unit leaf area.
 (iii) Increased activity of the photosynthetic apparatus.

Changes in morphology and anatomy found in sun and shade leaves can certainly affect internal CO_2 diffusion (Holmgren, 1968; see also p. 50), but there is doubt as to the extent that this limits photosynthesis. Changes in resistance to CO_2 movement between leaf and air, however, may be important. *Atriplex triangularis* showed a three-fold increase in stomatal conductance for CO_2 when grown at 920 as compared to 92 $\mu E\ m^{-2}\ s^{-1}$, but at any given conductance, the high irradiance plants photosynthesized much faster (Björkman *et al.* 1972), so that the difference must be due to changes in photosynthetic efficiency.

Shaded leaves generally have enhanced chlorophyll levels per unit weight, though often not per unit area (Shirley, 1929). Absolute amounts of chlorophyll are usually greater at high irradiances, but it does not seem that any good generalization can be made about the relationship between chlorophyll and irradiance.

Björkman and Holmgren (1963) grew plants of *Solidago virgaurea* from shaded and exposed habitats in low (30 W $m^{-2} = 6\cdot6$ E $m^{-2}d^{-1}$) and high (150 W $m^{-2} = 33$ E $m^{-2}d^{-1}$) irradiances and found that P_{max} was generally higher in the plants from exposed habitats, an indication of genetic differentiation (Table 2.6). In these plants P_{max} was reduced if the plants were grown in low light, showing physiological plasticity. The plants from shaded habitats, by contrast, had the same maximum photosynthetic rate irrespective of the irradiance at which they were grown; but the quantum efficiency was increased by growth in low irradiance in these shade plants, while remaining unchanged in the exposed plants. The two sets of populations are therefore physiologically distinct, but both are capable of plastic modification, which occurs by raising P_{max} in the sun plants and by raising the quantum of efficiency in the shade plants. These adaptations are appropriate to the situations likely to be encountered by the plants. Similar effects have been shown for arctic and alpine populations of *Oxyria digyna* (Mooney and Billings, 1961), and *Thalictrum alpinum* (Mooney and Johnson,

TABLE 2.6   Differences in photosynthetic characteristics and plasticity in clones of goldenrod *Solidago virgaurea* from shaded and exposed habitats (Björkman and Holmgren, 1963).

| | *Habitat of plants* | |
	Shaded	Exposed
1. Light-saturated photosynthetic rate (P_{max}, μmol CO_2 m^{-2} s^{-1}); plants grown at 33 E m^{-2} d^{-1}	11·3	16·1
2. Ratios of P_{max} for plants grown at 33 E m^{-2} d^{-1} to that of plants grown at 6·6 E m^{-2} d^{-1}	0·97	1·84
3. Ratios of quantum efficiency for plants grown at 33 E m^{-2} d^{-1} to that of those grown at 6·6 E m^{-2} d^{-1}	0·64	1·04

1965), and for shaded and exposed races of *Solanum dulcamara* (Gauhl, 1976), though these may be as much to do with water deficits as photosynthetic energy capture. These effects are probably widespread, although it may be that plants with high leaf turn-over may be able to adjust by producing new leaves with altered physiology.

These responses are linked to enzymic changes. In *Solidago virgaurea* Björkman (1968) found that ribulose bis-phosphate carboxylase (RuBP) activity, expressed in μmol CO_2 mg^{-1} protein min^{-1}, was 0·25 for exposed habitat plants grown in strong light and only 0·21 for those grown in weak light. The shaded habitat plants gave values of 0·15 and 0·12 for strong and weak light respectively. Once again both genetic differentiation and plasticity are apparent. On the other hand, although Hiesey *et al.* (1971), found a strong correlation between RuBP carboxylase activity and radiant flux density in two species of *Mimulus*, when expressed on a fresh weight basis (4·4 μmol CO_2 g^{-1} min^{-1} at 18 W m^{-2} to 13·5 at 106 W m^{-2}), this was associated with a parallel increase in protein; all activities were around 0·5 when expressed per unit protein. The increase in RuBP carboxylase activity may not, therefore, always be specific.

Generally, therefore, shade plants appear not to have the ability to respond to increases in irradiance by significant increases in enzyme activity. Since RuBP carboxylase is the largest single component of leaf protein, this must represent a major saving in terms of the amount of nitrogen that shade plants must invest in their leaf tissues, especially as changes in CO_2-fixing capacity are paralleled by changes in electron transport capacity. Not all plants, however, show the same forms of adaptation. In *Solidago* the sun and shade races are clearly differentiated in their response to irradiance, but in *Solanum dulcamara* at least one shade race had a "sun" physiology and was more sensitive to water status (Gauhl, 1969, 1979), and two species of *Mimulus*, the coastal *M. cardinalis* and the subalpine *M. lewisii*, were able to grow equally well in both low (25 W m^{-2}) and high (120 W m^{-2}) irradiance (Hiesey *et al.*, 1971).

C. Photosynthesis at High Irradiance

High irradiance is a relative term. Shade plants suffer reversible damage when grown in full sunshine (Björkman and Holmgren, 1963): shade-adapted plants of *Solidago virgaurea*, grown for one week at a high irradiance, had a much poorer response to increase in flux density than a control plant, but after a week at low irradiance this damage was repaired. The cause of the damage lies in the inability of the chloroplasts to dissipate the excitation energy of the absorbed radiation (Foyer and Hall 1980). Similar effects have been shown for seedlings of *Quercus petraea* (Jarvis, 1964): in sun leaves, photosynthesis at 40 W m^{-2} was reduced by 12% by a prior exposure for 2 h to 400 W m^{-2}, whereas the corresponding reduction for shade leaves was 45%.

In 1965 Hatch and Slack demonstrated the existence of a new photosynthetic pathway in sugar cane, *Saccharum officinarum*, known as the C_4 pathway, and initially thought to be quite distinct from the C_3 Calvin pathway. The C_4 pathway uses the enzyme PEP carboxylase for the primary fixation of CO_2, and the fixed carbon travels initially through various 4–carbon dicarboxylic acids—oxaloacetate, malate, and aspartate—but the CO_2 is eventually released and re-fixed by the universal carboxylating enzyme—RuBP carboxylase. It is best thought of therefore as a CO_2 concentrating mechanism. Plants possessing this pathway are usually anatomically distinct as well, having "Krantz" anatomy, with a well-marked bundle sheath surrounding the vascular bundles and the chlorolplasts concentrated in a ring of mesophyll cells radiating out from the sheath (Fig. 2.12), and in the sheath itself. There are also ultrastructural differences in the chloroplasts. The major distinctions between C_3 and C_4 plants are listed in Table 2.7.

1. C₄ and CAM
Carbon dioxide fixation by PEP carboxylase is not in itself novel; it is well known as part of crassulacean acid metabolism (CAM), characteristic of desert succulents (Ting, 1985). In CAM, however, fixation occurs at night and CO_2 is released again in the daytime for photosyn-

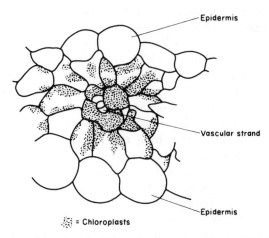

= Chloroplasts

FIG. 2.12. Cross-section of a single bundle-trace in a leaf of *Digitaria sanguinalis*, showing Krantz anatomy. Chloroplasts are stippled (redrawn from Black, 1971; × 340).

TABLE 2.7   Characteristics of C_3 and C_4 photosynthesis.

Characteristic	C_3	C_4
Initial CO_2-fixing enzyme	RuBP carboxylase	PEP carboxylase
Location of RuBP carboxylase		
Operating internal CO_2 concentration $\mu l\ l^{-1}$	220–260	100–150
Effect of O_2	Inhibitory (photorespiration)	None in range 2–21 kPa
Temperature response, 20–40°C at 330 μl CO_2 l^{-1}	Usually slight	Strong
Water use efficiency	Low	High

thesis; the malic acid produced by fixation is probably transported into the vacuole—concentrations achieved would depress cytoplasmic pH to below 3—and then shipped back and decarboxylated in daylight (Lüttge and Smith, 1982). In C_4 plants, however, the PEP carboxylase is active at the same time as RuBP carboxylase. The exact physical movements of carbon in the C_4 system are unclear, for it is both biochemically and anatomically complex, but the result is a physiology with very high *PAR* saturation, high temperature optima, and the ability to reduce mesophyll space CO_2 concentrations to very low levels, so increasing CO_2 diffusion rates and alleviating one of the limiting factors in photosynthesis at high photon flux densities. C_3 plants cannot lower internal CO_2 concentrations further because of the oxygenase capacity of RuBP carboxylase, which results in the phenomenon known as photorespiration—effectively reverse photosynthesis. In C_4 plants PEP carboxylase acts as a CO_2 scavenging system, and as a result photosynthetic rate is independent of O_2 concentration, in contrast to C_3 plants in which the rate increases markedly with declining O_2 concentrations. It has been suggested that the principal function of the C_4 pathway is the re-assimilation of CO_2 produced by photorespiration (Nasyrov, 1978).

C_4 plants are most frequent in hot, dry sunny environments and are largely confined to low latitudes, but even there they are rarely more abundant than C_3 plants (see pp. 171 and 172). A few temperate plants are apparently C_4 species, such as *Spartina anglica* (Mallot *et al.*, 1975) and *Euphorbia peplis* (Webster *et al.*, 1975); both are coastal species where irradiance is high and salinity may affect water relations. Other temperate species, such as *Jovibarba sobolifera*, *Sedum acre* and several *Sempervivum* species have CAM (Osmond *et al.*, 1975); all are succulent. CAM differs from C_4 photosynthesis principally in the temporal separation of PEP carboxylase and RuBP carboxylase fixation steps in CAM. The morphological consequences include succulence, since a large vacuolar volume is required to store the malic acid produced at night. In strict CAM plants, stomata are closed in daytime and open at night, so minimizing water loss: again succulence is a necessary feature since evaporative cooling cannot occur and the plant relies on the high specific heat of the stored water acting as a heat sink.

In extreme situations, where water supply is so limited that all gas exchange ceases, some CAM plants continue to display acidity fluctuations apparently because they re-utilize CO_2 produced by respiration. This behaviour has been termed CAM-idling (Rayder and Ting, 1983) and is distinct from CAM-cycling (Ting, 1985), where organic acid levels fluctuate although normal daytime C_3 fixation continues, as in many bromeliads. This is possibly a "tick-over" mechanism, that allows facultative CAM plants to respond rapidly to environmental changes.

Since C_4 photosynthesis was initially discovered in and is largely confined to tropical plants, it was at first thought to be a distinct evolutionary line. Björkman and his co-workers have demonstrated that both C_3 and C_4 species occur in *Atriplex* (Björkman *et al.*, 1971), and the same is true in *Euphorbia* (Webster *et al.*, 1975). Two species of

Atriplex, A. patula (C_3) and *A. rosea* (C_4), form fertile hybrids, which are intermediate in all respects (leaf anatomy, enzyme activities, photosynthetic characteristics) between the parents. PEP carboxylase activity was closer in the hybrids to that of *A. patula*, and RuBP carboxylase activity was the same in *A. patula*, the F1 and the F2; *A. rosea* had significantly less (Björkman *et al.*, 1971; Table 2.8). Strikingly, there was little correlation in the F2 between the anatomical characteristics of C_4 plants and PEP carboxylase activity, suggesting the genetic links between the various C_4 characters were easily broken. The hybrids showed poorer CO_2 assimilation rates at all irradiances than either parent and at light saturation were operating at about 50% of the rate of *A. patula* (Table 2.8). Lacking photorespiration, *A. rosea* was not stimulated photosynthetically by low oxygen, in contrast to *A. patula*; in this respect the hybrids were all characteristic of C_3 plants.

TABLE 2.8   Enzyme activity and photosynthetic efficiency in *Atriplex rosea* (C_4), *A. patula* (C_3), and their hybrids (Björkman *et al.*, 1971).

	A. rosea	A. patula	*F1*	*F2*
PEP carboxylase activity μmol CO_2 m^{-2} s^{-1}	111	6	24	23
RuBP carboxylase activity μmol CO_2 m^{-2} s^{-1}	18	44	45	45
Light-saturated photosynthetic rate 27°C; 300 μl l^{-1} CO_2; 21% O_2	13·8	7·4	4·1	4·3

All this evidence points to C_3 and C_4 photosynthesis being closely related in an evolutionary sense, and it is likely that the divergence has occurred frequently in response to the selective pressure of hot, dry environments. Both C_3 and C_4 plants possess the enzymes of the other pathway (cf. Table 2.8) although the PEP carboxylase involved in C_4 photosynthesis is in a distinct molecular form, and in C_4 plants both are simultaneously active—as was apparently true also for Björkman's F1 *Atriplex* hybrids. The classification of a plant as C_3, C_4 or CAM therefore depends on the relative activity of the two carboxylating enzymes, and so, possibly, on environmental conditions. Epiphytes, for example, even in wet forests, may suffer water deficits because they have very limited access to water supplies. Of 27 epiphytic bromeliads examined for CAM by Medina and Troughton (1974), 13 showed no dark CO_2 fixation and discriminated strongly against the stable isotope ^{13}C (a characteristic of initial RuBP carboxylase fixation); these were considered to be C_3 plants. A further 13 exhibited dark fixation and had low discrimination, and so were classified as C_4 or CAM plants. One species, *Guzmania monostachya*, had pronounced dark fixation (accumulating 27·5 μmol malate g^{-1} in 12 h in the dark) but strong discrimination, implying that both enzyme systems were operating. CAM plants are often found in dry habitats where dark fixation is an advantage, as stomata can be closed in daytime and so conserve water; *Guzmania monostachya* is an epiphyte in sunny, humid habitats, where possibly both pathways can be valuable, but bromeliads in general have very flexible photosynthetic systems, many having CAM-cycling.

As more and more carbon isotope discrimination analyses are carried out, it is increasingly clear that many species show the ability to switch from one pathway to the other, at least in part (see for example, Troughton *et al.*, 1977), and in one species, *Frerea indica*, Lange and Zuber (1977) have shown that the perennial succulent stems have CAM and the seasonal leaves, only produced in the wet season, have C_3 photosynthesis. Even more striking evidence of the adaptability of these various CO_2 fixation pathways is the inducibility of CAM metabolism in the succulent *Mesembryanthemum crystallinum* (Winter, 1974). Within 7 to 14 days of the onset of water stress the activity of PEP carboxylase rises and dark fixation starts, which can be shown to be due to *de novo* synthesis of the enzyme (Von Willert *et al.*, 1976; Queiroz, 1977), and there is a general trend from C_3 to CAM as the season progresses and water becomes scarcer (Fig. 2.13; Winter *et al.*, 1978). In fact many CAM species, particularly those in the Crassulaceae (e.g. stonecrops), and including the only commercially significant CAM plant, pineapple, which is a bromeliad, show a mixture of C_3 photosynthesis and CAM, largely controlled by their water status. It would be wrong, however, to see CAM solely in terms of water stress. The important feature is the flexibility of time of CO_2 fixation it confers. The aquatic pteridophyte *Isoetes* has CAM (Keeley, 1983) and clearly does not suffer water deficits. The postulated explanation is that it is unable to compete for CO_2 (as dissolved HCO_3^-) with other aquatic plants in the daytime, but can utilize HCO_3^- at night by a CAM mechanism.

2. *C_4 and C_3 Photosynthesis Compared*
As soon as the intrinsically greater potential photosynthesis of C_4 plants

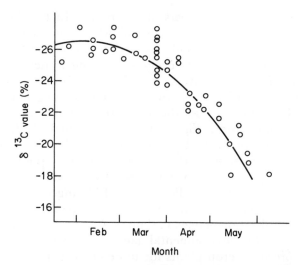

FIG. 2.13. Induction of CAM in *Mesembryanthemum crystallinum* growing in the wild on rocky cliffs by the Mediterranean in Israel. The declining $\delta^{13}C$ value shows decreasing discrimination against the heavy carbon isotope in CO_2, which indicates that RuBP carboxylase was no longer the primary CO_2-fixing enzyme. Redrawn from Winter *et al.* (1978).

was recognized, much interest centred on the implications for plant productivity, for example by breeding C_4 characters such as lack of photorespiration into C_3 crop plants, such as wheat. The ecological implications of an apparent division of plants into classes of high and low productivity also excited interest (Black, 1971). The basis for these comparisons was, however, often inappropriate and it is now known that many C_3 plants have a high photosynthetic capacity and, under favourable conditions, can achieve photosynthetic rates comparable to those of C_4 plants (Pearcy and Ehleringer, 1984).

The C_4 pathway represents a distinctive adaptation. Under conditions of high irradiance, high temperature, and long growing seasons, and particularly where water deficits play a greater role in determining plant distribution, PEP carboxylase is commonly the primary CO_2-fixing enzyme in grasses but less so in dicots. Even in environments of this type, however, C_3 plants are usually more frequent, and at high latitudes C_4 plants are rare (Fig. 4.11).

The large differences in photosynthetic rates often reported have as much to do with the environmental conditions as with photosynthetic capacity. Hot deserts are one of the classic C_4 environments, and yet they also contain large numbers of C_3 species. In the flora of short lived, ephemeral species that develop following sporadic rainfall in the Sonoran desert, there are both C_3 and C_4 species. C_3 ephemerals predominate in the flora that develops following winter rainfall when daily maximum air temperatures are 15–30°C. In summer these temperatures increase to 35–45°C, and the corresponding flora is almost exclusively C_4. Nevertheless, the two groups have similar P_{max} at appropriate temperatures (Fig. 2.14). The perennials *Atriplex hymenelytra* (C_4) and *Larrea divaricata* (C_3), which grow in the same habitat, also give no evidence to support the contention that C_4 photosynthesis is inherently more productive, even in this extreme habitat, for they have almost identical P_{max} values (Pearcy and Ehleringer, 1984). Since in less extreme environments, photosynthesis is probably as much limited by the rate of utilization of fixed carbon in the rest of the plant (i.e. it is sink-limited) as by environmental factors, it is probably futile to attempt to breed C_3 crop plants for more efficient photosynthesis. It may be significant that leaves of C_3 plants are more palatable than those of C_4 plants to grasshoppers (Caswell *et al.*, 1973) and to sucking insects (Kroh and Beaver, 1978), since very productive plants are normally considered most palatable to insects (Harper, 1969). Possibly herbivore pressure in the tropics has been so great that extra photosynthate in C_4 plants is diverted into chemical defence (see Chapter

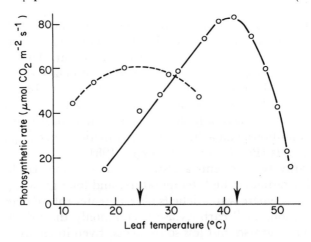

Fig. 2.14. The relationship of photosynthetic rate to leaf temperature in two desert ephemerals, the cool-season C_3 *Camissonia claviformis* (– – – –) and the hot-season C_4 *Amaranthus palmeri* (——). The temperature optimum for *A. palmeri* is almost 20°C higher, but the maximum photosynthetic rates are very similar, and at 25°C, roughly the temperature optimum for *C. claviformis* (P_{max} about 60 μmol CO_2 m^{-2} s^{-1}), the photosynthetic rate of *A. palmeri* is only about 30 μmol CO_2 m^{-2} s^{-1}.

8).The critical variables however are probably water use efficiency and temperature. The balance between the carboxylation and oxygenation activities of RuBP carboxylase is temperature sensitive, the latter being progressively more important as temperature rises. As a result the quantum yield of C_3 photosynthesis (that is the number of moles of CO_2 fixed per Einstein (mole of photons) absorbed) declines by about 35% over the range 15–40°C (Ehleringer and Werk, 1986). In contrast the quantum yield of C_4 photosynthesis is more or less constant, because the transfer of CO_2 from PEP carboxylase means that RuBP carboxylase is operating at a high enough CO_2 concentration to inhibit the oxygenase reaction. The quantum yield of C_3 photosynthesis is greater than that of C_4 photosynthesis at temperatures below about 30°C and less above that. As much as anything else, this may explain the general distribution of C_3 and C_4 plants, although sensitivity to low night-time temperatures is also considered important. Several studies (Ehleringer, 1978; Vogel et al., 1986) have shown distribution patterns of C_4 plants which correspond to temperature or rainfall gradients (see Fig. 4.11), but it is not possible to draw firm conclusions about causation from such correlative data.

3. Mineral Nutrients

I. INTRODUCTION

All green plants require the same basic set of mineral nutrients and the various elements are used by different plants for essentially similar ends. One or two groups have specialized requirements, such as that for cobalt by legumes co-existing symbiotically with nitrogen-fixing *Rhizobium* bacteria, or for sodium by plants utilizing the C_4 photosynthetic pathway (Brownell and Crosland, 1972), and by some salt-marsh and salt-desert plants. Species differ, however, in both their absolute ion concentrations and the relative proportions of different ions they contain. Some species are typically found on nitrogen-rich soils, for example *Chenopodium rubrum* which is abundant on manure heaps, and others, such as *Urtica dioica*, on phosphate-rich soils. The same is true for other nutrients.

All plants possess uptake mechanisms capable of moving ions across their cell membranes, chiefly protons and hydroxyl ions, and ions of nitrate and ammonium, phosphate, potassium, calcium, sulphate, magnesium, iron, manganese, copper, boron, zinc, and molybdenum. In addition an enormous range of other elements, ranging from the abundant (aluminium, sodium, chlorine) to the most obscure (zirconium, titanium, and similar elements), are accumulated by plants. Sometimes this accumulation is characteristic of a particular group: selenium is accumulated by some *Astragalus* species and silicon may be a major component of grass stems. As far as is known these elements have no metabolic function, though they may be ecologically important in providing support (silicon) or protection from grazing animals (silicon and selenium) for example.

In a range of common lowland grassland plants, concentrations of the macronutrients N, P and K may vary by a factor of 5 or more (Table 3.1), and greater differences can be found if more disparate species are compared. Herbs tend to have high potassium contents and legumes high nitrogen, the latter an obvious result of their symbiotic

TABLE 3.1   Nutrient concentrations (mg g^{-1} dry leaf tissue) of some common British herbaceous species (from various sources).

Species	Habitat	N	P	K	Ca
Juncus squarrosus	Acid heath	7·9	0·9	15·0	0·6
Nardus stricta	Acid heath	10·2	1·0	7·9	1·3
Festuca ovina	Acid grassland	14·0	1·5	14·3	2·0
	Calcareous grassland	13·6	1·2	13·6	3·3
Dactylis glomerata	Neutral grassland	14·7	1·8	19·1	3·7
Centaurea jacea	Neutral grassland	18·7	2·0	23·9	11·5
Vicia sepium	Neutral grassland	36·6	2·6	17·5	8·8
Urtica dioica	Disturbed ground	39·0	4·4	27·0	71·0
Range (highest/lowest)		4·9	4·9	3·4	118

nitrogen-fixers. For Ca the range is greater, reflecting the enormous differences in soil calcium between soils formed over calcareous (chalk, limestone) and non-calcareous substrata.

Soil calcium content is one of the major determinants of soil pH, since Ca^{2+} ions occupy the exchange sites on the soil minerals and act as a buffer system, and pH is intimately involved in the availability of many nutrients. At low pH microbial activity is inhibited, so that decomposition and the release of ions from organic matter are slowed; phosphate reacts with aluminium hydroxides, which are highly active below pH 4·5, and may become unavailable; iron is present as ferric ions; molybdenum is unavailable, and so on. Many of these effects, particularly where Al^{3+} and Mn^{2+} are involved, are of toxicity rather than nutrient availability and are considered in Chapter 6, but even from its effects on the major nutrients it can be seen why pH has such a powerful influence on community composition in the field. It has long been known that limestone added to acid soils has beneficial effects on crop yield—Pliny noted that the Belgae used chalk on their fields. Where limestone is added to natural vegetation, as observed by Hope-Simpson (1938) on a partly-limed field on acid greensand, marked changes in species composition may ensue; in this case, after about 50 years and with a pH change from under 5 to over 7, 15 of the original species had been replaced by 41 new invaders.

Similarly addition of nutrients to soil can transform vegetation. The finest example of this is on the Park Grass plots at Rothamsted Experimental Station, a field that was divided up in 1856 into plots which have received, in most cases, the same controlled nutrient application ever since. On an unfertilized plot from which a hay crop has been removed each year, representing a substantial loss of nutrients, there are now about 40 species (including many herbs) present producing less than 1500 kg of hay per hectare per year. By contrast a plot receiving a complete fertilizer and limestone dressing annually has a yield about four times as great, but with only 10–15 species, two or three of which are clearly dominant. Other treatments produce even more extreme results: the grass *Holcus lanatus*, for example, comprises almost 100% of the biomass on unlimed plots that have received heavy doses of ammonium, P, K, and Mg, producing a nutrient-rich but very acid (pH 4·0) soil. A similar treatment omitting P, K, and Mg, however, permits *Agrostis tenuis (capillaris)* to dominate, and if the soil is limed, *Alopecurus pratensis* is the major species. Such experiments provide valuable ecological information on the nutrient requirements of individual species, and Richards (1972) points out that a microspecies of dandelion, *Taraxacum pallescens*, thrives best on the Park Grass plots at a pH greater than 6·5 in soils of high potassium status.

Such effects can readily be demonstrated. Willis (1963) added a complete fertilizer (N, P, K, S, Mg, and micronutrients) to short dune turf and found that after two years the turf was completely dominated by *Festuca rubra*, with slow-growing plants such as *Thymus drucei* and short plants such as mosses being almost eliminated. Similarly, small dune annuals were crowded out when nitrogen was added to sand-dunes at Holkham in Norfolk by Boorman and Fuller (1982), and *Kobresia simpliciuscula*, a small, arctic-alpine sedge which survives, in a small relict outpost from its main distribution, in Upper Teesdale in northern England, was crowded out by *Festuca ovina* if phosphate was added to the soil (Jeffrey and Pigott, 1973).

Field experiments like these have the advantage of ensuring that the response measured is ecologically real, operating in natural vegetation, but suffer from the laxity of the connection between cause and effects that is their unavoidable characteristic. When phosphate is added to soil, for example, it might benefit a species in several ways:

(i) simple growth stimulation;
(ii) depression by toxicity of the growth of a competitor (Foote and Howell, 1964; Green and Warder, 1973);
(iii) precipitation of Al^{3+} (toxic) or Fe^{3+} (a nutrient) at the root surface (Brown, 1972; Wright, 1943);
(iv) stimulation of N-fixing activity of associated legumes;

quite apart from any effects it may have on the soil microflora (cf. p.

107). To make the link between addition and response, one needs also results under controlled conditions, in growth room or greenhouse. Such studies as have been made tend to confirm the simple response hypothesis, particularly those of Bradshaw and co-workers on the growth of several grasses at various levels of Ca, N, and P in solution (Bradshaw *et al.*, 1960a, b, 1964), though species vary greatly in their ability to respond, and growth depressions also certainly occur (Fig. 3.1).

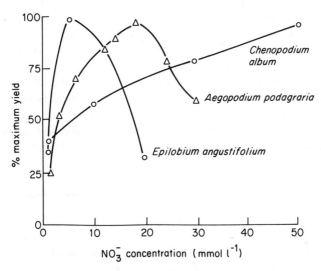

FIG. 3.1. Response of three contrasting plant species to nitrate concentration in solution culture (redrawn from Kinzel, H. (1982). "*Pflanzenökologie und Mineralstoffwechsel.*" Ulmer, Stuttgart).

Often differences in response to nutrients accord with the habitat preferences of the species. For example, Clarkson (1967) studied the responses of three species of bent grass *Agrostis* to phosphate. At low rates of phosphate supply (33 µg $H_2PO_4^-$/plant/week) *A. setacea*, a plant of acid, nutrient-poor heaths, maintained exponential growth for 10 weeks, whereas both *A. canina* and *A. stolonifera*, though initially growing faster, were slowed by phosphorus deficiency after seven weeks. The optimum rate of supply for both *A. setacea* and *A. stolonifera* was 66 µg/plant/week, when both contained about 0·3% phosphate in the shoots, so that the success of *A. setacea* at low P supply seems to lie in its ability to continue to grow, if very slowly, at low internal P concentrations.

It seems that some species utilize nutrients differently from others. Rorison (1968) found that four species from very different habitats had appropriate differences in their growth response to phosphate— *Urtica dioica* giving an almost linear response, *Deschampsia flexuosa* almost none (Fig. 3.2). Two species, *Rumex acetosa* and *Scabiosa columbaria*, were intermediate and similar, though they come from very different habitats and *Rumex* has a much faster growth rate. It may be relevant that *Scabiosa* stores much of its phosphate as inorganic P (up to 80%) in the root (Nassery, 1971). Similar uptake in excess of requirements, perhaps allowing storage against times of shortage, occurs in *Banksia ornata*, growing on extremely P-deficient soils in Australia (Jeffrey, 1964).

Such differences also occur between populations of many species. Snaydon and Bradshaw (1961, 1969) have shown for both *Festuca ovina* and *Trifolium repens* that the response of populations to a range of nutrients is related to the nutrient status of the soil on which the populations were collected, and even more strikingly, Davies and

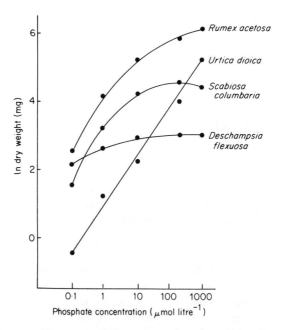

FIG. 3.2.   Response of four ecologically contrasted species to phosphate concentration in solution culture, after 6 weeks (from Rorison, 1968).

Snaydon (1973a, b, 1974) have found that plants of *Anthoxanthum odoratum* from various parts of the Park Grass plots, in some cases separated by only a few metres, show the same correlation of response to environment. These differences have arisen in some cases in less than 50 years, and have a genetic basis (Crossley and Bradshaw, 1968; Ferrari and Renosto, 1972); they are not just examples of plasticity in response to environment. That such differences in response to nutrients are hereditary should not be a surprise to anyone who considers the achievements of plant breeders in moulding the fertilizer responses of crop plants.

II. NUTRIENTS IN THE SOIL SYSTEM

A. Soil Diversity

Soils vary in their ability to supply nutrients, and the factors which give rise to this variation are those which determine the development of soils, viz: parent material, climate, topography, age, and vegetation. *Parent material* normally sets the upper limits on nutrient content of soil (except for N which can enter by fixation), so that soils formed on granite are base-poor, for example. Soils formed from limestones, on the other hand, initially are very base-rich—pure chalk is 100% $CaCO_3$—and typically lose Ca^{2+} ions through the action of rainfall (climate), leaching out carbonates with dissolved carbon dioxide (carbonic acid), a process dependent on *topography*. Such effects are clearly time-dependent, hence the *age* factor. Topography may reverse the usual situation by allowing accumulation of base-rich water: fen peats are wholly organic, and typically have no native inorganic calcium, but are base-rich from continual flushing by ground-water. *Climate*, of course, has more general weathering effects and in some areas nutrient inputs in rain can be important, particularly of Na and K in coastal areas (Etherington, 1967).

The effect of *vegetation* is more complex, involving ecological succession. Soil formation does not take place without vegetation so that these two time-sequences may be considered as essentially interlinked, but at any stage in either process the vegetation both reflects existing soil conditions and influences the future course of pedogenesis, thereby secondarily determining succession. Excellent instances of this can be seen in the work of Grubb and Suter (1971) on the acidification of soil by *Ulex europaeus* and in the effect of nitrogen-fixing alders (*Alnus crispa*) on glacial moraines in Alaska (Crocker and Major, 1955).

Any of the factors producing large-scale variation in soil nutrients can also be active on a small scale. Leaching can cause local acidification of hummocks in raised bog (Pearsall, 1938) and in fen (Godwin *et al.*, 1974), outcropping of calcareous parent material in otherwise acid turf allows calcicole species to invade, and in fact the details of the vegetation mosaic of most communities can be related to variation on the appropriate scale. The mechanisms are various but the following occur:

(i) litter fall: the litter of plant species varies greatly in nutrient content, with species of infertile soils generally returning least to the soil (Small, 1972). Most species withdraw the bulk of the N, P, and K from their leaves before abscission, but Ca concentrations normally build up (Fig. 3.3). Nutrients in

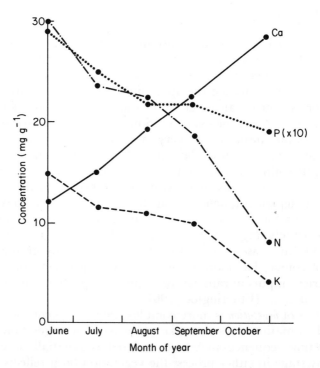

FIG. 3.3.   Mean mineral concentration of leaves of nine broad-leaved deciduous trees as a function of time of year. Note the large increase in Ca concentration. The scale for P has been expanded by a factor of 10. Data from Alway, F. J., Malli, T. E. and Methley, W. J. (1934). *Soil Survey Assoc. Bull.* **15**, 81.

organic matter cycled through herbivores may be returned to the soil, for example as urea, to give even more concentrated pockets of high fertility, as a glance at a cattle field will show.

(ii) acidification: many species produce acid litter which can have marked effects on gross soil pH. More subtle influences arise from the secretion by roots of protons, normally when cation exceeds anion uptake. This occurs, for example, when NH_4^+ rather than NO_3^- is the N source (Riley and Barber, 1971), and various species differ markedly in their preference for these two ions (Gigon and Rorison, 1972). Equally the oxidation of NH_4^+ to NO_3^- in soil can cause acidification.

(iii) depletion: plant uptake causes depletion of the soil, yet plant roots are relatively short-lived and so for immobile nutrients this will produce a mosaic of depleted, recharging, and unexploited soil zones.

(iv) fixation: some Cyanobacteria and bacteria such as *Azotobacter* can fix atmospheric nitrogen to NH_4^+ either independently or symbiotically with fungi (lichens) or higher plants. Where these organisms are active local areas of high NH_4^+ or NO_3^- concentration may occur. *Azotobacter* tends to be specifically associated with roots (Brown, 1975). In agricultural soils an analogous effect is produced by the addition of granular fertilizers.

These mechanisms will tend to produce a random patchwork of low and high nutrient concentrations, but the effects of vegetation and leaching normally interact to give a more ordered, and very widespread variation pattern—soil zonation. Except in soils where intense earthworm activity causes continuous mixing (for example brown earths), plant roots draw up minerals from deep layers which are then deposited on the surface as litter. Rainfall moves these down the profile again, but since the soil is not chemically inert, it reacts with some more than others, giving characteristic horizons. These may vary widely in pH, nutrient content, and organic matter, and provide distinct soil environments within a single soil area, with important consequences for species co-existence.

Soil formation is a time-dependent process and so long-term changes may be brought about by any of the factors discussed above, but of more importance to most plants are effects on a seasonal or shorter cycle. In natural, as opposed to agricultural, soils there is well-marked seasonal nutrient variation related to the amounts passing through the organic cycle. The main input occurs as leaf-fall, in the autumn in temperate climates; in moist tropical and sub-arctic evergreen forests this seasonality is less clear, and in the former may be related to rainfall. In the typical temperate cycle, however, the input coincides with the slowing down of microbial activity caused by falling temperatures, and so though there may be a small autumn peak, particularly of elements easily washed out of leaves, such as potassium, the bulk of the release occurs the following spring. This spring peak is followed by a summer fall, the result of root and microbial activity, although further differences may arise from the differential uptake activity of the roots of various species.

B. Concentrations

The concentrations of nutrients in solutions expressed from natural soils are typically very low (Table 3.2). Even in fertilized soils the values may be well below those that would saturate ion uptake systems. Nevertheless, plants will grow well in solution culture at these or sometimes much lower concentrations. Asher and Loneragan (1967) grew several species in water culture and found that some, such as *Erodium botrys*, *Bromus rigidus*, and *Trifolium subterraneum*, showed growth reponses up to 5 μM P, while one, *Vulpia myuros*, reached a peak at 1 μM. For four species good growth was made at less than 1 μM P. It might, therefore, be a cause for surprise that conventional nutrient solution recipes recommend P levels around 1 mM, apparently three orders of magnitude more than required. The difference is that these are for traditional static culture, whereas Asher and Loneragan used a sophisticated flowing culture system, so that the supply of 1 μM was maintained. This concept of supply is fundamental and is covered below briefly, and in more detail by Nye and Tinker (1977).

TABLE 3.2. Mean values and range of soil solution concentrations for seven ions (mM).

	Ca^{2+}	Mg^{2+}	K^+	Ion NO_3^-	NH_4^+	$H_2PO_4^-$	SO_4^{2-}
n	9	7	9	4	3	6	5
Mean	8·9	3·7	1·7	9·1	5·7	0·02	1·6
Min	1·7	0·3	0·1	1·0	5·0	0·00	0·3
Max	19·6	10·3	6·8	27·6	6·1	0·09	3·5

Most available data are from agricultural soils. Uncultivated soils will have lower values but few data are available. These data are from ADAMS, F. (1974). The soil solution. *In* "The Plant Root and Its Environment" (Ed. E. W. Carson). University Press of Virginia, Charlottesville.

Ion concentrations vary greatly among soils and spatially within them, depending on factors involved in pedogenesis (parent material, climate, topography, vegetation, and the age of soil), but of more importance to the individual plant is variation in time. The process of uptake in static solution culture reduces concentration (a fact, incidentally, that often complicates the interpretation of physiological experiments), and in soil the same will occur, its extent depending upon how well the soil is buffered against depletion.

Soil buffering powers vary greatly for different ions, depending on the extent to which they are adsorbed by soil. Strongly sorbed ions such as phosphate are powerfully buffered, since the bulk of the ion is in an insoluble state and acts as a reservoir. By contrast, nitrates are very soluble and so are not stored in soil to any extent; the reservoir for nitrate in soil comprises N in organic matter and adsorbed NH_4^+, both of which require microbial conversion to NO_3^-, and indirectly atmospheric N_2 which is fixed by various bacteria and cyanobacteria, to give NH_4^+.

C. Ion Exchange

The bulk of most ions in soil occurs in the solid phase, either as a component of soil minerals, or as part of the soil organic matter. Because of their reactivity, the most important inorganic components of typical soils are clay minerals, which are predominantly aluminosilicates. They consist of alumino-silicate sheets, bound together by numerous cations such as Al^{3+}, Mg^{2+}, K^+ and Fe^{3+} (Fig. 3.4). Where they fracture they expose surfaces which have a fixed negative charge. This charge is balanced by cations which are therefore held exchangeably on such surfaces, giving rise to the cation exchange capacity of the soil. In practice, diffusive forces cause cations held adjacent to the fixed negative charges to migrate away from them, so that the cation layer around such a surface is more diffuse than would otherwise be the case, giving a distribution known as a Gouy double layer (White, 1979).

In soils of pH greater than 5, the principal cations found in this layer are Ca^{2+}, Mg^{2+}, K^+ and Na^+, and the concentrations of these ions in solution are largely controlled by equilibria set up between surfaces and solution. If these cations are lost from soil, for example by leaching, the unsatisfied fixed negative charge is increasingly counteracted by protons or by Al^{3+} ions derived from the weathering of the clay minerals. Such soils suffer progressive acidification. One of the main problems caused by acid deposition from power stations and other solid fuel combustion sources, is this leaching of basic cations,

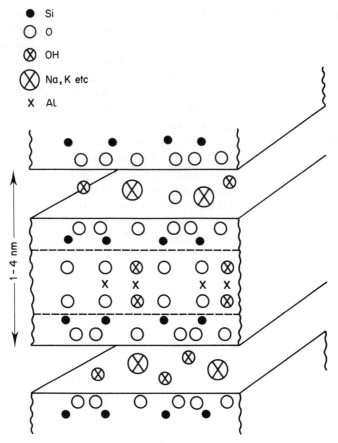

Fig. 3.4. Diagrammatic representation of the structure of a typical clay mineral, montmorillonite, to show the position of the exchangeable ions.

and their consequent replacement on the exchange sites by Al^{3+}. This leads to a rapid decline in soil fertility.

The supply of most cations and anions depends largely on inorganic equilibria, best illustrated by phosphorus. A typical soil may contain anything from 0·05 to 1·0 g P per kg of soil, but the concentration in soil solution is likely to be between 0·1 and 10 μM, equivalent to less than 60 μg P per kg of a moist soil, or about 0·005% of the total amount. The remaining 99·995% is in the solid phase, in a bewildering variety of compounds, most of which are not simply characterizable chemically. Solid phase phosphate is best classified according to its relationship with the solution phase, producing a labile pool in equilibrium with the solution, and a non-labile pool.

The non-labile pool is usually much the largest (Larsen, 1964), but the distinction is empirical. The labile pool is that part that will exchange with phosphate in solution, determined experimentally with $H_2{}^{32}PO_4^-$. Its size depends on the time allowed for equilibriation, as over long periods less labile ions will become exchangeable. Solid phase phosphate does not therefore exist as two discrete phases, but in a continuous spectrum of lability, depending on the nature of the binding to calcium, aluminium, ferric, and also organic compounds in the soil.

The value of this model is that it enables rates of supply to be quantified. A soil solution at 10^{-6} M phosphate represents about 10 g phosphate in solution per hectare; assuming crop uptake of 20 kg ha^{-1} and a growing season of 2000 hours, this solution would be wholly depleted in 1 hour, and long before that uptake would have been greatly reduced. Necessarily labile phosphate must come into solution to maintain the supply, and so the labile pool–solution equilibrium is of crucial importance. This equilibrium can be measured by taking advantage of the fact that if phosphate is added to soil in solution , the bulk is immediately adsorbed. By varying the amount added an adsorption isotherm can be constructed, relating the amount adsorbed to the concentration in solution after a given time, and indicating the equilibrium between labile and solution P. Equally, as phosphate is removed from solution by plant uptake or artificially, ions come into solution from the labile phase by desorption, and an analogous desorption isotherm exists. In practice such isotherms exhibit hysteresis and the desorption isotherm, which controls the supply of phosphate to absorbing roots , cannot be predicted from the adsorption isotherm.

Figure 3.5 shows three adsorption isotherms for contrasting soils, one steep, one gradual, and one almost flat. For the clay soil with a steep isotherm, large changes in the quantity of adsorbed phosphate are needed to bring about a small change in solution concentration; for the sandy soil the reverse is true. For a given amount of sorbed phosphate the clay soil will therefore show a lower concentration in solution, but will maintain this concentration better against depletion; the clay soil is therefore more strongly buffered than the sandy soil. The buffer power of a soil for an ion represents its ability to maintain solution concentrations in response to the removal of the ion from solution.

Buffer power measurements can be used to predict fertilizer additions required to maintain plant growth (Ozanne and Shaw, 1967;

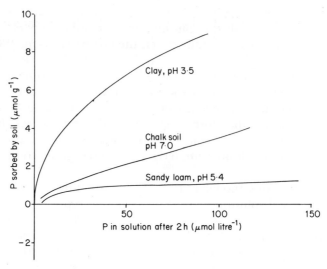

Fig. 3.5. Adsorption isotherms for phosphate of three contrasting soils (Fitter, unpublished).

Fitter and Bradshaw, 1974; Holford, 1976), but increasing the labile pool by fertilization will also alter the equilibrium between labile and non-labile pools (Fig. 3.6), and some labile will be immobilized as non-labile (Larsen, 1967). This description for phosphate applies equally to other adsorbed ions such as potassium (Beckett, 1964; Fergus *et al.*, 1972), molybdenum (Barrow, 1973), manganese (Lamm *et al.*, 1969), sulphate (Fox *et al.*, 1971), and cadmium (John, 1972).

The supply of adsorbed ions in soil tends therefore to be less variable seasonally than that of ions for which organic matter is a primary source, in particular nitrate. The dynamic balance of the inorganic

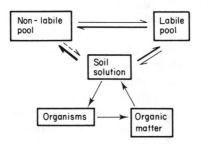

Fig. 3.6. Model of compartmentation of adsorbed ions in soil. The labile pool is the proportion of non-dissolved ion that is readily exchangeable, in the case of phosphate for ^{32}P. The thickness of the arrows indicates the balance of the equilibrium.

adsorption system provides a buffer against the fluctuations exhibited by nitrate in soil. Nevertheless, phosphate is supplied from organic matter as well, and in some soils quite marked seasonal variation can occur (Gupta and Rorison, 1975).

D. Cycles

The supply of ions from soil can be viewed in terms of mineral cycles, with inputs to and losses from the ecosystem and rates of transfer between components of the system. This approach is valuable for nitrogen, where inputs due to rainfall and fixation and losses due to leaching and denitrification represent a large proportion of the total amount cycling through the system. For adsorbed ions, however, these inputs are usually trivial in comparison with the total amounts present, although leaching losses in fertilized soil may be important for some.

The nitrogen cycle is complex (Fig. 3.7) and the organic matter compartment is dominant. Several bacteria are involved in the conversion of ammonium to nitrate (*Nitrobacter, Nitrosomonas,* and *Nitrococcus* are the most important), but both forms can in fact be taken up by

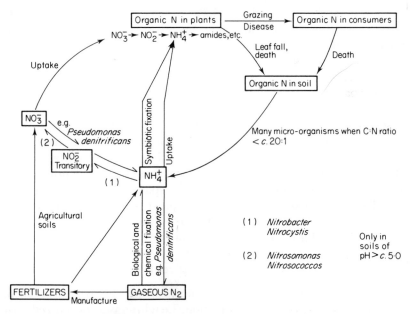

FIG. 3.7. Simplified scheme of nitrogen cycle, showing major sources and pools, and primary processes involved.

most plants with equal facility. More significant is NH_4^+ production from organic matter, brought about by various fungi and bacteria. These decomposers also require N, however, and if the material is low in N, it will be incorporated into their biomass and not liberated until carbon supply is reduced. There appears to be a critical value of between 1·2 and 1·8% (corresponding to C:N ratios of between 30:1 and 20:1) N in litter, below which little or no NH_4^+ is released. Since most plants conserve N by recovering it before leaf-fall, and in extreme cases may short-circuit the N cycle by retaining almost all in the perennating organs (Small, 1972), freshly deposited leaves tend to have a high C:N ratio (up to or exceeding 100:1), and the more slowly N is made available to plants, the higher will be the C:N ratio of the litter, which accentuates the infertility.

Nitrogen is not the only element which cycles through the organic matter. Freshly fallen leaves contain a wide range of elements, particularly calcium, since it is immobile within the plant and so not withdrawn from senescing leaves, unlike most other nutrient minerals. Some, such as potassium, tend to wash straight out into the soil and so enter the inorganic cycle; others, such as phosphorus, are mainly in organic compounds in the leaf, but there is little evidence that organic P is of great significance in soils, partly because to become available to plants it must be released as $H_2PO_4^-$ ions, whereupon it will be promptly precipitated or adsorbed by iron, aluminium, or calcium surfaces in soil.

E. Transport

The supply parameters considered above adequately describe nutrient availability at a fixed point in soil, but absorption by a root necessarily involves movement of the ion from the soil to the root surface. Since roots are scattered in soil and even in a densely-rooted horizon are unlikely to occupy more than 10% of the space, with 1% being more usual (Dittmer, 1940), such movement must occur over considerable distances if the plant is to exploit more than a fraction of the soil volume. The mobility of ions in soil has now been comprehensively discussed by Nye and Tinker (1977). In practice there are two ways in which an ion can move towards a plant root:

   (i) by mass flow or convective flow down the water potential gradient caused by transpiration. Leaching is a specialized form of mass flow in which the water flow is gravitational, down the soil profile. It is of great importance in agriculture but is little influenced by roots and will not be discussed here.

(ii) by diffusion down concentration gradients created by the uptake of ions at the root surface.

The speed at which ions can move by mass flow is vastly greater than by the very slow process of diffusion, but in practice it is only of importance for some ions. The movement of ions by mass flow can be represented as follows (Tinker, 1969):

$$F = V \times C_1$$

where F is the flux across the root surface in moles $cm^{-2} s^{-1}$, V is the water flux in g $cm^{-2} s^{-1}$ ($= cm^3$ $cm^{-2} s^{-1}$ or cm s^{-1}), C is the concentration of ions in moles cm^{-3}, and the suffix 1 represents liquid phase. When a single root grows into a volume of soil, the entire soil volume can in theory be exploited by mass flow; the adequacy of the supply, however, depends entirely on whether the product of water flux and bulk soil solution concentration satisfies the plants requirements.

At the root surface uptake lowers the concentration; if the uptake rate is greater than the rate of supply by mass flow, either because V is inadequate, as for example at night or under daytime water stress when transpiration ceases, or because C_1 is too low, a concentration gradient will develop from the bulk soil down to the root. Diffusion will then occur down this concentration gradient. Conversely ions may arrive faster than they are taken up, so that concentrations build up at the root surface and diffusion will then operate in the opposite direction. This may happen for calcium (Barber, 1974). The full equation for the flow of nutrient ions through soil to a plant root must therefore include both mass flow and diffusive components:

$$F = VC_1 + (-D dC/dx)$$

where D is the diffusion coefficient in soil and dC/dx the concentration gradient.

Several predictive models have been constructed to examine the effects of mass flow and diffusion in the supply of ions to plant roots (Passioura, 1963; Nye, 1966; Olsen and Kemper, 1968; Nye and Marriott, 1969; Baldwin, 1975; see Nye and Tinker (1977) for a full discussion). Typical results are illustrated in Fig. 3.8: as the water flux increases the zone of depletion around the root becomes a zone of accumulation. Where the line lies above the dotted line in Fig. 3.8, mass flow would be the sole source of supply, and this can therefore be

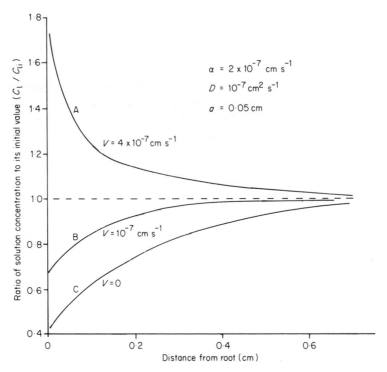

Fig. 3.8. Changes in ion concentrations around a root after about 12 days (10^6 s) at different mass flow rates, for given conditions. (After Nye and Marriott, 1969.)

viewed, hypothetically, as a relationship between ion flux and water flux, as in Fig. 3.9. Here the line through the origin shows what would occur with mass flow only (i.e. $F = VC_1$); the dotted line represents the actual situation for magnesium—at values of V less than about 2×10^{-6} cm s^{-1} ion flux across the root surface is greater than mass flow would allow, so that a concentration gradient downwards towards the root must exist and the diffusive contribution must be positive. Above this critical value of V, mass flow supplies ions faster than they are taken up and diffusion operates away from the root, lowering the fl ʌx into the root below the figure that could theoretically be supplied by mass flow (cf. A in Fig. 3.8).

Using these theoretical considerations and an increasing number of experimental confirmations, the relative importance of mass flow and diffusion for all important nutrient ions can be approximately stated. From data of available P, K, Ca, and Mg and known values of water use and nutrient uptake by a maize crop, Barber (1974) has estimated

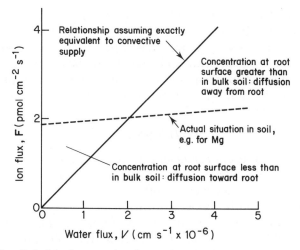

Fig. 3.9. Relations between ion flux and water flux (from Tinker, 1969).

the relative contributions of mass flow, and (by difference) diffusion (Table 3.3). For potassium this suggests that 11% was supplied by mass flow, which agrees well with data obtained experimentally by Tinker (1969) by growing leek seedlings in soil and measuring water and ion uptake, ion status of the soil, and the relevant plant parameters. He found that VC_1 (the mass flow contribution to flux) represented between 4 and 13% of the total flux, F, depending on harvest interval.

Mass flow is a rapid process if water flux and soil solution concentrations are great enough. By contrast, diffusion is a slow process in soil, measurable in mm per day. Where mass flow is insufficient to satisfy plant demand, therefore, ion concentrations at the root surface are

TABLE 3.3. Estimated contributions of root interception, mass flow, and diffusion during one season's ion uptake by a maize crop (Barber, 1974).

Ion	Uptake by crop ($kg\ ha^{-1}$)	$kg\ ha^{-1}$ supplied by	
		Mass flow	Diffusion
Ca^{2+}	45	90 (200)	negative (—)
Mg^{2+}	35	75 (214)	negative (—)
K^+	110	12 (11)	95 (89)
$H_2PO_4^-$	30	0·12 (0·4)	29 (>99)

Figures in parentheses are percentages of uptake for each ion.

reduced below that of the bulk soil solution, and marked zones of depletion occur. In such cases only a part of the soil volume is exploited, except for very mobile ions. The extent and development of these depletion zones are controlled partly by the plant, in creating the concentration gradient seen as a zone of depletion in autoradiographs, and partly by the soil, via the diffusion coefficient, D. This varies as follows (Nye, 1966):

$$D = D_1 \times \theta \times f \times 1/b$$

where D_1 is the diffusion coefficient in free solution,
   θ is the volumetric water content of the soil,
   f is an impedance factor, and
   b is the soil buffer power (dC/dC_1).

The drier and the more strongly buffered the soil, the lower the diffusion coefficient will be; clearly in a dry soil diffusion will be negligible and the wetter it is the nearer it approaches the free solution state. Impedance is a measure of the length of the diffusion path, which is affected by both moisture content and degree of compaction of the soil. The exact effect of these two interacting factors is complex (Fig. 3.10). The buffer power term accounts for sorption of ions by soil and explains why the diffusion coefficients of non-adsorbed ions are relatively little affected by soil. Typical values of D in soil are shown in

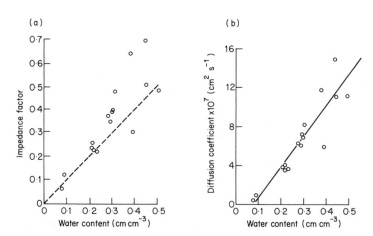

FIG. 3.10. (a) Relationship between water content of a silt-loam soil and the impedance factor (f) for potassium; the dashed line is the 1:1 relationship. (b) The relation between potassium diffusion coefficient in the same silt-loam soil and soil water content; the line is fitted by eye. Redrawn from Kuchenbuch, K., Claassen, N. and Jungk, A. (1986) *Plant and Soil* **95**, 221–231.

Table 3.4; the extremely low values for phosphate permit negligible movement and may severely limit supply. Even in poorly buffered soils phosphate diffusion is slow and pronounced, though narrow, depletion zones develop (Fig. 3.11).

An important consequence of this, then, is that for ions for which the soil solution concentration is great enough to permit considerable

TABLE 3.4. Range of measured values for the effective diffusion coefficient $(cm^2 s^{-1})$ of various ions in soil (summarized from several authors by Barber, 1974; Fried and Broeshart, 1967).

	Ion	Minimum	Maximum
Cations	Na^+	1×10^{-7}	1×10^{-5}
	K^+	2×10^{-7}	2×10^{-6}
	Rb^+	6×10^{-12}	7×10^{-6}
	NH_4^+	4×10^{-8}	1×10^{-6}
	Ca^{2+}	3×10^{-8}	3×10^{-7}
	Zn^{2+}	3×10^{-10}	2×10^{-7}
	Mn^{2+}	3×10^{-8}	2×10^{-7}
Anions	NO_3^-	5×10^{-7}	1×10^{-5}
	Cl^-	3×10^{-7}	1×10^{-5}
	$H_2PO_4^-$	1×10^{-14}	4×10^{-9}

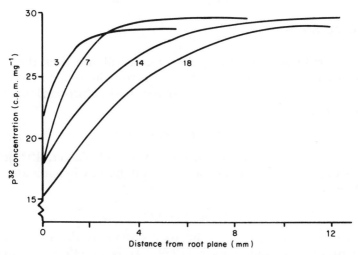

FIG. 3.11. Development of depletion zones for phosphate around an actively absorbing root (figures on curves represent days from start of the experiment). (From Bagshaw *et al.*, 1972).

movement by mass flow, little depletion occurs at the root surface, though the radius of the zone of influence is large; in contrast, ions supplied largely by diffusion typically have narrow (often less than 1 mm), but well-marked depletion zones. However, some ions with high diffusion coefficients, such as nitrate, are present at low concentrations in most soil solutions, and diffusive supply is important. Since the rate of diffusive supply is much greater for these ions than for ions such as phosphate, the depletion zones that develop will be wide and shallow, and this will mean that the whole soil volume will rapidly be exploited. In other words, the entire soil pool of nitrate is available to the plant, and where several roots are involved, competition will inevitably ensue. For phosphate, on the other hand, with an effective diffusion coefficient several orders of magnitude less than that for nitrate, the immediately available amount is that within a small radius of each root, and the chances of depletion zones overlapping and so reducing the concentration gradient from bulk soil is much less at any root density. The extent of competition for ions in soil therefore depends in large measure on the diffusion coefficient, and for phosphate it will only occur at unusually high root densities or in soils with very low buffering capacities, and hence higher diffusion coefficients. Figure 3.12 shows that ion uptake per unit amount of root is much more sensitive to root density for K^+ (D usually between 10^{-6} and 10^{-8}) than for $H_2PO_4^-$ (D usually 10^{-9} and 10^{-11}).

F. Limiting Steps

Ion transport in soil can be very slow, but it will only limit the uptake process where it is slower than the flux across the root surface. That flux (F), expressed on a surface area of root basis, is proportional to ion concentration at the root surface:

$$F = \alpha \, C_{lr} \; (\text{mol cm}^{-2} \, \text{s}^{-1})$$

where α is the root absorbing power (Nye and Tinker, 1977). The uptake can also be expressed on a root length basis, when it is known as an inflow (I):

$$I = 2\pi \overline{\alpha a} \, C_{lr} \; (\text{mol cm}^{-1} \, \text{s}^{-1})$$

The term $\overline{\alpha a}$, averaged over the root system, is the root demand coefficient of Nye and Tinker (1969). It is in effect a transport coefficient, in the same units ($\text{cm}^2 \, \text{s}^{-1}$) as a diffusion coefficient (D). Since the two processes—diffusion through soil and transport across the

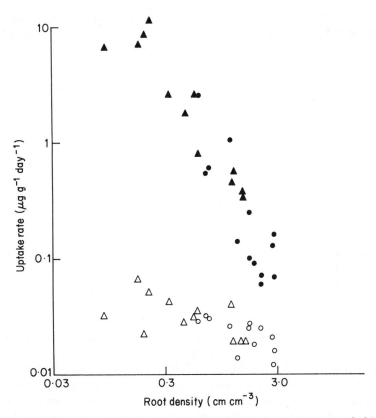

FIG. 3.12. Effect of root density on uptake of phosphorus (open symbols) and potassium (closed symbols) by *Lolium perenne* (circles) and *Agrostis tenuis* (triangles). (From Fitter, 1976).

root surface—are in series, the slower of the two will limit the rate of uptake. Values of $\overline{\alpha a}$ can be determined experimentally (Table 3.5) and are found to vary with internal and external ion concentrations, plant growth rate and root morphology, as predicted by Nye and Tinker (1969). Where $\overline{\alpha a}$ is less than D, transport across the root surface will be limiting, and this will often be the case for ions such as NO_3^- and Ca^{2+}. Conversely the values of $\overline{\alpha a}$ for K^+ in Table 3.5 are very much greater than typical values of D for K^+ (around 10^{-7} cm^{-2} s^{-1}: see Table 3.4), so that diffusion will normally be the rate-limiting step here.

TABLE 3.5.   Root demand coefficients $(\overline{\alpha\, a})$ for four species grown in flowing culture at different potassium concentrations over 42 days (Wild *et al.*, 1974).

		$\overline{\alpha\, a} \times 10^6 \ cm^2 \ s^{-1}$		
[K], μM	1	3	10	33
Species				
Dactylis glomerata	11·9	8·0[a]	3·2	1·0
Anthoxanthum odoratum	7·9[a]	4·5	1·1	0·4
Trifolium pratense	7·2	5·2[a]	2·1	0·6
Medicago lupulina	30·1	19·9[a]	6·9	2·3

[a]Concentration above which no further growth increases occurred.

III. PHYSIOLOGY OF ION UPTAKE

A. Kinetics

Plant root cells generally contain much higher ionic concentrations than their surrounding medium and are normally electrically negative with respect to it. This potential, of the order of -60 to -250 mV, is created by the activity of proton pumps, which transport H^+ across the membrane. These pumps are ATPases, and so the maintenance of the potential difference across the membrane involves energy expenditure. As a consequence most cations appear to be able to enter the cell passively, down the electrochemical gradient, although usually against the chemical concentration gradient for that ion; in some cases, certainly for Na^+, cations are actively exported back into the medium in exchange for protons. By contrast, anions must be transported actively into root cells in company with protons, but against the electrochemical gradient and in most cases the concentration gradient too. Whether active or passive, however, transport is energy-dependent and so necessarily influenced by a number of external and internal factors, in particular the concentration of the ionic substrate and variables such as temperature that affect respiration. When energy supply is adequate, the dominant influence on the rate of ion uptake is the concentration gradient.

Many experiments have been performed on ion uptake, some using excised roots, some whole plants, but almost all in non- equilibrium conditions, typically using roots previously grown in 0·2 or 0·5 mM

$CaSO_4$ and measuring uptake over very short periods. They have shown that there is a characteristic relationship between uptake rate and concentration which, because of its apparent similarity to that between the rate of an enzyme-mediated reaction and its substrate concentration, has been analysed in terms of Michaelis–Menten kinetics. The typical hyperbolic curve is shown in Fig. 3.13 together with two plots based on enzyme kinetics which produce linear plots. The enzyme analogy implies that there are molecules ("carriers") in the cell membrane, more or less specific for the ion, which can transport ions at a particular rate (v) under given conditions, so that at extremely high substrate concentrations all carriers are operating at maximum rate and saturation uptake rates (V_{max}) are reached. The Michaelis–Menten equation can then be applied:

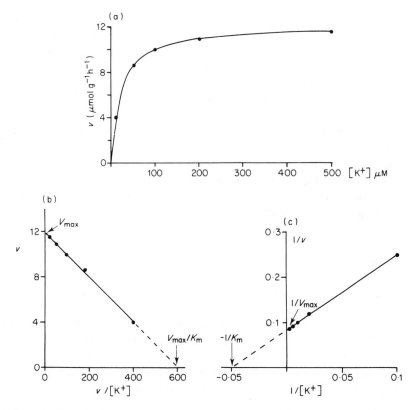

FIG. 3.13. Absorption isotherm for potassium ions (a) and two plots to obtain linearity: (b) Hofstee plot and (c) Lineweaver-Burke plot. $V_{max} = 12\ \mu mol\ g^{-1}$ FW h^{-1}; $K_m = 20\ \mu M$. Hypothetical data.

$$v = (V_{max} \times C_{ext})/(K_m + C_{ext})$$

where K_m represents that external concentration (C_{ext}) at which v is half V_{max}. This Michaelis constant, K_m, therefore describes a property we may call the "affinity" of the uptake system; a low K_m implies high affinity, the system becoming half-saturated at low concentrations.

At higher concentrations a distinct isotherm can be observed (Epstein, 1973), and some workers believe there is a series of consecutive isotherms which come into play as the external concentration rises ("multiphasic isotherms"; Nissen, 1974). These phenomena are, however, only observed in highly artificial systems, and it is becoming increasingly clear that the uptake of ions by plant roots growing in soil is rather little influenced by uptake kinetics (Clarkson, 1985), being more responsive to transport processes in soil and internal demand for the ion. Many physiological studies *in vitro* are therefore of limited value in understanding ion uptake in an ecological context. For example the classic material for studies of ion uptake has been a short length of root tip cut from a barley plant that has been nutrient-starved by being grown in 0·2 mM $CaSO_4$ for 5 days (Epstein, 1961). This has the advantage of ensuring responsiveness, and the disadvantage that it is a system very far from equilibrium. Starvation may alter the uptake characteristics of a root; Leigh *et al.* (1973) showed profound differences between maize grown simply on 0·5 mM $CaCl_2$ and that grown on 7·5 mM NaCl with 2·5 mM KCl and 0·5 mM $CaCl_2$. Removing the shoot, the sink to which ions are transported and the source of fixed carbon, also has profound effects, since there is a direct relationship between K^+ transport to the shoot, for example, and shoot relative growth rate (Pitman and Cram, 1973; Fig. 3.14).

B. Interactions

Typically physiological experiments are performed in very simple solutions. In their classic experiments Epstein and Hagen (1952) used single-salt (2 ion) solutions for simple uptake studies and 3 ion solutions (e.g. Na^+, Rb^+, Cl^-) for studies of interactions. Most workers have followed suit. For an understanding of the uptake mechanism this is a valid approach, but if more complex solutions are used many interactions are found, both stimulatory and inhibitory:

  (i) inhibition of the uptake of one ion by another can result either in an increased K_m with no change in V_{max} (known as competi-

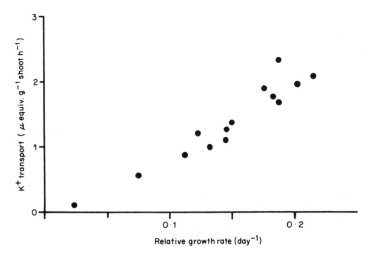

FIG. 3.14. Transport of potassium to the shoot of barley plants as a function of shoot growth rate (from Pitman and Cram, 1973).

tive inhibition) or in a reduction in V_{max} with no change in K_m (non-competitive inhibition). Competitive inhibition occurs for example between K^+ and Rb^+ or Ca^{2+} and Sr^{2+}, and depends on the relative frequency of the two ion species. It can therefore be overcome by raising the concentration of the test ion; this is not true of non-competitive inhibition, where the system is actually incapacitated.

(ii) stimulation of uptake is most often associated with changes in Ca^{2+} concentration, which appears to affect membrane function (Hepler and Wayne, 1985).

Clearly in complex soil solutions the relationship may be radically different to that determined in experiments. Even in more realistic work using whole plants growing in nutrient culture, only essential elements are normally used (Hewitt, 1967), and very little is known of the effects of non-essential ions (with the notable exception of Al^{3+} — see Chapter 6). In many soils silicate ions are dominant and these can interact with manganese (Rorison, 1971), and in calcareous soils bicarbonate ions are abundant and powerfully inhibit iron uptake by non-adapted species, such as *Deschampsia flexuosa* (Woolhouse, 1966). To complicate matters further many soil solutions contain a wide variety of organic ions, such as phenolic acids derived from the breakdown of lignin, which are present in some quantity in the soil around roots, and these have been shown dramatically to inhibit phosphate and potassium uptake by barley roots (Glass, 1973, 1974; Chapters 6 and 8).

C. Internal Control

It is a simple matter to characterize the kinetics of ion transport into roots using the Michaelis–Menten model, but the values of V_{max} and K_m obtained are highly dependent on the conditions of measurement. Barley has been a favourite material for experimenters (Table 3.6), and a wide range of responses to nutrient deprivation have been found. Either V_{max} or K_m may be affected, and though the commonest response is for V_{max} to increase and/or K_m to decline, this pattern is not invariable.

TABLE 3.6.   Alterations to uptake kinetics of barley roots brought about by starving plants of various ions.

| Ion | Effect on | | Authors |
	V_{max}	K_m	
$H_2PO_4^-$	Increased	No change	Drew et al. (1984)
	Increased	No change	Lee (1982)
	No change	Reduced	Cartwright (1972)
SO_4^{2-}	Increased· ·	No change	Lee (1982)
K^+	Increased	Reduced	Drew et al. (1984)
	No change	Reduced	Glass (1978)
NO_3^-	Reduced	Increased	Smith (1973)

If either K_m falls or V_{max} rises, the rate of uptake of the ion from very low external concentrations will be increased. These kinetic changes can occur very rapidly—after 1 day of K deprivation, 14 day-old barley plants experienced a 5-fold reduction in K_m from 53 to 11 µM (Drew et al., 1984). Similar effects are found for other species, such as the crucifer *Arabidopsis thaliana* (Doddema et al., 1979), and in maize, the uptake of K^+ is inversely related to shoot K concentration (and hence presumably to root K as well) over a very wide range from 20 to 100 mg Kg^{-1} (Fig. 3.15). Many plants can probably therefore react to short-term fluctuations in the root surface ion concentration (ion

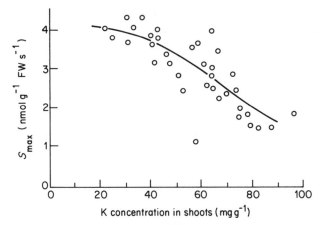

FIG. 3.15. Relationship between maximum specific absorption rate (S_{max}) and shoot K concentration for maize *Zea mays*. Redrawn from Claassen, N. and Barber, S. A. (1977). *Agronomy J.* **69**, 860–864.

"availability") by this sort of kinetic change, probably the result of synthesis of new carriers in the membrane or possibly of allosteric changes in existing carriers.

Earlier evidence suggested that these changes might be determined by fluctuations in cytoplasmic ion concentrations, but this was based on whole root ion concentrations (Glass, 1978). Recently the development of NMR (neutron magnetic resonance) techniques has allowed the measurement of ion concentrations in specific cell compartments, and it now appears that over a wide range of external P concentrations, the inorganic P (P_i) concentration of the cytoplasm is almost unaltered, while that of the vacuole changes markedly (Lee and Ratcliffe, 1983). In the light of this, it is thought that the flux across the tonoplast may regulate carrier behaviour.

D. Demand

Whatever the external concentration and the rate of uptake, the internal concentration of the ion will depend on the rate at which it is incorporated metabolically (except for K^+ and a few others which are not metabolized) and the rate at which it is diluted by growth. Growth rate is therefore an important variable in the determination of ion uptake. When Drew *et al.* (1984) deprived 14 day-old barley plants of potassium for periods of up to 7 d, their growth rate was unaffected and the internal K^+ concentration therefore fell rapidly (Table 3.7). Equally, if growth rates are inhibited by some other factor, uptake may continue and internal concentrations increase, leading to luxury accumulation. The regulating mechanism will then act to limit these changes in internal concentration by reducing either ion uptake or growth rate.

TABLE 3.7   Changes in internal potassium concentration of barley plants as a result of K-deprivation (Drew *et al.*, 1984).

		Days of deprivation		
K concentration ($\mu mol\ g^{-1}$)	0	1	4	7
Shoots	1690	1490	1030	737
Roots	1370	960	454	332

Inhibition of metabolic activity will lead, therefore, to a build-up of unutilized ion which will feedback on uptake rates. Tobacco plants supplied with both N and K had incorporated 40% of the N into protein after 5 h, with only 30% remaining as nitrate, whereas those supplied with N alone had put less than 20% into protein and over 40% was still present as nitrate (Koch and Mengel, 1974). Here K limitation had reduced the amount of nitrate that could be utilized. Similarly plants supplied with high N levels often have more rapid P uptake rates (Blair *et al.*, 1972), and White (1973) suggested that in lucerne this was due to increased metabolic activity in the root permitting greater incorporation of P_i into organic forms (Loughman, 1969).

Such a response pattern will ensure a well-balanced uptake of essential ions in accordance with the plant's immediate requirements, and is characteristic of fast-growing plants in disturbed environments, such as crop plants and weeds. Plants growing in less productive environments, however, experience a supply pattern for which such a response could be inappropriate. If nutrient availability varies during the growing season, or if the time of availability for different nutrients is not synchronized, then growth and uptake would always be inhibited. Two responses appear to have evolved to cope with this.

1. Storage
Many species contain larger amounts of particular nutrients than they can utilize immediately, apparently stored and used when other conditions permit. Two alpine grasses from New Zealand grow in

habitats differing in soil phosphate status: *Chionochloa pallens* favours more P-rich soils than *C. crassiuscula* (Chapin *et al.*, 1982). When grown on a range of P concentrations in solution culture, *C. pallens* grew much faster than *C. crassiuscula* at high P supply rates (Fig. 3.16) and took up more than twice as much P. The two species had, however, very similar P absorption rates, and in fact *C. crassiuscula* had a higher uptake rate at high external P concentrations; consequently it also had higher internal P concentrations at all supply rates, though the P was present largely as P_i (Fig. 3.16). Thus the low-P adapted species takes up P at

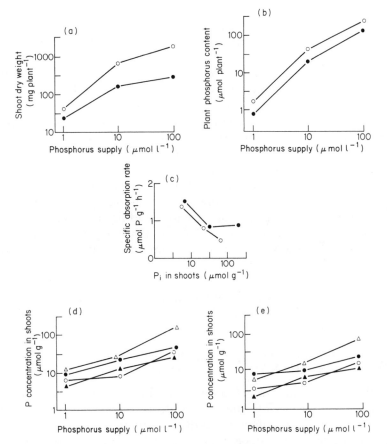

FIG. 3.16. Responses of two grasses, *Chionochloa crassiuscula* (● in a, b, c; and d) and *C. pallens* (O in a, b, c; and e) to phosphate supply. (a) and (b) show shoot dry weight and whole plant phosphorous content as a function of P supply. (c) shows P uptake rate as a function of shoot P_i concentration. (d) and (e) show P fractions: △ P_i; O phosphate esters; ● nucleic acids; ▲ phospholipids. Redrawn from Chapin *et al.* (1982).

the same rate as or faster than *C. pallens*, but stores it as P_i rather than using it for growth and metabolism, whereas the high-P adapted species does use it for growth and as a result becomes larger, and takes up more P in total but at a lower internal concentration.

A similar effect was shown by Barrow (1975), who found that *Lolium rigidum* transports more phosphate to the shoots than *Trifolium subterraneum*, and that at the high shoot phosphate concentrations so attained, produces more photosynthate and transports more of that to the roots, so maintaining high uptake rates. Consequently the grass is more successful than the clover at high phosphate supply.

2. Flexibility

Most nutrients are taken up by plants in a single ionic form. Phosphate may occur as $H_2PO_4^-$ or HPO_4^{2-}, depending on pH, but the uptake kinetics are the same. Nitrogen differs from other minerals both in having no substantial mineral reservoir in soil and in being present in most soils in two interconvertible forms—NO_3^- and NH_4^+. The conversion is by bacteria and the relative abundance of the two ions depends upon the rate of NH_4^+ production from organic matter, the adsorption of NH_4^+ by the soil exchange sites, the rate of bacterial conversion and the rate of loss of nitrate by leaching, denitrification and uptake. Plants may therefore experience wide fluctuations, both in space and time.

Most plant species can utilize either ion, although if one or other is the sole N source, large changes in root-zone pH result. When ammonium is taken up, protons are lost leading to acidification, and the converse applies to nitrate. As a result plants adapted to calcareous soils normally grow very poorly on ammonium-based nutrient solutions unless the pH is maintained above 6.

If ammonium is taken up, it is incorporated directly into organic compounds, initially via glutamate and glutamine. If nitrate is the source, however, it must first be reduced to ammonium by two linked enzymes:

$$\text{nitrate} \xrightarrow[\text{NADH}\rightarrow\text{NAD}]{\text{nitrate reductase}} \text{nitrite} \xrightarrow[\text{red. ferredoxin}]{\text{nitrite reductase}} \text{ammonium}$$

Both enzymes are substrate-induced, but the more important is nitrate reductase (NR), since it controls the level of nitrite, the substrate for nitrite reductase (NiR). Reduction may take place in either roots or leaves, depending on species. Several other enzymes (glutamine synthetase, glutamate synthase and glutamate dehydrogenase) are also part of the complex involved in N assimilation.

When a plant absorbs nitrate the level of NR in the tissue is proportional to the concentration of nitrate in the soil solution, and if the soil is fertilized with nitrate and then leached to produce rapid fluctuations in soil nitrate, the NR activity in the plant changes concomitantly in less than 24 h (Long and Woltz, 1972), since in the absence of the substrate existing enzyme is rapidly degraded or reversibly inactivated. The plant does not therefore retain non-operational enzyme, but the fluctuations in NR level may, by controlling internal nitrate or amino-acid levels, affect nitrate uptake (Oaks and Hirel, 1985). The involvement of ferredoxins as reductant for NiR probably explains the evolution of rhythmic behaviour of NR activity in some plants, such as *Anthriscus sylvestris* (Janiesch, 1973), where activity peaks at mid-day. Nitrate uptake may, therefore, be controlled by several environmental factors.

The feature of major significance, however, is the response to soil nitrate. Stewart *et al.* (1972) found a 50-fold range of NR activity in plants of *Suaeda maritima* on a salt-marsh, which was correlated with a 3-fold range in the total N content of the plants and a 20-fold range in tissue nitrate concentrations. Later (1973) they showed that NR activity in 18 salt-marsh species ranged from a mean of $0 \cdot 23 \, \mu mol$ NO_2^- produced g^{-1} fresh weight h^{-1} in *Triglochin maritima* to $3 \cdot 60$ in *Beta vulgaris* ssp *maritima*. The variation within each species was almost as great, however, with *Suaeda* for example ranging from $0 \cdot 06$ to $1 \cdot 93 \, \mu mol \, NO_2^- \, g^{-1} \, h^{-1}$, and when nitrate was added to the vegetation the NR activity of all 18 species was increased by anything up to 9-fold 72 h later.

Nevertheless, some species showed consistently high activities (*Beta, Atriplex hastata*) and some consistently low (*Festuca rubra, Triglochin maritima*), raising the possibility of ecological adaptation being based on difference in NR activity. Some species, particularly in the Ericaceae (heaths), characteristic of acid soils where nitrification is inhibited and NH_4^+ the major N source, do not appear to produce NR at all (Dirr *et al.*, 1973). It appears to be often true that plants from acid soils have lower NR activities in the field (Havill *et al.*, 1974), probably because of lower rates of nitrification, but they are generally capable of much higher rates if supplied with nitrate, and it is coming to be realized that even many acid, waterlogged soils still contain appreciable nitrate levels. Typical NR activities for plants from a variety of habitats are shown in Table 3.8, and the species seem to fall into three groups:

(i) those of high nitrate sites, characterized by flushes of nitrate availability, such as nettle *Urtica dioica*;

TABLE 3.8.   Nitrate reductase activity in the field before and after induction by added nitrate, in species from various soil types (Havill *et al.*, 1974).

Soil	Species	*Nitrate reductase activity* (μmol NO$_2^-$ g^{-1} FW h^{-1})	
		Before induction	After induction
Calcareous	*Poterium sanguisorba*	1·11	3·80
	Scabiosa columbaria	1·26	2·12
Neutral	*Urtica dioica*	7·93	16·10
	Poa annua	4·05	8·40
Acid	*Molinia caerulea*	0·52	1·50
	Galium saxatile	0·72	3·06
Acid peat	*Vaccinium myrtillus*	<0·1	<0·1
	Drosera rotundifolia	<0·1	<0·1

(ii) an intermediate group of plants of both acid, neutral and calcareous soils, among which plants of acid soils tend to have the smallest activities in the field; and

(iii) those of acid peat soils, typically raised bog where soil nitrate levels are negligible, which seem to lack the enzyme altogether.

Although the large differences in NR activity found between these extremes can easily be seen as adaptive, the majority of plants fall into the second group, where few differences can be discerned. Such plants grow in habitats whose nitrogen supply regime is extremely hard to characterize. Woldendorp (1983) compared two plantains, *Plantago major* and *P. lanceolata*, which grow in such soils, in this case sand-dune grassland in Holland. Soil around *P. lanceolata* roots had lower ammonium (6·2 μg g^{-1}) and nitrate (0·5 μg g^{-1}) concentrations and fewer nitrifying bacteria (10 mg^{-1}) than soil from *P. major* roots (17·5, 1·2 and 109 respectively). Leaf NR activity was also much higher for *P. major* when measured in the field (1·02 as against 0·38 μmol NO$_2^-$ produced g^{-1} fresh weight h^{-1}). He postulated that *P. major*, a plant characteristic of rather open ground, occurred on sites with a higher and more reliable nitrate supply, and showed that it exhibited a very constant NR activity in culture (around 20 μmol NO$_2^-$ g^{-1} h^{-1}), even when the nitrate supply was varied 50-fold. In contrast, *P. lanceolata*, which inhabits sites of more variable and unpredictable nitrate supply, had NR activity ranging from <10 to 20 μmol NO$_2^-$ g^{-1} h^{-1} in response to a change in supply (Fig. 3.17).

The utilization of nitrate thus illustrates both the plasticity of

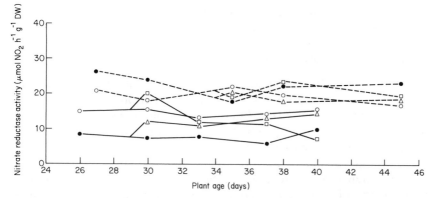

FIG. 3.17. Effect of changing the nitrate supply on nitrate reductase activity in *Plantago lanceolata* (——) and *P. major* (– – – –). Treatments were: ●, continuous low supply (75 μM); ○, continuous high supply (3·75 mM); △, supply increased from low to high; □, supply reduced from high to low. Redrawn from Woldendorp (1983).

response conferred by an inducible enzyme system in fluctuating environments,and genetic differences between species exposed to stable environmental differences. Even within a species there can be heritable differences in NR activity, as shown by Goodman *et al.* (1974) for *Lolium perenne* and *L. multiflorum*.

IV. MORPHOLOGICAL RESPONSES

For most ions the rate-limiting step in uptake is transport through soil and not across the root (see p. 80). In such circumstances, changes in uptake kinetics will have little effect on uptake, and the only useful response a plant can make is to alter the morphological properties of the roots. In other words to have greater unit activity on the same amount of root will not increase uptake, but to have more root will.

A. Root:Shoot Ratio and Root Weight Ratio

The simplest index of allocation to roots is the root:shoot ratio (W_R/W_S) or the root weight ratio (W_R/W_{RS}). The latter, though less widely used, is in many ways a more useful variable as it represents the proportion of total plant biomass allocated to roots. Both are very plastic and generally increase with:

   (i)   low soil water supply (Chapter 4);
   (ii) low nutrient supply (Fitter, 1985);
  (iii) low soil oxygen;
  (iv) low soil temperature (Davidson, 1969);

and may show less predictable responses to other environmental variables such as light intensity and photoperiod. Such effects are summarized by Aung (1974). On the whole, plants seem to put more of their resources into root production in environments where growth is limited by soil-derived resources or factors.

A corollary of this is that one might expect plants characteristic of stressful environments to have higher and less variable root weight ratios. Certainly three species of low growth rate associated with poor soils, *Deschampsia flexuosa*, *Carex flacca*, and *Scabiosa columbaria* examined by Hunt (1970), bore this out in contrast with species such as *Urtica dioica*. On the other hand the data of Bradshaw *et al.* (1964) show a clear increase with nitrogen deficiency, but no evidence of any correlation with habitat, *Nardus stricta* being the most plastic and *Lolium perenne* the least plastic of the species studied. This is confirmed by Higgs and James (1969) who also show that $W_R:W_S$ tends to be lowest in species of high growth rate.

This apparent contradiction is probably due to the complex nature of plant weight as a character. Plants in stressful conditions frequently allocate a large proportion of their photosynthate to storage organs which are often subterranean (cf. Chapters 4 and 5). For example, *Plantago lanceolata* growing in stressful habitats (sea-cliffs, toxic soils) has much higher root weight ratios than meadow plants, and this is due to the development of a massive root stock and not to large numbers of absorbing roots. Root weight ratio is therefore a good guide to the stressfulness of the environment, but bears little relation to nutrient absorption.

B. Root Diameter and Root Hairs

All except the oldest roots possess some uptake ability. The segment behind the root tip, however, possesses a special importance by virtue of its root hairs, except in a few families (e.g. Alliaceae). The effect of root hairs is to increase the volume of soil exploited, which will have little effect on the uptake of mobile ions such as nitrate, but which one might expect to promote uptake where diffusive supply is important, as for phosphate. In the latter case depletion at the root surface will be extensive and the whole root hair zone may be depleted, as shown by autoradiographs (Bhat and Nye, 1973). Further, in field conditions where soil moisture varies greatly and the solution contact with the root may be broken, root hairs can penetrate very small pores (diameter < 5 μm) and by secreting mucilage maintain a liquid junction between root and soil without which both ion and water uptake would cease (Barley, 1970). Experimentally, however, the role of root hairs is not fully confirmed; Barley and Rovira (1970) showed that whereas "hairy" root zones are no more active in P uptake in stirred solution culture where transport is not limiting, in slightly compressed clay (which hairs cannot penetrate) phosphate transfer to pea roots was reduced by 78%, while transfer to anion exchange paper was increased by 29%, as compared with the uncompressed soil. Similarly Caradus (1981) showed that P uptake by white clover was directly proportional to root hair length in a range of genotypes. On the other hand Bole (1973), using chromosome substitution lines of wheat differing in root "hairiness", could find no relationship between P uptake from soil and root-hair development. Most probably root hairs become increasingly important as the supply of an ion becomes more limited by diffusion.

The main value of root hairs is likely to be their ability to penetrate soil pores too small for roots, but root diameter itself can also vary and may be an important factor when depletion is less and the concentration in the root free space is high, so that uptake occurs into all the cortical cells, not just the epidermis. In this case uptake will be related more to root volume than to surface area, as found experimentally by Russell and Clarkson (1976) and shown theoretically by Nye (1973). Mean root diameter is certainly very variable for some plants, being lower in arctic than in temperate species (Chapin, 1974) but increased by low temperature in the latter. Wild *et al.* (1974) found the radius of "coarse" roots to vary between 3 species, but that of the "fine" roots to be more or less constant, while Christie and Moorby (1975) subjected three Australian grasses to P concentrations ranging from 0·003 to 30 mg l^{-1} and recorded increases in mean root diameter of 17, 22, and 41%.

C. Root Density and Distribution

When a single root grows in an unlimited volume of soil, uptake is influenced by soil factors and plant demand. In real situations, however, the depletion zones around adjacent roots will frequently overlap. There will then be interference to supply patterns by one root upon another—in other words competition for nutrients (Nye, 1969). This interference will result from the overlapping of the depletion zones resulting in the diminution of the diffusion gradient.

Competition will only appear, however, at high root densities for ions of low diffusion coefficients (such as phosphate), because the depletion zones are narrow and so unlikely to overlap. Figure 3.12 shows the different effect of root density on uptake of K^+ and $H_2PO_4^-$ by monospecific stands of *Lolium perenne* and *Agrostis tenuis* (Fitter, 1976). These effects are typical and can be shown equally with a single plant (Newman and Andrews, 1973). As a result an increase in local root density is an effective way to increase uptake of immobile ions such as $H_2PO_4^-$ or NH_4^+, but not of mobile ions such as NO_3^- (Robinson and Rorison, 1983).

When the roots of more than one species (or genotype) are interacting in this way, the result may not be just a general lowering in uptake rates for all roots, for the roots of one species may be able to lower the ionic concentration at the root surface more than the other. In this case nutrient ions will move down the resulting steeper concentration gradient more rapidly to the root surface of the former species (cf. Chapter 8, Fig. 8.1), which will therefore effectively exploit a greater soil volume. There is very little experimental evidence on this point, though it seems certain that this is what occurs in soil and it is well established that competitive ability for nutrients varies between species, with species balance dependent on nutrient levels. Welbank (1961), for example, showed that added N had a greater effect on the growth of *Impatiens parviflora* when its roots were competing with those of *Agropyron repens* than when they were in monoculture. Similarly, van den Bergh (1969), in a replacement series experiment in which the density of the two species was varied, found that at high soil fertility *Dactylis glomerata* always replaced *Alopecurus pratensis*, at pH 4·2, 6·2, or 6·7; at low fertility, however, *Alopecurus* was favoured at low pH and an equilibrium was established at the two higher pH values.

The mechanism of competition for nutrients and its interaction with competition for light and water, is discussed fully in Chapter 8 , but clearly the form, extent, and distribution of the root system will be critical. All these features are plastic characters and respond to various soil environmental factors such as texture (Veihmeyer and Hendrickson, 1948; Kochenderfer, 1973) and pH (Mukerji, 1936; Fitter and Bradshaw, 1974). There are also dramatic responses to nutrient concentrations (Duncan and Ohlrogge, 1958; see review by Viets, 1965), a fact exploited in agriculture by the practice of banding fertilizer so that roots can grow preferentially in the fertile zone. Such effects can be mimicked in water culture (Fig. 3.18) and increases in

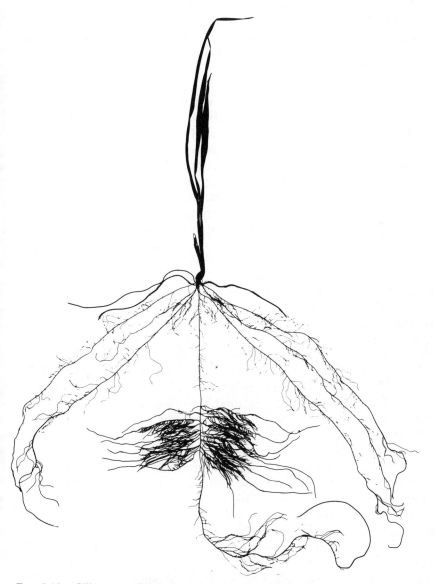

Fig. 3.18. Silhouette of a barley plant grown with its root system in a nutrient solution containing 0·01 mM nitrate, except for the central section which received 1·0 mM nitrate. Note the proliferation of secondary laterals in this zone and their absence elsewhere. (Courtesy Dr M. C. Drew).

root growth in response to local concentrations of, in particular, N and P, can be ascribed to the fact that lateral root growth only occurs at a point on the root if adequate external nutrient concentrations exist there (Drew et al., 1973). Growth of the main root axes of cereals is not concentration-dependent in this way, so that in infertile soil these will continue to grow, but lateral production will be inhibited until a fertile zone is reached. The root is thus adapted to explore poor soil volumes but to exploit fertile ones.

These striking and dramatic changes are not accompanied by changes in shoot growth (Drew, 1975), and nutrient uptake of the whole plant is scarcely affected in a plant only part of whose root system is exposed to high nutrient concentrations, as compared to one wholly in the high concentration, as a result of a compensatory increase in uptake in the favourably placed segment (Drew and Saker, 1975; Drew and Nye, 1969). In other words, the plasticity of uptake kinetics observed between individuals in many species, for example, can also exist simultaneously between different parts of a root system.

What is remarkable is how little attention has been given to the possibility that differences in this response might exist between species, in other words that some species might be inherently more able to respond to variation in nutrient supply than others. Grime et al. (1986) have grown plants in complicated chambers in which parts of the root system could be exposed to different nutrient regimes. They found that *Agrostis stolonifera*, a grass of nutrient-rich sites, was very much more plastic than *Scirpus sylvaticus*, a sedge found in less fertile sites. When offered pulses of nutrients that lasted only 24 h, the sedge simply failed to respond at all, and though the grass did show increased root growth in the pulsed chambers, it was unable to obtain sufficient additional nutrients as a result to make the investment worthwhile. When the pulses were longer, however, the plasticity of the grass root system allowed it to obtain very much more N than the sedge. In environments where such fluctuations are normal, then, this plasticity could confer considerable advantages, but in more infertile habitats the rather rigid growth pattern of the sedge is probably favoured.

Such effects also occur in soil (Fitter, 1976), with the precise distribution and growth of the root system being a response to variations in soil nutrient concentrations, so that one can assume that the highest root densities will occur in fertile soil. In this case, competition for immobile ions such as phosphate, which involves closely adjacent roots so as to obtain overlap of narrow depletion zones, will occur mostly in fertile soils or soil zones, and it may be that plants that are poor competitors for phosphate rely on exploiting less phosphate-rich soil volumes where root densities will be lower.

One final aspect of root systems which has largely been overlooked is architecture, or the precise distribution, angles and sizes of branches within the system. These differ markedly between root systems, both within and between species. Herringbone-like patterns (Fig. 3.19(a)) are more efficient in terms of the volume of soil explored for a given construction cost, but diffuse-branched systems (Fig. 3.19(b)) are more transport-efficient and generally minimize construction costs. It seems

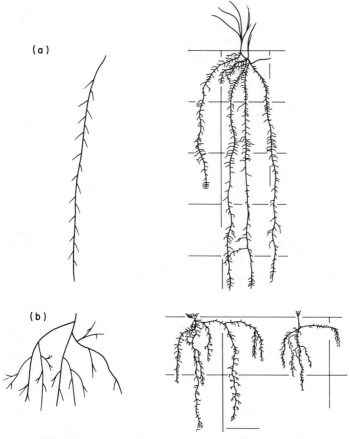

FIG. 3.19. Idealized branching patterns (left, a and b) and actual examples which approach each pattern (right, a and b). (a) Herringbone pattern, *Calamovilfa longifolia*; (b) dichotomous pattern, *Phlox hoodii*. Root systems reproduced from Coupland and Johnson (1965), with permission from Blackwell Scientific Publications Ltd.

that where acquisition of nutrients is limiting growth, herringbones are more likely to be adopted (Fitter, 1987).

D. Turn-over

Ion uptake is proportional to surface area, though not necessarily the external surface, owing to diffusion into the cell walls. Not all parts of the root system are equally active in uptake, however, since old roots may in some case become completely suberized , and Ca^{2+} in particular is only taken up by the youngest roots in which the endodermis does not constitute a barrier (Harrison-Murray and Clarkson, 1973). Evidence for other ions, such as $H_2PO_4^-$ and K^+, suggests that earlier assumptions that old roots are inactive are incorrect. Phosphate is readily taken up by barley roots 50 cm from the tip and can be translocated from there to the shoot (Clarkson et al., 1968). At this distance from the meristem the endodermis is massively thickened and has a pronounced suberin layer (the Casparian band); this prohibits movement of ions such as Ca^{2+} beyond the cortex, since they travel primarily in the free space of the cell walls (apoplastically), but provides no barrier to $H_2PO_4^-$ and K^+, which travel symplastically in the cortical cytoplasm, and the endodermal cells are traversed by many plasmodesmata (Clarkson et al., 1971).

In ideal conditions, roots can therefore continue absorbing ions for long periods, but in time that rate will diminish because of the depletion of the surrounding soil (i.e. transport through soil will become limiting) and because of the damage inflicted by pathogens and grazing soil animals such as nematodes. If nutrient levels elsewhere in soil are higher, there might be a benefit to the plant in allowing existing fine roots to senesce, and to grow new roots in unexploited soil zones. This is what happens above ground: deciduous trees are characteristic of productive environments in which the carbon investment in a leaf is rapidly repaid by photosynthesis. The leaves are not retained over unfavourable seasons (dry or cold) and senescence here is a controlled event, with resources, especially N and P, being withdrawn and used for new growth. On infertile soils, however, and in other unproductive environments (boreal and high-altitude forests, bogs, semi-deserts), many plants have small, resistant evergreen leaves, which can continue to produce, albeit at a slower rate, for several seasons (Small, 1972; Goldberg, 1982; Gray, 1983; cf. Chapters 4 and 5).

Whether root senescence is a controlled phenomenon as for leaves is apparently unknown, but turn-over of fine roots can be considerable and it seems likely that it will one day be shown to involve recycling rather than simple resource loss. If so, one would expect to find root longevity to be related to soil nutrient status.

V. SOIL MICRO-ORGANISMS

A. The Nature of the Rhizosphere

The root is not the only living component of the root–soil system; indeed soil ecosystems contain remarkable numbers of bacteria, fungi, protozoa and small invertebrates. One estimate (Brookes et al., 1985) suggests that the microbial biomass below ground in the Broadbalk wheat plots at Rothamsted is around 5 Mg fresh weight ha^{-1}, the equivalent of around 100 sheep! The supply of many nutrients depends on microbial degradation of organic matter, and soil animals play critical initial roles in this process, fragmenting the litter into pieces suitable for microbial attack. Standard procedures of sampling for soil microbes involve taking soil samples, suspending them in water, and plating them on agar to detect microbial growth. This gives a general view of the soil flora, though tending to emphasize the importance of the spore population, and indeed the activity of much of this population is very low. More precise work shows that these microbes are not uniformly distributed; rather they are highly aggregated around energy sources, with the mineral matrix representing a microbial desert. Plant roots represent one of the major sources of energy for soil microbes and their influence on the soil microflora was first studied systematically by Starkey (1929), though the term "rhizosphere", to cover the volume of soil in which the microbial populations are influenced by the proximity of a root, was coined by Hiltner in 1904.

The most striking feature of the rhizosphere is the stimulation of bacterial numbers and activity. The effect is normally quoted in the form of an R (root):S (soil) ratio (Katznelson, 1946), and typical R:S values range from 2 to 100, though figures as high as 2000 are on record. The R:S ratio is based on the numbers of micro-organisms in the soil adhering to the roots relative to the numbers in bulk soil, and is valuable for unicells such as bacteria (though spores cause problems), but difficult to interpret for filamentous fungi. Rovira and co-workers have approached the problem directly, using scanning electron micrographs of root surfaces to examine rhizosphere (or strictly rhizoplane) populations (e.g. Rovira *et al.*, 1974), and several techniques are now available (see, for example, Polonenko and Mayfield, 1979).

The microbes in the rhizosphere are stimulated by increased concentrations of various chemicals that act as energy sources, and these are deposited there from a number of sources (Newman, 1985):

(i) by the sloughing of root cap cells as roots grow through soil. The few figures available here suggest that this is a surprisingly minor source of material, offering less than 10 mg g^{-1} root according to Newman's (1985) calculations.

(ii) by the production of mucigel by the root cap, which facilitates its passage through the soil (Juniper and Roberts, 1966; Greaves and Darbyshire, 1972). Newman (1985) suggests that this may account for up to 50 mg g^{-1} root.

(iii) by exudation or secretion of compounds from intact cells. Rovira (1969) suggested that in general exudation amounted to about 0·1% of carbon fixed by photosynthesis, but species differ in amount and type of exudate, and individuals vary with age, temperature, and nutritional status. An enormous range of compounds is exuded—from wheat alone 10 sugars, 20 amino acids, 10 organic acids and several other compounds have been reported. Using radiocarbon tracers Whipps and Lynch (1983) and Whipps (1984) measured both soluble and insoluble carbon lost to the rhizosphere of wheat plants over a three week period. Their data and others suggest that not more than 100 mg of soluble carbon g^{-1} root finds its way into the rhizosphere by this means (Newman, 1985).

(iv) by the death of root cells. There is often very much more insoluble material in the rhizosphere than can be accounted for by the sources listed so far, and the most likely explanation for the discrepancy is in the death of root hairs and epidermal and cortical cells.

The total effect of these losses can be considerable, but even so cannot account for figures such as those of Martin (1977), who found that of the carbon transported to wheat roots over 23 days at 10°C, an average of nearly 40% was lost to the soil, although much of this may be by respiration. In the experiments of Whipps and Lynch (1983) a large part of the ^{14}C lost to soil was not recovered in the rhizosphere, but in the bulk soil. This material is almost certainly dead roots. It seems likely that root turn-over can be very rapid indeed (see p. 106), and that dead roots are an extremely important food source in soil, particularly for filamentous microbes such as fungi which can move from site to site by growth, rather than migration.

In the light of such complexity and enrichment it is not surprising that the rhizosphere flora is qualitatively as well as quantitatively different from that of the bulk soil (Rovira and Davey, 1974). Bacteria requiring amino acids are abundant (Lochhead and Rouatt, 1955) and root-inhabiting microbes, both symbiotic bacteria (nitrogen-fixing *Rhizobium*) and pathogenic fungi (such as *Fusarium*), may show increased populations.

B. Effects on Nutrient Uptake

This immensely complex microbial population might influence nutrient uptake by roots in four ways:

(i) by altering the supply at the root surface—competition;
(ii) by altering root or shoot growth, or by directly damaging the root;
(iii) by interference with nutrient uptake itself—inhibition or stimulation;
(iv) by mineralization of organic or dissolution of insoluble ions.

Evidence on all these points is conflicting, and early work which suggested that roots stimulate the activity of phosphate (apatite)-dissolving bacteria and hence improve P supply, is now interpreted in terms of root growth stimulation by gibberellic acid produced by rhizosphere microbes (Gerretsen, 1948; Brown, 1975). Certainly soil microbes influence uptake. Barber *et al.* (1976) have clarified earlier reports and showed that microbes generally stimulate uptake and transport of phosphate in young plants and in short-term experiments, but depress both in older plants and over longer periods (Fig. 3.20). There appears to be a balance between the stimulation due to growth substances released by micro-organisms and only effective on young roots, and competition for phosphate between the microbes and the roots. In older root systems, the proportion of young root segments declines, and competition becomes the dominant force. In addition to these direct effects, soil micro-organisms may indirectly reduce the ability of roots to take up ions. Thus many fungi break down lignin in soil, releasing phenolic acids, particularly in the rhizosphere (Pareek and Gaur, 1973), and these acids markedly inhibit ion uptake (Glass, 1973; Fig. 6.2).

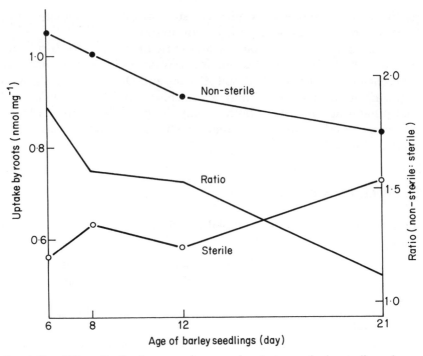

F<small>IG</small>. 3.20.   Effect of soil micro-organisms on phosphate uptake by sterile and non-sterile roots of barley plants of different ages (from Barber *et al.*, 1976).

C. Nitrogen Fixation

Among the many bacteria found in the rhizosphere are those capable of both free-living and asymbiotic fixation of gaseous nitrogen. Several common genera of soil bacteria, such as *Azotobacter, Klebsiella* and *Clostridium*, can fix N_2 in the rhizosphere using energy obtained from root-derived materials. High levels of nitrogenase (the enzyme responsible for fixation) activity have been found in the rhizosphere of some tropical grasses, particularly *Paspalum notatum* (Döbereiner and Day, 1976). The bacterium most commonly responsible is *Azospirillum brasilense*, which may actually invade dead cortical cells, producing what has been called an endorhizosphere. For some time there was great excitement about the possibility that this associative symbiosis could be exploited agriculturally to produce strains of crop plants which would have very low fertilizer nitrogen requirements. Unfortunately, the energy costs of N_2 fixation are high (estimates range from 4–

174 g C g^{-1} N fixed) and the energy supply in the rhizosphere is quite inadequate (Giller and Day, 1985). On the most optimistic assumptions, such associations could only supply a few per cent of the N needs of a crop. In most natural ecosystems where there is a net input of N from fixation, this derives either from symbiotic fixation or from photosynthetic, N$_2$-fixing cyanobacteria, which have their own energy source.

In symbiotic fixation, the energy supply is direct from the host. The bacteria develop in a nodule on the host root which is supplied by the host vascular system. For unknown, but probably profound biochemical reasons only prokaryotic organisms have the ability to fix nitrogen—which probably explains why attempts to transfer *nif* (*nitrogen-fixing*) genes for N$_2$-fixing ability from bacteria to higher plants have been unsuccessful—and one bacterial genus, *Rhizobium*, has proliferated strains or species capable of forming nodules with a single plant family, the Leguminosae. Depending on the criteria used there are from three species upwards of *Rhizobium* (Sprent, 1983), but infection of a legume species is specific to a particular strain. Most species in the two sub-families Papilionoideae and Mimosoideae form nodules, but only a third of the more primitive and largely tropical Caesalpinoideae. In other families nodule formation is known in 158 species in 14 genera, including *Alnus* (alder), *Myrica* (bog myrtle), *Casuarina* and *Hippophae* (sea buckthorn). All bar one are infected by a different symbiont, the actinomycete *Frankia*, and form larger, clustered nodules (Bond, 1976). The exception is *Trema cannabina* (Ulmaceae) from New Guinea, which is symbiotic with *Rhizobium* (Trinick, 1973).

The nodule acts as an important sink for fixed carbon. Over a 24 h period a pea plant (*Pisum sativum*) transported 11·02 mg of carbon to its roots, of which 6.29 mg (57%) were used as carbon skeletons for fixed nitrogen, and 4.34 (39%) for respiration (Minchin and Pate, 1974). Only as long as the extra nitrogen supplied to the plant gives it some ecological advantage will this outweigh the carbon loss, since that must affect the potential for growth. Many legumes are pioneer species on N-deficient soils, as are many of the non-legumes capable of fixation. In the classic successional studies at Glacier Bay, Alaska (Crocker and Major, 1955), it was N-fixation by alders that facilitated the invasion of spruce and the furtherance of the succession. A similar role is played on abandoned china-clay waste heaps in Cornwall by gorse *Ulex europaeus* and tree lupins *Lupinus arboreus* (Marrs *et al.*, 1983).

Where soil nitrogen levels are higher, however, the competitive advantage derived from fixation is lost, and carbon used for growth lends more competitive ability than that used for fixing nitrogen.

Legumes and other N-fixers are generally rare in mature communities and where they are found, as trees in tropical forest for example, they belong typically to the little-nodulated Caesalpinoideae (Sprent, 1983).

D. Mycorrhizas

The rhizosphere is a quantitative phenomenon, defined by the increased microbial population. There is consequently a gradient of influence away from the root with the most enhanced area being the root surface, the rhizoplane. The cover and frequency of bacteria at the rhizoplane vary between species; for example *Hypochaeris radicata* may have twice as many bacteria as *Plantago lanceolata* (Rovira *et al.*, 1974). Generally the area of root covered by bacteria seems to vary little amongst species, but the number of bacterial cells varies greatly and can be altered by the proximity of the roots of other species (Christie *et al.*, 1974). In view of the known effects of the rhizosphere flora, discussed above, such interactions could have important implications for plant competition.

Mycorrhizas, symbiotic associations of a fungus and a plant root, can be viewed as a highly specialized development of a rhizoplane association, that has become at least partly invasive. Ectomycorrhizas (sheathing mycorrhizas) involve usually a Basidiomycete fungus and a tree; the fungus forms a mantle of hyphae around the root, a network of intercellular hyphae in the cortex and a ramifying mycelium in the soil. No intracellular connections are made and the roots are generally stunted. Endomycorrhizas are of three main types, two of which are very specialized, occurring respectively in the order Ericales and the family Orchidaceae, and will not be further discussed. The third type, formed with a small group of Phycomycetous fungi, is known as the vesicular-arbuscular (VA) mycorrhiza. In addition there is a range of puzzling, intermediate forms known as ectendomycorrhizas; some host species can form more than one type. The whole field is comprehensively reviewed by Harley and Smith (1983).

VA mycorrhizas occur in most families of higher plants, as well as some Bryophytes (liverworts but not mosses) and Pteridophytes. The relationship is intracellular, the fungus producing haustoria that invaginate the plasmalemma and produce arbuscules ("little trees"), masses of minute hyphae less than 1 µm in diameter, and vesicles which are storage bodies. The fungi belong to several genera in the Endogonaceae, principally *Glomus* and *Gigaspora*. They are exceedingly widespread and have been reviewed by Powell and Bagyaraj (1984).

The primary interest of both types of mycorrhizas lies in their ability to improve plant growth by enhancing P uptake. In P-deficient soils mycorrhizal plants typically grow markedly better than non-mycorrhizal ones (Fig. 3.21), but the reverse may be true in soils well supplied with phosphate (Fig. 3.22). Indeed in such soils plants normally show very low levels of infection. The advantage of the mycorrhizal plants cannot be explained on the basis of root morphology, since they take up phosphate faster per unit root length than non-mycorrhizal ones (Table 3.9). In fact mycorrhizal plants often have shorter root systems, visibly so in the case of ectomycorrhizas; it is possible that reported effects of rhizosphere microbes reducing root length may in part be due to endomycorrhizal infection (Crush, 1974; Fitter, 1977).

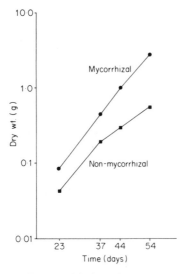

FIG. 3.21. Growth rate of mycorrhizal and non-mycorrhizal onions (data from Sanders and Tinker, 1973).

For both ectomycorrhizas (Harley and Lewis, 1969), orchid mycorrhizas (Alexander and Hadley, 1984) and VA mycorrhizas (Rhodes and Gerdemann, 1975)—but not for ericaceous mycorrhizas—it is well established that phosphate travels from fungus to root, so the enhanced uptake must be due to the greater ability of the fungus to obtain phosphate. It has been suggested that the fungus might utilize sources of P not available to plants, but when Hayman and Mosse (1972) grew VA mycorrhizal plants on soil enriched with ^{32}P, they found that infected and uninfected plants had the same specific activity as the soil,

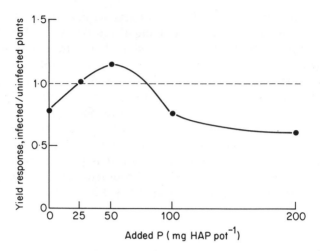

Fig. 3.22. Yield response of soybeans *Glycine max* to infection by the VAM fungus *Glomus fasciculatum* in sand culture amended with hydroxyapatite (HAP) as a phosphorus source. Note the marked depression of yield at high P supply rates. From Bethlenfalvay, G. J., Bayne, H. G. and Pacovsky, R. S. (1983). *Physiol. Plant.* **57**, 543–548.

Table 3.9. Phosphate inflows of mycorrhizal and non-mycorrhizal onion plants (Sanders and Tinker, 1973).

| | *Harvest interval* | | *Inflow* (pmol $cm^{-1} s^{-1}$) | |
No.	Duration (days)		Mycorrhizal	Non-mycorrhizal
1	14		0·17	0·050
2	7		0·22	0·016
3	10		0·13	0·042
		Means:	0·17	0·036

though the mycorrhizal plants had absorbed up to 30 times as much phosphate (Table 3.10). If they had been using unavailable phosphate, the extra P they took up would not have been labelled and they would have shown lower specific activities.

An alternative explanation would therefore seem to be that the mycorrhizal root has different phosphate uptake kinetics, perhaps a lower K_m, but though changes in kinetics have been reported, they are never great enough to account for the increased uptake, even for ectomycorrhizas, where the root is enclosed in the fungal sheath, making such an explanation more attractive. For VA mycorrhizas,

TABLE 3.10.   Phosphorus content and specific activity of mycorrhizal (Myc) and non-mycorrhizal (Non-myc) onion plants in relation to specific activity of P in soil (Hayman and Mosse, 1972).

Soil	P content of plants (roots and shoots (mg))		Specific activity of ^{32}P (c.p.m. mg^{-1} (roots))		(c.p.m. mg^{-1} (soil))
	Myc	Non-myc	Myc	Non-myc	Soil
7	1·13	0·04	4·4	3·7	6·7
8	0·33	0·26	9·7	10·1	12·9
10	0·21	0·10	24·7	20·1	24·8
11	1·33	0·13	2·1	1·9	3·0
12	1·22	0·48	2·1	2·0	3·5
15	0·14	0·08	26·1	23·8	29·5
16	1·29	1·03	1·7	1·9	2·5

however, Sanders and Tinker (1973), in an experiment with onions, calculated that the maximum rate of diffusive supply, irrespective of the uptake kinetics, would be 0·035 pmol cm^{-2} s^{-1}, achieved when the root surface concentration would be effectively zero, and this was approximately the rate achieved by uninfected roots (Table 3.9). Mycorrhizal roots had higher rates, by a factor of about 5, an impossible result unless the effective surface area for uptake had been incorrectly determined, which would be true if the fungal hyphae in the soil, attached to the mycorrhizal root, were exploring unexploited soil outside the depletion zone of the root, and translocating phosphate from there to the root. They found about 80 cm of hypha per cm of root in the soil, which adequately explains the superior uptake of mycorrhizal roots.

It is generally accepted that VA mycorrhizas act in this way, exploring soil inaccessible to the plant root, because of the limitations of diffusion, and Rhodes and Gerdemann (1975) have shown that mycorrhizal roots can exploit soil for phosphate up to 70 mm from the root, whereas a figure of 10 mm or less would be more typical for uninfected roots. All this has been established in pots in growth rooms or glasshouses, and many fresh demonstrations of such a mycorrhizal response are published each year. The significance of VA mycorrhizas in the field is still much less clear.

Many strains and species of mycorrhizal fungi exist, and ectomycorrhizal associations may be specific. Species of the genera *Boletus*, *Suillus*, *Amanita*, and *Russula*, and many other Homobasidiomycetes, are normally found only under particular tree species—the fly agaric, *Amanita muscaria*, for example is characteristic of birch woods. By contrast over 100 different fungi can form associations with *Pinus sylvestris*, and these in turn with more than 700 other tree species (Trappe, 1962). Very many fewer fungal species are involved in the endomycorrhizal association. A taxonomic hierarchy was proposed by Gerdemann and Trappe (1975) and Hall (1984) has updated this and published keys. Descriptions of new species are published frequently, but many of these seem likely only to be local types. These strains and species vary greatly in their effectiveness at increasing P uptake and growth and in their behaviour on different soils (Mosse, 1972). Table 3.11 shows that onion growth was improved by a factor of between 2 and 30 times by different strains, though heavy infection occurred in all but the two least effective, and that on two soils the order of effectiveness of eight strains was not the same. Although, therefore, there is little or no specificity of infection, a given species or strain may well differentially benefit various plant species, depending on soil type.

Indeed in the field, mycorrhizal fungi indigenous to a soil are often found to give no benefit to introduced crop plants, which require inoculation with cultured strains to show a response; this is certainly true of *Trifolium repens* in hill pastures in New Zealand (Powell, 1976). Field studies of native communities suggest that there is widespread infection of the roots of co-existing species (Read *et al.*, 1976), but that

TABLE 3.11.   Effect of six strains of VAM fungi on the growth of onions (dry wt. (mg)) over 12 weeks, and the influence of bacterial washings and soil type. The two soils used are selected from those in Table 3.9. (From Mosse, 1972.)

Strain	Soil Leachings[a]	Inoculated	Soil 11	Soil 7
Laminate				
(*Glomus caledonium*)[b]	48	714	737	370
E$_3$ (*Glomus* sp.)	22	640	708	441
Honey-coloured				
(*Acaulospora laevis*)	40	525	689	475
SBC				
(*Glomus ?fasciculatum*)	41	397	649	631
Glomus microcarpa	31	61	151	134
Bulbous reticulate				
(*?Gigaspora calospora*)	25	43	139	153

[a]Material obtained by washing roots and filtering so as to exclude fungal spores.
[b]We are grateful to Dr D. P. Stribley for suggesting probable identification.

considerable variation exists between species. It is not clear whether this reflects the chance of colonization, differences in foraging ability of uninfected root systems, or differential P demand. There is little evidence that depriving plants in the field of infection is deleterious, and indeed it may give some benefit (Fitter, 1986). Similarly, on Pennine grassland soils Sparling and Tinker (1975) could show no benefit of infection to the native grasses.

VA mycorrhizal infection therefore appears to be a mechanism which enables plants to overcome limitations of diffusion in P uptake, but because for much of a plant's life P may not be limiting growth, and because in any case the external hyphae which obtain the phosphate are very susceptible to grazing damage by soil animals (Finlay, 1985), they may only offer a benefit at, for example, particular times in the life-cycle such as seedling establishment. The failure of seedlings to establish is often due to root competition with mature plants (see Chapter 8, p. 303); infection might obviate the need to establish an early root system, with the obvious resource costs that entails. Alternatively, since soil moisture content directly affects diffusion coefficients in soil (see p. 84), it might be only in dry periods that mycorrhizal transport of P was beneficial.

Some plants, notably sedges and rushes (Powell, 1975), normally exhibit low levels of mycorrhizal infection, while others seem almost obligately mycorrhizal. Baylis (1975) has related such variation to variation in root morphology. Plant families and groups which on floral characters are considered primitive, such as the Magnoliales, have very coarse root systems with all roots more than 0·5 mm in diameter; they have no root hairs and apparently are heavily dependent on infection. At the other extreme the grasses, generally accepted as an advanced family, have finely branched roots, a dense root hair zone, and are often only mycorrhizal in soils of low P status. Rushes and sedges have similar roots, but even longer root hairs. Since hyphae are very narrow, with a radius usually less than 5 μm for VA fungi, their cost (in carbon or other resource terms) is certainly no greater than root hairs, which typically have a radius of 5–10 μm, and they can extend much further into soil. Nevertheless it seems that VA infection may consume 5–10% of total photosynthate (Snellgrove *et al.*, 1982; Koch and Johnson, 1984) and ectomycorrhizas may be even more expensive (Fogel, 1985).

VA mycorrhizas seem then to be more or less facultative associations, in which the flows of carbon and phosphate between plant and fungus vary depending upon season, conditions and life-cycle stage. The dependence of the host on the association is related to root morphology and the whole symbiosis appears to be very flexible. Ectomycorrhizas, in contrast, seem to act as more permanent extensions of the root system, which is often severely reduced. The mycelium contains some hyphae with a very large cross-section and there is evidence that they may be responsible for the acquisition of water and other ions as well as phosphate. Certainly they tend to be more abundant than VA forms on the least fertile soils, both in temperate (Malloch *et al.*, 1981) and tropical (Alexander and Hogberg, 1986) forests.

VI. GENERAL PATTERNS OF RESPONSE TO SOIL NUTRIENTS

Until recently, concepts of mineral acquisition have been dominated by ion transport physiology. It is becoming apparent that this emphasis has prevented recognition of more general patterns (Chapin, 1980; Clarkson, 1985). Variations in uptake kinetics, though they certainly occur, are probably only of significance as short-term responses to local fluctuations in ion concentrations. Long-term adaptation is achieved by changes in demand for and use of nutrients, and by changes in root morphology (including symbionts) and distribution.

In effect, reducing demand for nutrients involves a reduction in growth rate and so implies that longer life-cycles will be favoured: nutrient-poor soils tend to be dominated by perennial vegetation. Annuals are of course favoured by disturbance, which physically removes perennial vegetation, and where nutrient-poor soils are frequently disturbed (for example on mobile sand-dunes), few species can persist and a community of low diversity is found. Another consequence is the greater allocation of resources to unproductive fibrous tissue, which has a low nutrient content, as opposed to photosynthetic tissue. This in turn reduces palatability and provides protection from grazing, so avoiding one possible loss of resources which might stimulate demand. An analogous response is an increase in the root weight ratio, which again reduces growth rate by diverting resources from photosynthetic tissue, and makes possible changes in root morphology which may improve nutrient acquisition.

Nutrient-poor environments are then characterized by high levels of allocation of resources to roots—from 40 to 85% in a range of natural ecosystems (Fogel, 1985). Nutrient deprivation also changes root morphology, with P-starved grass plants, for example, having main axis diameters of around 0·5 mm as compared to 0·7–1·0 mm in high P conditions (Christie and Moorby, 1975). Root diameter can be integrated for a whole root system by measuring the length per unit root weight or specific root length (SRL, cm mg^{-1}). SRL is nearly always increased by nutrient deprivation (Fitter, 1985)

In field soils nutrient supply varies as much in time as in space, even or perhaps especially in fertile soils. Species capable of a good growth response to nutrients also have very much more plastic root morphology, including SRL (Christie and Moorby, 1975; Table 3.12). The experiments of Grime et al. (1986), described on p. 104, using growth chambers to which pulses of nutrients were supplied, vividly demonstrate the greater root plasticity of the more nutrient-responsive species.

TABLE 3.12.　Responsiveness to phosphate supply, specific root length and root diameter of two arid-zone grasses (Christie and Moorby, 1975).

	Thyridolepis mitchelliana	Cenchrus ciliaris
Response to P[a]	3·0	29·6
Specific root length (cm mg^{-1})		
3 mg l^{-1}	16·2	13·6
0·003 mg l^{-1}	19·2	27·8
ratio	1·10	2·04
Root diameter (μm) Nodal axes		
3 mg l^{-1}	704	968
0·003 mg l^{-1}	552	460

[a]Response to P is the quotient of total dry weight of plants grown at 3 mg P l^{-1} to those grown at 0·003 mg P l^{-1}.

Plants of infertile soils seem then to have a less flexible morphology and physiology. They are slow-growing, unpalatable, devote much photosynthate to root growth, are usually mycorrhizal and tend not to respond morphologically to changes in nutrient concentration. These characteristics are well suited to the efficient acquisition and utilization of nutrients (and in many cases of water too—see Chapter 4) when they are scarce. In another context they are recognizable, too, as the characteristics of stress-tolerance (Grime, 1979). As a consequence, such species are readily displaced from more fertile soils by species with opposite characteristics, which imply greater photosynthetic activity, shoot growth and competitive ability for light.

The emphasis on research in this field is now shifting from studies of uptake kinetics, often at high external concentrations, to those in which the whole soil–plant system is considered. The recognition of the importance of allocation patterns, morphology and growth physiology may prove valuable in many fields, such as the breeding of crop plants for low-input agriculture.

4. WATER

I. PROPERTIES OF WATER

Water is the major component of green plants, accounting for 70–90% of the fresh weight of most non-woody species. Most of this water is contained in the cell contents (85–90% water) where it provides a suitable medium for many biochemical reactions. However, water has many other roles to play in the physiology of plants and it is uniquely fitted, by its physical and chemical properties, to fulfil these roles. An early, but still very useful review of these properties can be found in Henderson (1913).

Because of the unusually strong hydrogen bonds between its molecules, water tends to behave as if its molecules were very much larger (Bernal, 1965; Davis and Day, 1961). For example, the melting point (0°C) and boiling point (100°C) of water (molecular weight 18) are anomalously high when compared with the values ($-86°C$ and $-61°C$, respectively) for the closely-related compound, hydrogen sulphide (H_2S, molecular weight 34)(Hay, 1981a). Thus, unlike all other substances made up of small molecules, water normally remains in the liquid state under terrestrial conditions, although problems for living organisms do result from the freezing of water. For the same reason, the specific heat ($4·2 \, J \, g^{-1}$), the latent heat of melting ($333·6 \, J \, g^{-1}$) and the latent heat of vaporization ($2441 \, J \, g^{-1}$ at 25°C) of water are all unusually high, with important implications for the thermal economy of plants. The high specific heat of water buffers the plant body against rapid fluctuations in temperature, whereas the high latent heat of vaporization provides an effective means of cooling leaves by the evaporation of water (Gates, 1976). The same properties are responsible for moderating the temperatures of moist soils, lakes and the ocean.

Water is an excellent solvent for three groups of biologically important solutes, namely:

(i) Organic solutes with which water can form hydrogen bonds, including amino acids and low molecular weight carbohydrates and proteins, which contain hydroxyl, amine or carboxylic acid functional groups. Water also forms colloidal dispersions with higher molecular weight carbohydrates and proteins (e.g. the cytoplasm itself).

(ii) Charged ions such as the major nutrient ions (K^+, Ca^{2+}, $H_2PO_4^-$, NO_3^-, etc.). Water molecules, which carry partial charges, orientate themselves round ions to give larger, but highly-soluble, hydrated ions. In the same way, water molecules become attached to fixed charges on the surfaces of plant cell walls, cell membranes and soil particles, giving tightly-bound layers of water a few molecules thick (Bernal, 1965).

(iii) Small molecules, such as the atmospheric gases (O_2, N_2), which, presumably, can fit into voids in the rather open structure of liquid water (Crafts, 1968a).

Thus, as well as being an ideal solvent for many biochemical reactions, water is also a suitable medium for the transport of organic molecules (e.g. sucrose in the phloem), inorganic ions (e.g. nutrients from root to leaf in the xylem) and atmospheric gases (e.g. diffusion of oxygen to sites of respiration).

Two other physical properties, tensile strength and viscosity, are highly important in the long-distance transport of water and dissolved solutes. In particular, the high tensile strength (cohesion) of liquid water columns in xylem elements (Table 4.8) means that water can be drawn to the tops of tall trees by transpirational pull alone (Oertli, 1971). This cohesion is another consequence of the unusual strength of the hydrogen bonds between water molecules; however, in spite of these bonds, water has a moderate viscosity (Meidner and Sheriff, 1976). This results in rapid flow in, for example, soil macropores, but may be less important in soil capillaries and cell walls, where the ratio of surface-bound to bulk water is high.

Water can move from the soil, through root and stem, to a transpiring leaf only if there is continuity of liquid throughout the pathway. Thus, in addition to continuous columns of water in the xylem, the plant also requires continuity of water in the capillaries of the soil and the apoplasts of both root and leaf. That this continuity does exist is the consequence of two properties of water. First, as explained above, continuous thin films form on the hydrophilic surfaces of soil and cell wall capillaries. Secondly, since the surface tension of water is very high (73.5×10^{-3} kg s^{-2} at 15°C, i.e. two to three times the value for most laboratory solvents), the filling of capillaries with water results in a large reduction in the energy (and increase in the stability) of the capillary/water system because of the reduction of water surface area in contact with air. These two forces retaining water in capillaries constitute the matric forces. For example, after the gravitational water has drained from a water-saturated soil, all the soil capillaries of diameter less than 60 μm are filled with water retained by matric forces (Russell, 1973). Both the distribution and the morphology of green plants are influenced by the fact that water absorbs specifically in the infra-red but is relatively transparent to short-wave radiation. Thus, aquatic plants can absorb photosynthetically active radiation (PAR) at considerable depths in clear water; on the other hand, in terrestrial plants, these optical properties have permitted the evolution of leaves in which the colourless outer layer of cells (the epidermis) permits the passage of PAR to the underlying layers of mesophyll cells which are active in photosynthesis. However, the epidermis does not appear to act as a heat filter, protecting the mesophyll; on the contrary, Sheriff (1977) has shown that, under moderate irradiances, the epidermis can be significantly cooler than underlying tissues owing to heat loss by transpiration.

As well as being a useful solvent, liquid water is also an important biochemical reagent in, for example, hydrolysis reactions. Much of the chemical activity of water is a consequence of its dissociation to give highly-reactive hydronium and hydroxide ions:

$$2H_2O \rightleftharpoons H_3O^+ + OH^-, \quad \text{where } K_w = 10^{-14} \tag{4.1}$$

Thus even pure water is a 10^{-7} M solution of hydronium ions.

Because transpiration involves the diffusion of water molecules in the gas phase, it is important to understand the different methods of expressing the water vapour content of air, and the inter-relationships between the different units which are used. The water vapour content of air in equilibrium with liquid water (expressed as the density of saturated water vapour or, more commonly, the saturation water vapour density, in g of water per m³ of moist air) rises sharply with increasing temperature, as does the corresponding partial pressure exerted by water vapour (i.e. the saturation water vapour pressure, in kPa) (Table 4.1). The water vapour content of a volume of *unsaturated* air can also be expressed in terms of the water vapour density (g m^{-3}), but since the air in the substomatal cavity is normally saturated, both the water vapour content of the bulk air and the driving force for the diffusion of water molecules from the substomatal cavity to the

TABLE 4.1. Methods of expressing the water vapour content of air at
            different temperatures.

Temperature (°C)	0	5	10	15	20	25	30	35
Saturation water[a] vapour density g m^{-3}	4·9	6·8	9·4	12·9	17·3	23·1	30·4	39·6
Saturation water vapour pressure kPa	0·6	0·9	1·2	1·7	2·3	3·2	4·2	5·6
Relative humidity[b]								
Water vapour pressure deficit (VPD) kPa								
0·1	83	89	92	94	96	97	98	98
0·2	67	78	83	88	91	94	95	96
0·3	50	67	75	82	87	91	93	95
0·4	33	56	67	77	83	88	91	93
0·5	16	44	58	71	78	84	88	91

[a]Note that, over the same temperature range, the *density of moist air* changes very little (from $1·3 \times 10^3$ to $1·1 \times 10^3$ g m^{-3}).
[b]The water vapour density of the air as a percentage of the saturation vapour density.

bulk air are commonly expressed by the water vapour pressure deficit or VPD (i.e. the difference between the actual vapour pressure of the bulk air and the saturated vapour pressure, in kPa). However, it must be stressed that a given VPD can correspond to a wide range of relative humidities, depending upon the temperature (Table 4.1).

II. THE WATER RELATIONS OF PLANTS AND SOILS

A. Water Potential

Water flowing down a hillside can turn a water wheel, yielding useful work. The water at the bottom of the slope has lost part of its capacity to do work and has a lower free energy content than the water at the top. Thus we have a flow of water whose driving force is the difference in free energy between the top and bottom of the slope. In a similar way, water moves in the soil–plant–atmosphere system in response to differences in the free energy content (capacity to do work) of water in different parts of the system. For example, in a well-watered plant which is transpiring, the free energy content of water decreases progressively from the soil, via the xylem and the leaf apoplast, to the bulk air, and water flows from the soil through the plant to the air in response to this gradient in free energy.

In plant physiology, it is customary to express the free energy content of water in terms of water potential (Ψ). Derivations of water potential from strict thermodynamic principles can be found in Slatyer (1967) and Meidner and Sheriff (1976); however, for the present purposes, it is sufficient to define water potential as the free energy per unit volume of water, assuming the potential of pure water to be zero under standard conditions (usually ambient temperature and atmospheric pressure). Since energy per unit volume has the same dimensions as pressure, plant and soil water potentials are expressed in pressure units, normally MPa, although the bar ($1 \text{ bar} = 10^5 \text{ Pa} = 0\cdot1 \text{ MPa}$) was widely used until recently.

Because water potential increases with temperature, it is important to maintain constant temperature during a series of measurements. In contrast, water potential is lowered below that of pure water by dissolved solutes and also by the binding of water to surfaces by matric forces. Since these effects are considered to be mutually independent, the water potential of a solution (Ψ) can be expressed as

$$\Psi = \psi_s + \psi_m \tag{4.2}$$

where ψ_s, the solute potential, is the lowering of water potential by dissolved solutes, ψ_m, the matric potential, is the lowering of water potential owing to matric forces, and Ψ will be negative, since the potential of pure water is set at zero.

Consequently, in the soil–plant–atmosphere system, water potentials are usually negative, water flows towards regions with more negative values, and the expressions "lower" and "higher" water potential indicate more and less negative values, respectively. For example, in rapidly-transpiring Sitka spruce trees, water flowed upwards, but down a water potential gradient, from the soil ($-0\cdot04 \text{ MPa}$) to the terminal shoots ($-1\cdot5 \text{ MPa}$) at 10 m above the soil surface (Hellkvist *et al.*, 1974).

Because the water potential in plant tissues can be raised by hydrostatic pressure, equation 4.2 must be modified to give

$$\Psi = \psi_s + \psi_m + \psi_p \tag{4.3}$$

where ψ_p, the pressure potential, is the increase in water potential owing to hydrostatic pressure. Since ψ_p is positive, the resulting water potential is less negative than would be the case if only the solute and matric effects were included. Occasionally, hydrostatic pressure within plants can result in zero, or even positive, water potentials, for example, in the xylem of guttating plants (Crafts, 1968b).

Methods for measuring water potential and its components are described in detail in Slavik (1974).

B. The Water Relations of Plant Cells

Most of the cells involved in plant–water relations are fully mature, with a large fraction of the cell water contained in a central vacuole. The thin layer of cytoplasm, together with its associated plasmalemma and tonoplast, can be treated as a complex semi-permeable membrane separating the vacuolar contents from the external medium.

In a leaf, the external medium is the water in the cell walls and intercellular spaces (the leaf apoplast), which is subject to atmospheric pressure (i.e. $\psi_p = 0$). Since the solute concentration is normally very low, ψ_s is small and, therefore, the water potential in the apoplast is determined by the matric forces exerted by the cell walls. Thus

$$\Psi_{apo} = \psi_m \tag{4.4}$$

where ψ_m will normally be high (> -0.1 MPa) in a leaf well-supplied with water but not transpiring (e.g. at night). In the vacuole, matric forces are much less important, and the water potential is determined largely by the solute concentration. Thus

$$\Psi_{vac} = \psi_s \tag{4.5}$$

where ψ_s can vary from -0.5 to -3.0 MPa and even lower in some cases, according to species and the degree of osmoregulation (see p. 132).

Since Ψ_{vac} is lower than Ψ_{apo}, water tends to flow inwards across the cytoplasm, raising the vacuolar water potential by diluting the vacuolar sap, but also increasing the volume of the vacuole. In a plant cell lacking cell walls, this influx of water would continue until either the cell burst or the difference in water potential was abolished. However, in a leaf, the volume of each mature cell is limited by its walls, and only a relatively small inflow of water can be accommodated by the elasticity of the walls (compare with the yielding of the walls of cells which are still expanding, see below). Consequently, hydrostatic pressure (turgor pressure) builds up within the vacuole, pressing the cytoplasm against the inner surface of the cell walls, thereby raising the vacuolar water potential. As turgor pressure rises, adjacent cells press against one another, with the result that the leaf, originally in a wilted, flaccid condition, becomes increasingly turgid. Ultimately, equilibrium is reached when the driving force for water influx (the difference in solute potential) equals the driving force for water efflux (cell turgor pressure), and there is no further movement of water into the vacuole. At this point, cell turgor pressure is at its maximum, and

$$\Psi_{apo} = \Psi_{vac} \tag{4.6}$$

i.e.

$$(\psi_m)_{apo} = (\psi_s + \psi_p)_{vac} \tag{4.7}$$

The continuous and progressive development of turgor pressure (pressure potential) caused by water influx into plant cells can be illustrated by modern versions of Höfler's diagram. For example, Fig. 4.1 shows a typical Höfler plot for leaf cells, in which, for clarity, the water potential of the bathing medium is fixed at zero. According to Fig. 4.1, a flaccid cell with a cell volume of $1 \cdot 0$ unit and a ψ_s of $-1 \cdot 6$ MPa will absorb water from the bathing medium, thus increasing cell volume and raising ψ_s by sap dilution. However, since the cell walls resist expansion, ψ_p rises at an increasing rate until maximum turgor pressure is achieved when $\psi_p = 1 \cdot 2$ MPa, whereas ψ_s has risen to $-1 \cdot 2$ MPa. At this point, since $\Psi_{vac} = \psi_s + \psi_p = 0$, the driving force for the entry of water has declined to zero and the cell has reached its maximum volume ($1 \cdot 3$ units).

The treatment so far has concentrated upon the water relations of cell vacuoles. Since virtually all the biochemical and physiological processes in a plant take place in the cytoplasm or cytoplasmic organelles, the water relations of the cytoplasm are of much greater interest. It is customary to assume that the cytoplasmic water is in thermodynamic equilibrium with the vacuolar water; consequently, under most conditions, the water potential in the cytoplasm (Ψ_{cyt}) will be identical to that in the vacuole (Ψ_{vac}), and subsequent discussion can be simplified by using the term cell water potential (Ψ_{cell}), where

$$\Psi_{cell} = \Psi_{cyt} = \Psi_{vac} \tag{4.8}$$

However, it must be stressed that although cytoplasmic and vacuolar potentials are equal, the relative contributions of the components of water potential will not normally be the same. In particular, because of high concentrations of colloidal particles, there may well be a significant matric potential component in the cytoplasm.

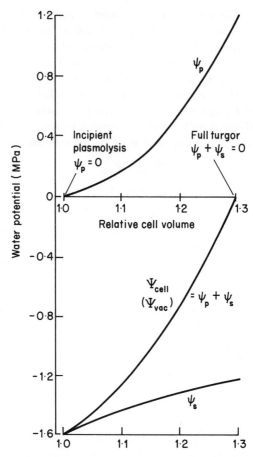

FIG. 4.1. "Höfler" diagram for an idealized leaf cell showing the relationship between vacuolar water potential (Ψ_{vac}) and its components, pressure potential (ψ_p) and solute potential (ψ_s), at different cell volumes (after Meidner and Sheriff, 1976).

C. The Influence of Water Stress on Plants and Plant Cells

Stress has been defined as "any environmental factor capable of inducing a potentially injurious strain in (plants)" (Levitt, 1980), where the "strain" can be reversible or irreversible (although it is difficult to apply these terms to plants which can sustain irreversible injury but replace damaged parts by resumed growth). It is relatively easy to quantify and measure certain stresses; for example, thermal and freezing stresses can be expressed in terms of the number of day degrees above or below a critical temperature for damage, and the degree of shading stress can be measured using appropriate radiation sensors. However, because of the complexity of plant/water relations, there is no single index of water supply by the environment (soil water content, bulk air humidity, etc.) which can be used to express the degree of water deficit stress (normally called water stress) to which a plant is subjected. As a result, it has become conventional to use plant rather than environmental indices of water stress (principally tissue water potential, but also relative water content or water deficit, each of which is a measure of the extent to which tissue water content has fallen below the maximum water content at full turgor (Fig. 4.1), at which conditions are optimum for growth and function). Since the photosynthetic uptake of CO_2 by the moist apoplast of the leaf mesophyll is inevitably associated with water loss to the atmosphere, and loss of turgor (see below), the leaves of plants growing in most environments are exposed to some degree of water stress throughout their lives during the daily periods of illumination. It is clear that coping with water stress is a routine aspect of the physiology of most plant species, not simply those growing in dry habitats.

In reviewing the effects of water stress on plant growth and function, Hsiao (1973) found it convenient to use three, rather loosely defined, degrees of water stress, in relation to a "typical mesophyte":

Mild stress: Ψ_{cell} slightly lowered, typically down to -0.5 MPa at most.

Moderate stress: Ψ_{cell} lowered to values in the range -0.5 to -1.2 or -1.5 MPa.

Severe stress: Ψ_{cell} below -1.5 MPa.

Figure 4.1 shows that the imposition of mild water stress is associated with a rapid loss of turgor. This loss of turgor continues, at a declining rate per unit of water potential, under moderate water stress, but severe water stress normally involves the complete loss of turgor $(\psi_p = 0)$, and leaf wilting. As the volume of the cell contents decreases, there is a tendency for the plasmalemma to shrink away from the cell wall (plasmolysis); however, according to Meidner and Sheriff (1976), plasmolysis of leaf cells is a rare event in nature since it is difficult for water or air to move inward to fill the space vacated between the plasmalemma and cell wall. Thus, exposure to severe water stress involves mechanical stress as well as serious dehydration.

Interpretation of the effects of different degrees of water stress on plant function can be complicated by the fact that responses can be evoked at both the tissue and the cell level. For example, the stomata of many mesophytic species close at leaf water potentials between -0.5 and -1.0 MPa (Fig. 4.2), thereby cutting off the supply of carbon dioxide to the mesophyll. Thus it is possible that the rate of photosynthesis of the leaf may be reduced considerably by levels of water stress which would have a smaller effect upon the photosynthetic apparatus of individual cells. In spite of such difficulties, there are many reports of the effects of different degrees of water stress on cell growth and function, as reviewed, for example, by Hsiao et $al.$ (1976)(Fig. 4.2). Cell and leaf expansion are the processes most sensitive to water stress because of their dependence upon turgor. The relative rate of volume increase of a cell can be described by the simplified equation:

$$\frac{1}{V} \cdot \frac{\mathrm{d}V}{\mathrm{d}t} = \varphi(P - \Upsilon)$$

(4.9)

where V is cell volume, φ is the (volumetric) extensibility of the cell

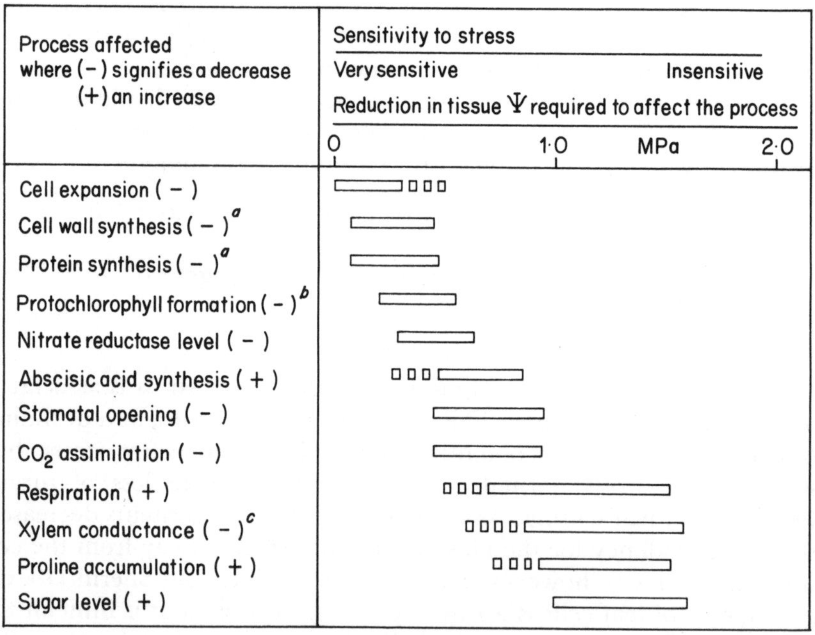

Process affected where ($-$) signifies a decrease ($+$) an increase	Sensitivity to stress		
	Very sensitive Insensitive Reduction in tissue Ψ required to affect the process		
	0	1·0 MPa	2·0
Cell expansion ($-$)			
Cell wall synthesis ($-$)[a]			
Protein synthesis ($-$)[a]			
Protochlorophyll formation ($-$)[b]			
Nitrate reductase level ($-$)			
Abscisic acid synthesis ($+$)			
Stomatal opening ($-$)			
CO_2 assimilation ($-$)			
Respiration ($+$)			
Xylem conductance ($-$)[c]			
Proline accumulation ($+$)			
Sugar level ($+$)			

[a]Rapidly-growing tissue; [b]etiolated leaves; [c]should depend on xylem dimensions

Fig. 4.2. The influence of water stress on the physiology of mesophytic plants. The continuous horizontal bars indicate the range of stress levels within which a process is *first* affected, whereas the broken bars refer to effects which have not yet been firmly established. The reductions in tissue Ψ used are in relation to the Ψ of well-watered plants under mild evaporative demand (after Hsiao *et al.*, 1976).

wall, P is cell turgor pressure and Υ is the yield threshold pressure (the minimum turgor pressure required for expansion to occur). Thus the rate of expansion of an immature cell can decrease as a result of a reduction in cell wall extensibility or in cell turgor, an increase in yield threshold, or a combination of these effects. Most studies of cell expansion (including the data reported in Fig. 4.2) have relied upon whole tissue measurements, although it is clear that each of these three characteristics (φ, P and Υ) can vary amongst the cells of a leaf, particularly between mature and expanding cells (Tyree and Jarvis, 1982). However, the *short-term* sensitivity to water stress of expanding plant cells appears to be mainly a consequence of changes in turgor rather than in the properties of the cell walls. For example, in expanding cells of the fifth leaf of maize which had a constant yield threshold of 0·4 MPa, a reduction in turgor pressure of only 0·25 MPa (corresponding to a change in water potential from $-0·55$ to $-0·9$ MPa) was sufficient to reduce the rate of expansion from its maximum to zero (Hsiao *et al.*, 1985). These values of cell water potential are lower than the range of values presented in Fig. 4.2, presumably because the latter are tissue values, averaged over expanding and mature zones of the leaf. Since water stress has less effect upon the duration (than the rate) of leaf expansion, long-term depression of the rate of expansion under water stress results in smaller leaves and less extensive leaf canopies. This will tend to reduce the quantity of solar radiation intercepted by the canopy, thereby reducing its potential for dry matter production. However, this need not be the case for short-term exposure to mild water stress because the expansion of leaves and other tissues can take place at night when stomatal closure permits the rehydration of above-ground tissues, and also because CO_2 assimilation and respiration are less sensitive to leaf water potential than is cell expansion (Fig. 4.2; Boyer, 1970). In general, as water stress increases from mild to moderate, cell biochemistry is increasingly disturbed. Thus protein and chlorophyll synthesis are reduced under mild stress, whereas, under moderate stress, nitrate reductase and the levels of growth substances begin to be affected. Moderate to severe stress is associated with serious disruption of metabolism as indicated by increases in respiration and the accumulation of proline and sugars (see also Chapter 6, p. 250).

Throughout this treatment of cell water relations, attention has been concentrated upon short-term effects on the leaf cells of a "typical mesophyte" with a solute potential of about $-1·5$ MPa and a response to water loss similar to that shown in Fig. 4.1. This has simplified the discussion of water stress and permitted the use of Hsiao's (1973) classification to review the effects of lowered water potential on growth and function (Fig. 4.2). However, there is considerable variation in vacuolar solute potential amongst plant species (normally within the range -0.5 to -3.0 MPa, the value being broadly related to the water relations of their natural environment), and also, the response to water stress over longer periods can be different from that shown in the Höfler diagram. In particular, plant species vary in their ability to maintain the solute potential in expanding cells; in some cases, the influx of water leads to dilution of the cell contents (Tomos, 1985), whereas, under water stress, other species are capable of lowering the solute potential of growing cells by secreting solutes into the vacuole, leading to increased movement of water into the cells, increased cell turgor and maintained cell expansion (i.e. osmoregulation; Morgan, 1984). Because of effects of this type, some prefer to indicate the severity of water stress by the fraction of cell water lost rather than the depression of water potential, although serious difficulties can be encountered using this approach (see p. 168).

Overall, it can been seen from Fig. 4.2 that exposure of plants to even mild water stress can affect growth, and lead to the disruption of metabolic processes. Depending upon their severity, these effects can reduce the ability of the plant to survive and reproduce. Consequently, it is crucially important for all terrestrial species either to avoid water stress or to evolve anatomical, morphological and biochemical adaptations which lead to the amelioration or tolerance of water stress. Much of the remainder of this chapter is devoted to the study of these adaptations, but, because of the great popularity of plant/water relations as a study area and the enormous quantity of information which is available from a range of dry environments, the treatment must be highly selective, omitting many important adaptations and habitats. A recent, more encyclopaedic review of plant water relations can be found in Lange *et al.* (1982).

D. The Supply of Water by the Soil

As we shall see later, the absorption of dew and rain by leaves can be an important factor in the survival of some plant species in arid zones. However, as far as most terrestrial plants are concerned, this mechanism is of negligible importance compared with the absorption of soil water through the root system.

The quantity of water held by a soil depends primarily on the climate and, in particular, on the excess of precipitation over evapotranspiration $(P-E)$. Thus, in "extremely humid" areas such as the northern and western parts of the UK (Geiger, 1965), where annual precipitation is at least twice evapotranspiration, it is unusual for the availability of soil water to be a factor in the survival and distribution of plant species. Exceptions to this rule occur in unusually dry habitats, for example on shallow or very coarse soils. In contrast, in "extremely arid" regions, where potential evapotranspiration is more than twice precipitation, soil moisture levels are normally so low that the vegetation is sparse.

Between these extremes, there is a spectrum of humid and arid climates, where the availability of soil water depends not only on the annual $P-E$ but also on the distribution of rainfall within each year. Many of the features of drier climates, where the risk of deficit is high, can be illustrated by Fig. 4.3, which compares the rainfall distribution for two contrasting, but not extreme, years in a tropical savanna region where virtually all of the annual rainfall is received during the months November to April. In 1970/71, the "wet season" began and ended decisively on 25 November and 8 March respectively, with a few scattered days of rain in early November, late March and April. During the wet season there were no prolonged dry spells when significant depletion of soil moisture could occur, and it can be assumed that the soil was near field capacity throughout. In contrast, in 1971/72, both the beginning and end of the wet season were ill-defined and a higher proportion of the annual precipitation fell in the first half of November and in April. In addition, there were at least two mid-season dry periods (26 December to 4 January and 26 January to 8 February) during which annual plants would have been exposed to water stress, according to agrometeorological calculations (Hay, 1981b).

To be successful under such conditions, a plant species must possess several characteristics, including the following abilities:

(i) to survive long periods without rain every year (six months in this case);

(ii) to make full use of the period of soil water availability for the completion of its annual life cycle without sustaining serious injury from water stress at the beginning and end of the wet season. Since the date of the beginning of the growing season is variable, the plant requires a mechanism to signal the earliest "safe" date for the start or resumption of growth. (Mechanisms of this kind are discussed below when considering the timing of germination of dormant seeds, and leaf production in deciduous perennials);

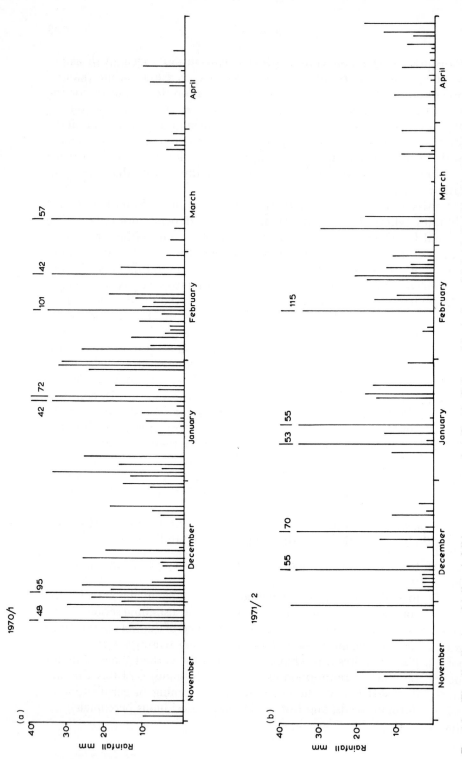

Fig. 4.3. The distribution of daily rainfall in a savanna region (Bunda College of Agriculture, Malawi, C. Africa) in (a) 1970/1971 and (b) 1971/1972. There was virtually no rainfall during the remaining months of each year.

(iii) to avoid, ameliorate or tolerate short periods of mild to severe water stress at different times and growth stages within the wet season;

(iv) to survive occasional years of drought when either the total amount of precipitation or the length of the wet season is greatly reduced.

In more humid and more temperate regions, where rainfall is less seasonal, (i) and (ii) become less important, but (iii) and (iv) remain essential characteristics for plant species growing in all but the most humid climates.

Within a given climatic zone, the availability of water for plant uptake depends upon the water-storing properties of the soil. Soils consist of mineral particles of varying diameter (sand 2–0·06 mm, silt 0·06–0·002 mm and clay 0·002 mm, Soil Survey of England and Wales classification; the USDA classification is identical except that the boundary between sand and silt is set at 0·05 mm), bound together into aggregates by organic matter and clay particles. Within and between these aggregates there is a network of interconnected spaces of diameter ranging from a few cm (drying cracks, earthworm or termite channels) through a few mm (between aggregates) down to a few μm or tenths of a μm (within aggregates). Although these spaces are irregular in shape, it is conventional to refer to the system of spaces as the pore space and to treat it as if it were made up of a set of regular pores. This simplification is used here but it can cause serious difficulties in more advanced treatments. When a soil is saturated with water after prolonged rainfall, the pore space becomes temporarily water-filled. However, a free-draining soil cannot hold all of this water for plant use.

Since the concentration of solutes in the soil water is generally very low, the major forces retaining water in soil pores are the matric forces, which increase as pore diameter (d) decreases. Consequently, the water potential in a water-filled pore is inversely related to pore diameter by the expression:

$$\Psi_{\text{pore water}} = \psi_m = -0\cdot3/d \qquad (4.10)$$

where water potential is in MPa and d is in μm (Russell, 1973). For example, pure water held in soil pores of diameter 10 μm will be at a potential of 0·03 MPa; to withdraw water from these pores requires a suction of at least 0·03 MPa. Similarly, since gravity exerts a suction equivalent to about 5 kPa in temperate regions (Webster and Beckett, 1972), all pores wider than 60 μm will tend to drain spontaneously after a soil has been saturated with water. In a freely-draining soil, this drainage can take two to three days, and once it is complete, the soil is at field capacity. At field capacity, the soil contains the maximum amount of water (normally expressed in g per 100 g oven-dry soil) that it can hold against gravity. The gravitational water normally lost in drainage is not available to plants unless drainage is impeded.

As indicated in an earlier section, transpirational loss of water from the leaves of a plant establishes a gradient of water potential in the plant/soil system. If, as a result, the water potential in the xylem of a root axis falls below the water potential in the soil pores adjacent to its surface, then water will flow into the root and pass via the xylem to the site of transpiration. Soil water is, therefore, available to plants only if the root xylem water potential (and, in turn, leaf water potential) can be lowered below the soil water potential. Significant quantities of water are retained in soil pores at potentials down to -3 MPa, but few mesophytes can tolerate the lowering of leaf water potential required to absorb all of this water.

Many of the species which have been studied in detail (principally crop plants) can withdraw water from pores wider than about $0 \cdot 2$ μm, corresponding to a water potential of $-1 \cdot 5$ MPa; once all the water in these pores has been exhausted, no more can be transported to the leaf and the plant will wilt permanently and ultimately die, unless the soil is recharged with water. Consequently, the property "permanent wilting point" has been defined as the moisture content (g per 100 g oven dry soil) after the soil has come to equilibrium under a suction of $1 \cdot 5$ MPa (at which there is relatively little water left in soils other than clays, Fig. 4.4). The available water content of the soil is then the difference between field capacity and permanent wilting point. Although this standard method of determining permanent wilting point and available water is invaluable in comparisons of the water relations of different soils (e.g. Fig. 4.4), it is no more than a laboratory convenience. In reality, a given soil will release differing quantities of water to different species (and even to plants of the same species grown under different conditions) according to the minimum leaf and xylem water potentials which can be tolerated (e.g. Table 4.7).

The amount of soil water which is available for uptake by a plant depends primarily upon the size distribution of the soil pores. This, in turn, is dependent upon both soil texture and structure (extent and type of aggregation), but, in general, medium to fine textured soils tend to hold more water for plant use than coarse textured soils (Fig. 4.4). Temperature also influences soil water availability through its effect upon the viscosity of water. In the foregoing discussion of field capacity, the conventional assumption has been made that only those

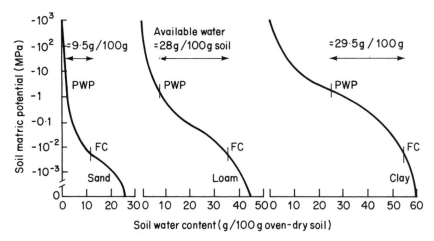

Fig. 4.4. Typical soil water release curves for coarse (sand), medium (loam) and fine (clay) soils. FC and PWP indicate field capacity and permanent wilting point, as defined in the text. Note that although the available water contents of the loam and clay soils are very similar, a larger proportion of this water is held at matric potentials above −0·1 MPa (i.e. in wider pores) in the loam soils (redrawn from Brady, 1974).

pores wider than 60 μm drain under the influence of gravity. However, this assumption is based on field measurements of soils in winter and spring in the UK (e.g. Webster and Beckett, 1972). Russell (1973) estimates that in sub-tropical zones, the lowering of water viscosity under warmer conditions, coupled with very free drainage, can lead to the drainage of pores as narrow as 10 μm. Depending upon pore size distribution, this can cause significant reductions in available water. Any increase in the solute concentration of soil water will result in a lowering of soil water potential and a reduction in the amount of water available for uptake; this effect is most pronounced in saline soils. Finally, in certain environments, soil texture and structure can have a large influence on the availability of water to plants through their influence upon soil surface properties and water infiltration. For example, in direct contrast to the normal conclusion, based upon soil water release curves (e.g. Fig. 4.4), coarse and poorly-textured soils of arid regions can sometimes hold a larger proportion of intense, episodic precipitation than can finer-textured soils, whose finer pores cannot accept water at a sufficient rate to avoid run-off.

Plants do not draw water only from the immediate vicinity of their actively-absorbing roots. As extraction of the water adjacent to a root proceeds, a depletion zone similar to those for mobile ions (see p. 85) develops (Hainsworth and Aylmore, 1986), causing water to flow over distances of at least several mm from the bulk soil to the root surface. However, at the same time, the withdrawal of water from the larger pores reduces the volume of soil through which flow can take place, and also increases the "tortuosity" or impedance of the pathway between the bulk soil and the root (see p. 84). These effects combine to give a progressive and drastic reduction in the hydraulic conductivity (reciprocal of resistance) of a soil as it dries. For example, a Pachappa sandy loam soil had a conductivity of 6 cm day^{-1} at a matric potential of -5 kPa (field capacity) which fell to less than 10^{-6} cm day^{-1} at -1.5 MPa (permanent wilting point) (Gardner, 1960). Consequently, the maintenance of a steady flow of water into a root from a drying soil requires:

(i) a progressive lowering of root xylem water potential to maintain the potential gradient between the xylem and the remaining soil water; and

(ii) a progressive increase in the steepness of the gradient to overcome the increasing resistance to water flow offered by the drying soil. This effect is illustrated in Fig. 4.7.

In conclusion, as we shall see in subsequent sections, it is common for the supply of water to the plant from the soil to be insufficient to meet demand during periods of rapid transpiration, with the result that the transpiring leaves are exposed to severe water stress long before the soil permanent wilting point has been reached.

E. The Loss of Water from Transpiring Leaves

Within a leaf, the walls of the mesophyll cells adjacent to sub-stomatal cavities (Fig. 4.5) must remain moist to permit the dissolution and uptake of carbon dioxide for photosynthesis. Consequently, as long as the stomata are fully closed, and the temperature is stable, the air contained in the leaf will tend to be saturated with water vapour (Table 4.1). Under these conditions, water can escape from the leaf to the surrounding air only by diffusing across the hydrophobic cuticle covering the epidermal cells. The rate of cuticular transpiration depends upon the thickness and composition of the cuticle, being low in all plants, and particularly low in young leaves with undamaged surfaces, and in many plants from dry environments.

When stomata begin to open in response to diurnal rhythms or environmental conditions (see below), the air outside the leaf is normally not saturated with water vapour (i.e. a vapour pressure

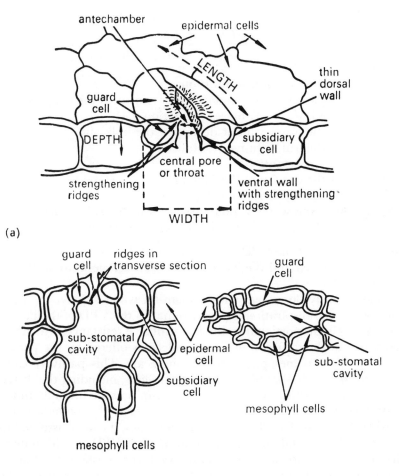

(a)

(b)   (c)

Fig. 4.5.   The structure and characteristics of stomata as illustrated by a typical elliptical stoma (as, for example, in *Vicia faba*). (a) Section shown as a perspective diagram, (b) transverse section including the substomatal cavity, and (c) longitudinal section through one guard cell and the substomatal cavity (from Meidner and Mansfield, 1968). Some idea of the considerable variation in the structure and characteristics of stomata amongst species and habitats can be gained from Table 4.2.

deficit, VPD, exists between the substomatal cavity and the bulk air). As a result of this VPD, water molecules diffuse out of the leaf air spaces into the bulk air via the stomatal pores, according to the equation:

$$F = \frac{d_1^s(T_1) - d_a}{R_1} = \frac{K\rho_{air}}{P} \cdot \frac{e_1 - e_a}{R_1}$$ (4.11)

where

F	is the rate of transpiration;
$d_1^s(T_1)$	is the saturation water vapour density at the temperature of the leaf (see Table 4.1, equation 5.4);
d_a	is the water vapour density of the bulk air;
R_1	is the leaf diffusive resistance;
K	is a constant;
ρ_{air}	is the density of the bulk air;
P	is atmospheric pressure;
e_1, e_a	are the water vapour pressures in the sub-stomatal cavity and bulk air, respectively (Kramer, 1969).

Note that $e_1 - e_a$ is the VPD of the bulk air if the air in the substomatal cavity is saturated and the leaf is at the same temperature as the bulk air (Table 4.1).

Consequently, for a leaf with a plentiful water supply, the rate of water loss by transpiration is proportional to the VPD of the bulk air, and inversely proportional to the leaf diffusive resistance, which is the resistance to the diffusion of water molecules offered by the pathway between the leaf mesophyll and the bulk air. This pathway can be resolved into a number of components in series, including the distance moved within the leaf air spaces; entry into, passage through and exit from the stomatal pore; and passage through the boundary layer outside the leaf surface. However, in most cases, it can be assumed that the rate of (stomatal) transpiration at a given VPD is determined by the sum of the stomatal and boundary layer resistances. (Note the common use of the reciprocal of resistance, conductance, whose dimensions are velocity, e.g. cm s^{-1}.) A more complete account of leaf diffusive resistance and its components is given in Chapter 7.

1. Stomatal Resistance

The resistance to the diffusion of water molecules offered by the stomata is proportional to the stomatal frequency (or density, i.e. number per unit area of leaf, Table 4.2) and inversely proportional to the diameter of the stomatal aperture, which, in turn, is dependent in a complex manner upon a number of environmental factors (Meidner

TABLE 4.2.   Stomatal frequencies, dimensions and pore areas on the upper and lower epidermis of fully-expanded leaves of a selection of plant species (from Meidner and Mansfield, 1968; Kramer and Kozlowski, 1979; see also Table 4.6).

Species	Frequencies (per mm²)		Length of stomatal pore (μm)	Pore area[a]
	Lower	Upper		
Herbaceous monocotyledons				
Allium cepa	175	175	24	2·0
Hordeum vulgare	85	70	17	0·7
Triticum aestivum	40	50	28	0·6
Zea mays	108	98	14	0·7
Herbaceous dicotyledons				
Vicia faba	75	65	28	1·0
Tradescantia virginiana	23	7	51	0·4
Helianthus annus	175	120	16	1·1
Sedum spectabilis	35	28	21	0·3
Angiosperm trees				
Carpinus betulus	170	0	13	0·6
Populus deltoides	226	187	30	3·0
Quercus robur	340	0	10	0·8
Tilia europea	0	370	10	0·9
Eucalyptus globulus	0	370	10	0·9
Gymnosperm trees				
Pinus sylvestris	120		20	1·2
Larix decidua	16	14	20	0·2

[a]Expressed as a percentage of total leaf area (both sides) with the pore fully open (width 6 μm). The value for hypostomatous tree species (stomata on lower side only), therefore, represent 50% of the pore area for the lower surface only.

and Mansfield, 1968; Willmer, 1983). The stomata of most species open in the light and close in the dark, either as a direct response to solar radiation or, more commonly, under the control of endogenous circadian rhythms. However, there are a number of important exceptions, such as those plants which employ the CAM pathway of photosynthesis (see p. 172) and the potato (see p. 287), whose stomata open or remain open at night. Stomata also tend to open if the CO_2 concentration in the sub-stomatal cavity falls below a critical level, whose magnitude is related to the photosynthetic pathway employed

(C_3, C_4 or CAM, see below, and Chapter 2). These responses to the environment are clearly related to the photosynthetic demand for CO_2 (Raschke, 1975).

Because of the experimental difficulties encountered in studying individual guard cells, the complex processes leading to the opening of stomata are not yet fully understood (Jarvis and Mansfield, 1981). However, it is known that exposure of leaves to either solar radiation or to lowered internal CO_2 concentrations results in the depression of the solute potential (ψ_s) of guard cells by the accumulation of solutes (principally potassium, chloride and/or organic acid ions) in the guard cell sap. Water is, therefore, drawn into the guard cells, their turgor pressure rises above that of the surrounding epidermal cells, and the stomata open as a consequence of the mechanical effects of this difference in turgor (Meidner and Mansfield, 1968; Willmer, 1983). The size of the stomatal aperture, and the resistance to gaseous diffusion through the pore, depends upon the magnitude of the difference in turgor pressure. Some examples of stomatal aperture size are given in Table 4.2.

The fundamental response of stomata to radiation or to circadian rhythms can be modified by several other factors. For example, increases in the rate of respiration at high temperatures can lead to closure in response to high levels of CO_2 in the leaf air spaces. This mechanism, which can lead to improvements in water conservation, is, presumably, the cause of mid-day closure of stomata under high evaporative demand and high leaf temperatures (e.g. Fig. 4.7), but it can also lead to longer-term, irreversible, closure even when an adequate supply of water is re-established. It may take several days before such stomata can again respond normally to the environment. As Meidner and Mansfield (1968) point out, such a delay in stomatal opening is particularly useful to mesophytes, which normally experience only short periods of drought, since by the time normal stomatal responses are re-established it is likely that the soil water will have been recharged with rain. Stomata also respond to overall leaf water relations, closing at relatively low thresholds of leaf water potential, the value of which is broadly related to the water supply in the natural range of the species (e.g. -1.0 MPa for *Vicia faba*; -1.8 MPa for *Zea mays*; -2.0 MPa for *Sorghum bicolor*; -4.3 MPa for *Gossypium hirsutum* (cotton); and -5.8 MPa for the desert evergreen *Larrea divaricata*; Ludlow, 1980; see p. 178 and Fig. 4.12). The plasticity of stomatal response to leaf water potential is discussed in more detail in Section IV.

Experimental evidence is accumulating that stomatal closure, lead-ing to the conservation of water and the avoidance of severe stress, can be caused by factors other than enhanced CO_2 levels. For example, under certain circumstances, stomatal aperture can respond directly to changes in the water vapour content of the air before changes in leaf water potential have had time to occur (Losch and Tenhunen, 1981). Such a response to humidity permits the leaf to control its water content very accurately. A further control of stomatal aperture is exercised by growth substances in the leaf (Mansfield, 1983); in particular, it has been established that abscisic acid, which is synthe-sized in the chloroplasts during water stress, can cause stomatal closure and delayed opening.

Stomatal resistance is, therefore, determined by the interplay of such a complex array of factors—irradiance, CO_2 level, water stress, humidity, wind, growth substances, endogenous rhythms—that it is not surprising that some consider stomata to be "miniature sense organs" (Mansfield and Davies, 1981). The success of a plant species in a given environment depends upon the manipulation of the responses to these factors to give a favourable balance between photosynthetic uptake of CO_2 and water conservation.

2. Boundary Layer Resistance

Even in the absence of wind, the bulk air surrounding a leaf is turbulent owing to convective heat exchange. Consequently the air is thoroughly mixed, and water molecules move rapidly from the trans-piring leaf into uniformly unsaturated air by mass flow rather than, much more slowly, by diffusion. However, at the leaf surface, there is a relatively undisturbed layer of air, the boundary layer, through which water vapour must diffuse before entering the turbulent bulk air (see Chapters 2 and 7). The thickness of the boundary layer depends upon wind velocity and upon leaf shape and size as shown in Fig. 4.10.

Using the data of Holmgren et al. (1965) and Meidner and Sheriff (1976) it is possible to compare the relative magnitudes of the resistances to the diffusion of water vapour across the cuticle, through stomatal pores and across the boundary layer in a typical mesophyte:

Cuticle (r_c) 20–80 s cm^{-1} (much higher values for certain tree species).
Stomata (r_s) 0·8 s cm^{-1} to 16 s cm^{-1} (dependent upon degree of opening).
Boundary Layer (r_a) 3·0 s cm^{-1} (at 0·1 m s^{-1}, Beaufort Scale Force 0); 0·35 s cm^{-1} (at 10 m s^{-1}, Force 6).

Note that r_s will be of the same magnitude as r_c when the stomata are fully closed.

Since the stomatal and cuticular pathways are in parallel, these high values of r_c indicate that cuticular transpiration will be a negligible fraction of total transpiration as long as the stomata remain open. The leaf diffusive resistance (R_l) of a transpiring plant is the sum of r_a and r_s; under most conditions, r_s is the dominant component of R_l and, therefore, stomatal aperture determines the rate of transpiration. However, as wind velocity falls, the thickness of the boundary layer increases and r_a becomes a larger fraction of R_l. At very low velocities or in still air (which are relatively uncommon occurrences), the boundary layer resistance can control transpiration over a range of stomatal apertures, as shown by the classic experiment of Bange (1953). In the same way, the rate of diffusion of CO_2 or of pollutant gases into the leaf can be very low if the measurement is carried out under controlled conditions in still air (see p. 285).

F. Water Movement in Whole Plants

In a freely-transpiring plant, water evaporates from the moist cell walls of epidermal and mesophyll cells in the interior of leaves and is lost to the atmosphere according to equation 4.11. As water loss proceeds, the water potential in the leaf apoplast falls below that of the leaf cells, and also below the water potential in the xylem and the soil. This results in the withdrawal of water from the leaf cells and a lowering of cell water potential (see Fig. 4.1). In contrast, although there is continuity of liquid water between leaf and soil via the xylem, rapid equalization of water potential throughout the plant by upward movement cannot occur because there is a resistance to hydraulic flow in the plant/soil system. As a result, the transpiration of water from the leaf sets up a gradient in water potential, down which water tends to flow from soil to leaf apoplast.

The pathway of water movement from the root surface to the site of evaporation in the leaf is predominantly extracellular (Spanswick, 1976). Many observations and theoretical calculations are consistent with the hypothesis that, in the apical 10–20 cm of young roots, water flows radially inwards through the cell walls and intercellular spaces of the root epidermis and cortex up to the endodermis, where further apoplastic movement is blocked by Casparian strips, in fully-differentiated regions. Thereafter, water passes through the cells of the endodermis before entering the lumina of the xylem elements by way of the stelar parenchyma apoplast (Anderson, 1976). The pathway then follows the stem xylem into the leaf where bundle sheaths and branching networks of veins deliver water to the apoplast within a few cells of the site of evaporation. Throughout its length, this pathway appears to cross membranes and pass through living cells only at the root endodermis (and, possibly in some species, at the leaf bundle mestome sheath which can develop suberized Casparian strips similar to those of the root endodermis; Esau, 1965). The water in most cells of the plant, including the leaf cells, is, therefore, not part of this pathway. The route of water flow is less clear for older and less permeable roots with a suberized exodermis (e.g. p. 267) or which have undergone secondary thickening; it may be that water enters the apoplast of such roots by cracks or lenticels.

Following van den Honert (1948), a number of workers have estimated the relative sizes of the resistances to water flow offered by different sections of the water pathway from root surface to leaf mesophyll, by treating the plant as a hydraulic system made up of a simple series, or catena, of hydraulic resistances through which water flows in response to a gradient in water potential. Because this flow is analogous to the flow of electrons in response to an electric potential difference, it can be described by an equation similar to Ohm's Law. Thus:

$$F = \frac{\Psi_{rs} - \Psi_{leaf}}{R} = \frac{\Delta\Psi_{root}}{r_{root}} = \frac{\Delta\Psi_{stem}}{r_{stem}} = \frac{\Delta\Psi_{leaf}}{r_{leaf}} \qquad (4.12)$$

$$\text{and } R = r_{root} + r_{stem} + r_{leaf} \qquad (4.13)$$

where

F is the steady state rate of flow of water through the plant

Ψ_{rs}, Ψ_{leaf} are the water potentials at the root surface and in the leaf mesophyll, respectively, and

$\Delta\Psi, r$ represent the difference in water potential, and the resistance to flow in different sections of the pathway, respectively.

Studies of this kind tend to show that the root resistance is slightly higher than that of stem or leaf, a finding which is consistent with the idea that the pathway crosses membranes only at the root endodermis. For example, Jensen et al. (1961) found the following ratios:

$$r_{root} : r_{stem} : r_{leaf} = 1 : 0\cdot42 : 0\cdot42 \text{ (sunflower)}$$
$$1 : 0\cdot42 : 0\cdot21 \text{ (tomato)}$$

Thus, in these species, the root resistance is the largest component but it accounts for no more than 54–61% of the total hydraulic resistance within the plant.

Although the catenary model has been useful in the assessment of hydraulic resistances within plants, and very influential in stimulating work on the factors limiting the rate of transpiration, there are a number of reasons why it should not be applied to plant/water relations in the field. First, plants do not consist of a single root, stem and leaf in series; they should rather be considered as a number of root axes, branches and leaves attached in parallel to a single, or multiple, stem. Thus, for example, the resistance of a complete root system could be evaluated by the expression:

$$\frac{1}{r_{root}} = \frac{1}{r_{x_1}} + \frac{1}{r_{x_2}} + \frac{1}{r_{x_3}} \dots \qquad (4.14)$$

where r_{x_1}, r_{x_2}, etc. are the resistances of individual root axes. Similar calculations could be performed for branches and leaves.

Secondly, at different points along the pathway within the plant, water can be withdrawn into, or released from, storage, thereby altering the rate of flow. For example, we have already seen that, although leaf cells do not fall on the direct route of water movement from soil to air, they supply water to the leaf apoplast at the beginning of rapid transpiration. In a similar way, the cells bordering the xylem in the root and stem can lose water during periods of high demand during the day, and absorb water during the night. The extent of such exchanges can be assessed by measuring diurnal fluctuations in leaf thickness, stem diameter and root diameter (Kozlowski, 1972). In the catenary model, this problem can be overcome by extending the electrical analogy to include both capacitance and resistance at different points in the system; the exchange of water with storage is then likened to the charging and discharge of electrons stored on the plates of a capacitor. However, this detracts from the essential simplicity of the catenary model; for example, Fig. 4.6 shows the complex array of resistances and capacitances required to describe the flow of water through a short branched axis of a root system.

Thirdly, any treatment of transpiration must include the entire pathway of water flow from bulk soil to bulk air, the so-called "soil–plant–atmosphere continuum". However, inclusion of the soil and gaseous components of the system effectively rules out the use of the Ohm's Law analogy because the basic conditions required for its use are not fulfilled:

(i) the system is only rarely at steady state because environmental factors affecting transpiration (irradiance, temperature, VPD,

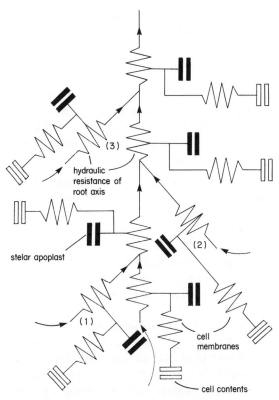

FIG. 4.6. Schematic diagram of the resistances and capacitances required for the description of water movement through a short root axis with three primary laterals (numbered 1 to 3). The filled capacitance symbols (—‖) represent storage adjacent to the "main pathways" of water movement (mainly water held in the stelar apoplast), whereas the unfilled symbols represent "off-line" storage (the water held within cell vacuoles). Water flow between these capacitances is controlled by the combined resistance (—/\/\/—) of the cell plasmalemma, cytoplasm and tonoplast. In practice, the system should include further orders of branching. The main pathways of water into and through the system are indicated by arrows (adapted from Meidner and Sheriff, 1976).

wind velocity etc.) are changing continually, in a regular diurnal pattern and in an irregular short-term fashion;

(ii) the system is not a set of (relatively) constant hydraulic resistances. On the one hand, the resistance offered by the soil increases progressively as water is withdrawn (both short-term, within hours, and longer-term effects), whereas the stomata represent a highly variable resistance under the active control of the leaf. There is also evidence of changes in root resistance as the rate of transpiration is altered;

(iii) the "continuum" involves a change of state from liquid water to water vapour, which ensures that the leaf diffusive resistance will always be very much larger than the resistances to liquid water movement within the plant. Furthermore, equation 4.12 cannot apply to the gas phase because the driving force for the diffusion of water vapour molecules is VPD and not water potential (which in the gas phase is logarithmically and not linearly related to water potential).

The many theoretical deficiencies of the Ohm's Law analogy are considered in detail in Passioura (1984). Overall, it can be concluded that the rate of transpiration at a given VPD is normally determined by leaf diffusive resistance and, more specifically, by stomatal resistance, except in still air. The hydraulic resistance to the flow of liquid water within the plant will not normally limit the rate of transpiration, but the resistance to water movement to the root from a drying soil can dominate plant water relations under certain circumstances, as outlined below.

The relative importance of the different resistances in the continuum can be illustrated by Fig. 4.7 which shows changes in leaf, root (surface) and soil water potential during a six day period of soil water depletion by an "ideal" plant. At the start of the first day, the stomata open progressively more widely over a period of several hours, in response to external stimuli (solar radiation, etc.) or endogenous rhythms, causing a progressive rise in the rate of transpiration. Because of the hydraulic resistances in the plant and soil, water begins to move from the soil to the leaf only after a gradient in water potential has been established. Since the soil is at field capacity ($\Psi_{soil} \simeq 0$), an adequate flow can be maintained without lowering leaf water potential below -0.6 MPa, thereby exposing the leaves to mild water stress for a few hours. Because of the low hydraulic resistance of the wet soil, water flows to the root in response to a very small difference in water potential (< 0.1 MPa).

In the evening, the rate of transpiration diminishes with progressive closure of the stomata and the upward movement of water to the leaf begins to exceed the rate of loss. Consequently, the leaf apoplast and cells are rehydrated, and the differences in water potential between soil and leaf are abolished overnight. However, because the stored soil water has been depleted by uptake during day 1, the equilibrium soil and plant water potential is now approximately -0.1 MPa. Plant/

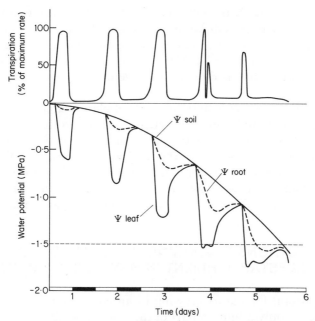

FIG. 4.7.   Schematic representation of the changes in leaf, root surface, and bulk soil water potentials, and in the rate of transpiration, associated with the exhaustion of the available soil water over a five day period. See text for a full description (adapted from Slatyer, 1967).

water relations during the second day are essentially similar to those of the first except that it is now necessary to lower leaf water potential to about -0.9 MPa to maintain the necessary gradient to ensure water flow. In addition, the soil hydraulic resistance is beginning to rise with soil drying, and a water potential difference of 0.1 to 0.2 MPa is now required to maintain water flow towards the root.

At the beginning of the third day, the equilibrium water potential in the soil, root and leaf has fallen to -0.4 MPa and it becomes necessary during the course of the day to lower leaf water potential to -1.2 MPa. The progressive increase in soil resistance has two effects: first, the potential difference between the root and soil required to ensure the same supply of water as in previous days has increased to 0.3 MPa but, what is more important, the slower movement of water through the soil is now delaying the overnight equilibration of water potential. Consequently, the leaves are exposed to mild to moderate water stress during most of the day. These developments become more serious on day 4 when leaf water potential falls below $-1 \cdot 5$ MPa and the stomata close for a short period at mid-day in response to high leaf temperatures or turgor effects (see above). Finally, by the end of day 5 when soil water potential has fallen to $-1 \cdot 5$ MPa (PWP), there is no available water left, the plant wilts on day six and eventually dies if the soil is not rewatered.

Although this is an idealized account of plant/water relations during a drying cycle, it underlines the general conclusion that *in the presence of available soil water*, the rate of transpiration is largely determined by the stomatal resistance, whereas the hydraulic resistance of the plant determines the leaf water potential lowering that is required for water to flow from a soil at a given matric potential, and the hydraulic resistance of the soil controls the rate of rehydration at night.

III. ADAPTATIONS FAVOURING GERMINATION AND SEEDLING ESTABLISHMENT IN DRY ENVIRONMENTS

It is important for plants growing in an environment where water is available for only a limited part of the year to be able to make full use of this period of favourable conditions for vegetative and reproductive growth. As a consequence, seeds or other propagules are normally ready for dispersal at the end of, or after, the wet season, at a time when conditions may still be favourable for germination but will rapidly become very unfavourable for the subsequent growth of seedlings. It is, therefore, not surprising to find that the seeds of many successful species from drier habitats, especially annuals, are innately dormant when first shed from the mother plant, and will not germinate until the start of the next prolonged period of favourable conditions (this is analogous to autumn dormancy which prevents the germination of the seeds of temperate species until after winter, p. 214).

For example, in tropical savanna regions, plant growth and development are not normally restricted by water supply from the beginning of the wet season up to 20–30 days after the end of the main rains (e.g. Fig. 4.3), but thereafter all the available water in the top 1–2 m of soil is progressively depleted by evapotranspiration. Seeds dispersed over the dry soil surface at the end of this period of favourable water supply can experience good conditions for germination during isolated periods of rain (e.g. in April 1972, Fig. 4.3), but the resulting seedlings would soon run out of water. Similarly in deserts, where rainfall is generally less predictable, it is essential that the seeds of ephemeral species do not germinate until the soil contains sufficient water to enable the resulting plants to complete their life-cycles. Seed dormancy is, therefore, not a reliable adaptation unless it is associated with a system which can signal the onset of conditions which are favourable for seedling establishment.

If the seasonal pattern of precipitation is reasonably predictable, then the system need not be elaborate. For example, the seeds of several tropical grassland species simply require a few weeks of after-ripening before they will germinate; in their natural range, the risk of isolated late rains will normally have passed before the after-ripening period has been completed (e.g. Fig. 4.3). In other cases, the hilum of the drying seed can act as a one-way valve, opening in dry air to permit the loss of water vapour but closing as the humidity of the air rises; the rehydration of the seed, leading to germination, can then occur only after prolonged immersion in liquid water, which would normally signal the start of the wet season (Koller and Hadas, 1982).

Alternatively, the need for environmental monitoring can be avoided by means of a "random strategy", which can also be effective in areas of irregular rainfall. For example, many legume species release soft, non-dormant, seeds and hard seeds whose coats are impermeable to water. In a given generation, the non-dormant seeds will germinate whenever conditions permit, whether subsequent conditions are favourable for seedling establishment or not. Since the removal of the impermeable seed coat by decay, abrasion or fire can take from a few months to several years, the dormant seeds of the same generation become ready for germination over a long period. To maintain the population of the species, it is sufficient for the germination of only one seed of this, or another, generation to coincide with the appropriate conditions for establishment (Koller and Negbi, 1966). For many species, the existence of an adequate water supply is not in itself a sufficient condition for germination, and other environmental factors must also be favourable. This can be clearly illustrated by the observation that, in many of the deserts of the world, rainfall at different times of the year causes the germination of different groups of ephemeral plants, even though the full range of species is represented in the seed bank (e.g. in the Sonoran Desert, Shreve and Wiggins, 1964). Work under controlled conditions, using plants from arid regions of Australia, has shown that this effect can be attributed to differing temperature requirements for germination (Mott, 1972; see also p. 178). Similarly, the seeds of other desert species require a specific light treatment before germination can begin (Koller and Negbi, 1966).

The complex inter-relationships between the plant and its

environment which can be required for successful reproduction in arid environments are well illustrated by *Mesembryanthemum nodiflorum*, an annual from the hottest and driest deserts of Israel (Gutterman, 1980/81). Although the mature seeds are released from the mother plant by the mechanical action of raindrops, germination cannot normally occur until the soil surface salinity has been reduced by leaching. Thereafter, germination is controlled by a regular seasonal variation in innate dormancy, which ensures that only a small fraction of the seeds germinates outside the winter wet season. This endogenous rhythm, which can persist for many years, does not appear to be synchronized with the seasonal variation in temperature (Table 4.3). Furthermore, there seems to be an additional variation in germinability, related to the time of seed dispersal, which acts to spread germination over a longer period. Finally, the mechanism of dispersal, which deposits the seeds close to the (now senescent) mother plant, ensures that any resulting seedlings establish in a microhabitat which has already proved suitable for the full development of the species.

With a few notable exceptions such as the groundnut and subterranean clover, the propagules of most terrestrial species are scattered over the soil surface, with the result that germination normally takes place within the surface layers of soil. Since these layers are subject to frequent and rapid cycles of wetting and drying, even under humid climates, it is important for seeds to lodge in "safe sites" where the water supply remains adequate throughout germination and seedling establishment. This is particularly important in those species which colonize disturbed ground (e.g. burned or cultivated), where bare soil is not protected from desiccation by vegetation. Many of these species display seed characteristics which favour germination in "safe sites".

TABLE 4.3. The germination of seeds of *Mesembryanthemum nodiflorum*, collected near the Dead Sea between 1972 and 74, and subjected to different temperature regimes under continuous light in the laboratory in 1978 (% of seeds germinated after 9 days) (Gutterman, 1980/81).

Month of germination test	Constant temperature			Alternating temperature	
	15°C	25°C	35°C	15–35°C	35–15°C
April	77	69	89	58	99
June	12	17	7	19	31
September	12	15	6	14	16
December	67	81	—	64	85

For example, many temperate weeds produce very small seeds which are able to lodge in cracks in the soil surface where better contact can be made with soil moisture, and evaporative loss from the seed is reduced by an undisturbed, humid, boundary layer (Harper *et al.*, 1965). In a number of other species from several families (e.g. Gramineae, Ranunculaceae, etc.), each dispersal unit is equipped with a hygroscopic awn (Fig. 4.8) whose twisting movements drive the seed into the soil during periods of fluctuating humidity (Darwin, 1876; Koller, 1972). Thus in *Avena fatua*, a common weed of temperate cereal crops, the awn first acts as a flight, enabling the seed to lodge upright in the soil, and then screws the seed into deeper layers of the soil where the water supply is less variable and evaporation is reduced (Thurston, 1960). In the dry tropics, this mechanism may have the additional advantage of placing seeds at a depth at which they cannot be damaged by grassland and bush fires (Lock and Milburn, 1971).

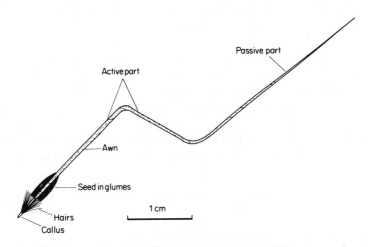

FIG. 4.8.   The dispersal unit of *Themeda triandra*, a perennial grass species native to savanna regions in sub-tropical Africa, Arabia and India. The "active part" of the awn twists and untwists in response to changes in humidity, thereby driving the seed into the soil surface (from Lock and Milburn, 1971).

In a study of several composites, Sheldon (1974) found that the pappus ("parachute") attached to the fruit not only aids the wind dispersal of the fruit, but also ensures that it lands, in soil cracks and crevices, with the attachment scar downwards. Consequently, good contact can be made between soil moisture and the micropile, which is adjacent to the scar, and through which water enters the fruit. In other species, contact with soil water is improved by the secretion of mucilage and by smooth seed coats (Koller and Hadas, 1982). The rate of germination of many temperate species can be reduced significantly if the soil is dried from saturation or field capacity to a matric potential of only −0·1 MPa (Harper and Benton, 1966), and the germination of the seeds of a given species is fully inhibited at a soil water potential which is broadly related to the water supply in its natural environment (Table 4.4). Thus the seeds of sorghum can germinate at lower matric potentials than can the seeds of maize, and halophytes germinate in soils and sediments with very low solute potentials. In general, these minimum values for germination are modest when compared with the matric potential of dry seeds (about −50 MPa; Hillel, 1972), and even at a soil water potential of −3 MPa (Table 4.4), there is still an enormous difference in water potential between seed and soil, tending to drive water into the seed, at least at the beginning of germination.

This apparent anomaly can be explained by the work of Owen (1952), who found that wheat seeds could absorb water and germinate in a gaseous environment corresponding to a water potential at least as low as −3·2 MPa (note that below this value, in terms of matric potential, there is very little capillary water left in most soils; Fig. 4.4). This is much lower than the corresponding value in Table 4.4 for two reasons. First, Owen's wheat seeds had access to an unlimited quantity of water, whereas in soil the moisture content falls rapidly with matric potential. Secondly, the resistance to the flow of water through the boundary layer round each seed in Owen's experiment would have been much lower than the resistance offered to the movement of liquid

TABLE 4.4. Minimum bulk soil water potentials at which the seeds of selection of plant species can germinate (from Levitt, 1980; Koller and Hadas, 1982).

Species	Water potential (MPa), mainly at 25°C
Trifolium repens	−0·35
Pisum sativum	−0·66
Triticum aestivum	−0·80
Cicer arietinum	−1·20
Zea mays	−1·25
Sorghum bicolor	−1·5 to −2·0
Salsola kali	−3·0
(Temperate halophyte)	

water through a drying soil (see above). Overall, it is likely that the rate of movement of water into seeds in soil is governed by the resistance to hydraulic flow and by the amount of water available in the soil (i.e. by soil and not seed characteristics); it is, therefore, surprising to find that pronounced differences in the ability to extract water do exist amongst the seeds of different plant species.

One possible solution to this problem has been proposed by Hegarty and Ross (1980/81) who found that the minimum water potential for the germination of a range of crop species was substantially higher than that for subsequent growth, in spite of the fact that much less water was required for germination. The mechanism of this differential sensitivity (between plant processes in the same plant) to water supply is not understood, although growth substances are implicated; however, its value to plants establishing in drier environments is clear since germination will not begin unless a substantial quantity of water is available for seedling growth.

IV. ADAPTATIONS FAVOURING SURVIVAL AND REPRODUCTION UNDER CONDITIONS OF WATER SHORTAGE

To grow and reproduce successfully in all but the most humid environments, plants must be able to survive periods of exposure to water stress varying in length from hours to years. In the driest zones (deserts, semi-deserts, etc., with discontinuous vegetation), the emphasis tends to be on survival, rather than growth and reproduction, since rainfall is episodic. For example, the herbaceous perennial, *Agave deserti*, sets seed once only at the end of its lifetime of 10–30 years (Schulze, 1982). Similarly conservative life cycles are found in very cold environments (p. 209). Under moister climates, such as tropical savannas, with more reliable seasonal patterns of precipitation and complete vegetative cover, survival and reproduction also involve inter-species competition. Under these conditions, the extreme sensitivity of cell and leaf expansion to water stress (Fig. 4.2) can have important implications for the interception of radiation during the period of favourable conditions. On the other hand, the life cycles of mesophytes, the typical plants of humid areas, do not normally need to be in phase with the seasonal variation of precipitation, and are more likely to be limited by temperature (Chapter 5); nevertheless, they have to endure unpredictable episodes of water stress (lasting from hours to weeks) during drought and in unusually dry habitats such as shallow soils over rock, and sand dunes. Note, however, that the terms xerophyte and mesophyte are not precisely defined and that, for example, there are few sharp discontinuities between mesophytic and xerophytic vegetation as we move from rainforest through savanna grassland to desert. In addition, apparently xeromorphic characteristics can turn up in some unexpected habitats (e.g. in the bog species *Erica tetralix*, p. 275).

Problems of water shortage and the maintenance of turgor are, therefore, universal amongst terrestrial plants. The physiological and morphological characteristics and life-cycles which have evolved in response to water deficit can be divided into three main classes:

(i) adaptations leading to the acquisition of the maximum amount of available water (avoidance of water stress, and the amelioration of its effects);

(ii) adaptations leading to the conservation and efficient use of the acquired water (amelioration and tolerance, but also avoidance in the case of those species which restrict their activities to periods of water availability);

(iii) adaptations (mainly biochemical and ultrastructural) which protect cells and tissues from injury and death during severe desiccation (tolerance).

Although it is convenient to discuss these classes separately in the following three sections, it must be stressed that some xerophytes can exhibit adaptations of all three types.

A. Acquisition of Water

1. Root System Morphology and Distribution
In humid zones, plants do not require deep and widely-spreading root systems for water uptake because soil water is plentiful and all the water required for transpiration can, in theory, be supplied by a relatively small volume of soil. Root:shoot ratios tend to be low; for example, it is estimated that roots account for only 21–25% of the total biomass of coniferous forests (Lange *et al.*, 1976). In drier, tropical savanna woodland, the proportion rises to 30–40%, whereas the root systems of some desert and prairie species, which grow to great depths, can represent 60–90%. However, conclusions based on simple ratios of this kind can be misleading because a variable fraction of the root system is active in water absorption, no account is taken of the contribution of mycorrhizal tissues, and the allocation of dry matter to

underground organs, in cold and arid environments, can represent the recharging of carbohydrate or lipid reserves, rather than a stimulation of root growth (p. 99). Furthermore, an extensive root system may be necessary for other purposes such as anchorage or uptake of phosphate (p. 101). The ratio:

$$\frac{\text{length (or surface area) of absorbing roots}}{\text{area of transpiring leaf}}$$

would be much more useful, but it is not practicable at present to discriminate between absorbing and non-absorbing roots in the field.

In contrast, there is a wealth of qualitative and semi-quantitative information on the distribution of plant roots in soil profiles, in relation to soil water availability. For example, much of the rain falling in deserts accumulates in depressions and river courses before draining into the soil to a depth of several metres (Walter, 1963); in this way the water table can remain stable from year to year, even in very arid regions (e.g. at about 2 m in the Sonoran Desert; Shreve and Wiggins, 1964). Ephemeral annual species can flourish briefly in moist depressions using shallow fibrous root systems, whereas perennials growing at low density have deep, unbranched root systems tapping the ground-water. For example, in a review of woody species native to mediterranean and desert areas of Israel and the Americas, Kummerow (1980) found maximum rooting depths of up to 9 m; other authors give much more extreme depths such as 18 m for *Welwitchia mirabilis* growing in South African deserts (cited by Kozlowski, 1964) and 30 m for an *Acacia* species near the Suez Canal (cited by Parker, 1968). The ability of such deep roots to conduct water to the shoot at a reasonable rate must depend upon the diameter of their xylem elements (Passioura, 1972). (This is analogous to the problem of raising water to the crown of a tall tree; p. 179).

In the North American prairies, where precipitation is higher, more seasonal and more readily accepted by the soil, the root systems of native grassland plants tend to be deep, but with profuse branching in the top 0.8 to 1 m of soil. For example, the maximum rooting depth of the dominant species *Agropyron smithii* varies from 1·5 to 3·6 m in the USA, and from 0·6 to 1·5 m in the moister Saskatchewan prairies (various reports, reviewed by Coupland and Johnson, 1965). This pattern of root distribution, which is also common in tropical savanna grasslands (Taerum, 1970), leads to efficient absorption of water and nutrients in the top layers of soil during wet periods, as well as the extraction of stored moisture from deeper horizons during drought or the dry season. Genera such as *Agropyron* can maintain their dominance by absorbing water from horizons below the maximum rooting depth of competing grasses. In the same environment, some shrubs and herbs have a root system morphology similar to that of desert species, with long, unbranched, tap roots absorbing ground-water below the grass roots (e.g. 1–3 m in Saskatchewan).

An important feature of prairie species is the morphological plasticity of their root systems in response to differences in soil conditions (Kummerow, 1980). Figure 4.9 illustrates variations in the spread and maximum depth of the roots of *Agropyron smithii* associated mainly with changes in soil moisture status. However, such results should be treated with caution since they may also reflect differences in above-ground growth. A much more remarkable example of root system plasticity is provided by the herbaceous perennial *Artemisia frigida* which relies upon a deep tap root in dry soil, but under moist conditions, the tap root senesces, to be succeeded by a mass of fibrous roots in the top metre of soil (Coupland and Johnson, 1965; Weaver, 1958).

Deep rooting is clearly an important feature of established perennial plants in desert, mediterranean, prairie and savanna regions, as well as other dry environments, such as sand dunes (Salisbury, 1952), where groundwater is a major source of supply. However, as pointed out by Parker (1968), it is difficult to understand how such species can become established. In some cases, this can be achieved by vegetative reproduction, producing rhizomes which can rely upon the parent root system until their own roots have extended sufficiently to reach groundwater. Alternatively, large quantities of seed can be released, a very few of which may germinate and establish under unusually wet conditions (see above). Even in this case, it is essential that the first roots extend rapidly to make contact with stored water while the surface layers remain moist. This explains the very high rates of root growth which have been recorded for some species in dry areas; for example, in their natural range in the Namibian Desert, seedlings of *W. mirabilis* can produce a tap root of 35 cm within ten weeks of germination (von Willert, 1985). Rapid root growth at the start of the growing season can also give a competitive advantage to annual plants. In particular, since the seminal roots of *Bromus tectorum* and *Taeniatherum asperum* can grow at lower temperatures than the roots of competing species in the western rangelands of the USA, these grasses absorb a disproportionate fraction of the limited supply of stored water in early spring by sending new roots into successive horizons ahead of their competitors (Harris and Wilson, 1970).

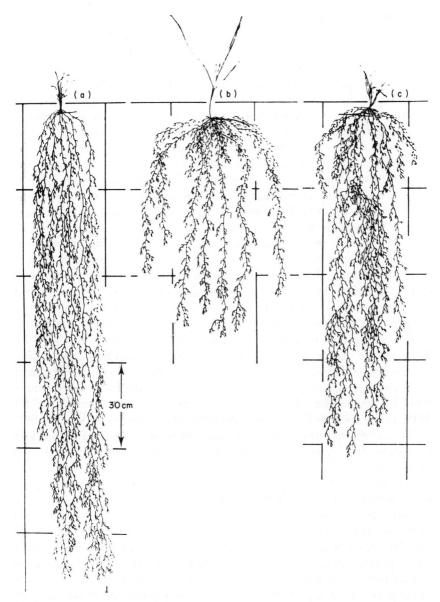

30 cm

FIG. 4.9.   Root systems of *Agropyron smithii* growing in the Saskatchewan Great Plains. Within the dark brown soil zone, rooting tends to be deeper on more xeric, south-facing slopes (a) than on level sites (b); rooting is also deeper in the drier climate of the brown soil zone (c) (from Coupland and Johnson, 1965).

2. *Leaf Characteristics*

Under conditions of low or irregular rainfall, the leaf structure of certain plant species has evolved to favour the absorption of dew or rain directly into the shoot. Thus *Chaetacme aristata* from the deserts of Natal bears specialized cells in the leaf epidermis which absorb water adhering to the leaf surface (Meidner, 1954). However, the collection of liquid water by leaves is not restricted to plants from arid regions; for example, the blade and sheath structure of the Gramineae, both temperate and tropical, leads to the accumulation of water within the leaf sheath. Although there has been considerable debate over the contribution of dew to the economy and distribution of xerophytes (e.g. Stone, 1957a), it does appear to prolong survival in certain cases, possibly as a result of the direct (temporary) suppression of transpiration (Stone, 1957b; Slatyer, 1967).

In addition to the enormous variation in their rooting characteristics, plant species can also vary in the ability of their roots to extract water from soils. The lowest matric potential at which a plant can extract water from a soil is determined primarily by the lowest leaf water potential which it can tolerate (in order to set up a sufficiently steep gradient in water potential) or, more precisely, by the lowest solute potential of its leaf cells (i.e. at incipient plasmolysis, Fig. 4.1). Because the solute potentials of mesophytes, including many crop species, fall within the range $-1 \cdot 0$ to $-2 \cdot 0$ MPa (Slatyer, 1963), their roots can dry soils down to matric potentials within the same range. This has led to the widespread use of the moisture content at $-1 \cdot 5$ MPa as the standard measure of permanent wilting point (see above).

However, since the solute potentials of xerophytes and halophytes tend to be much lower (commonly in the range $3 \cdot 0$–$4 \cdot 0$ MPa, but even lower for certain halophytes; Slatyer, 1963), the true permanent wilting point (which is really a characteristic of the plant and not the soil) for these plants will be much lower than $-1 \cdot 5$ MPa, and the quantity of additional water which is available for uptake will depend upon soil texture and structure (Fig. 4.4). As far as water uptake is concerned, there is little advantage to be gained from lowering plant water potentials much below -4 MPa because the minute quantities of capillary water left in the soil are distributed discontinuously in the finest pores. Where extremely low leaf water potentials do occur (e.g. down to -16 MPa in the xerophyte *Artemisia herba-alba*; Richter, 1976), they are the consequence of severe desiccation and salinity, and are not an adaptation for the acquisition of more water from the soil because the leaves cannot function at such low water potentials. In contrast, certain woodland herbs can absorb much less soil moisture, having permanent wilting points corresponding to soil matric potentials and cell solute potentials in the range -0.5 to -1.0 MPa.

Although low leaf water potentials may be necessary for the extraction of water from drying soils, there is also the need for the cells of expanding tissues to maintain turgor pressure above the yield threshold, if growth is to continue (equation 4.9). For a rapidly increasing number of species, it has been shown that turgor can be increased temporarily by osmoregulation, the lowering of the solute potential of the expanding cell by the accumulation of solutes (chiefly sugars, amino acids and potassium ions). For example, the long-term maintenance of full turgor in the presence of diurnal variations in cell water potential (similar to Fig. 4.7) has been demonstrated for a range of largely mesophytic crop plants; this can involve parallel diurnal variations in solute potential of up to 2.0 MPa (Morgan, 1984). This type of osmoregulation is induced by exposure of the plant to water deficit, although variation in irradiance and CO_2 concentration can affect the response, but the ability to osmoregulate normally declines once the tissue is fully expanded. It is still too early to judge whether mesophytes and xerophytes differ significantly in their patterns of osmoregulation, particularly since most of the work to date has concentrated upon crops. However, Bennert and Mooney (1979) found that for the desert halophyte *Atriplex hymenelytra* and the desert xerophyte *Larrea divaricata*, diurnal changes in solute potential gave full turgor maintenance at water potentials around -3 MPa, whereas the desert ephemeral *Camissonia claviformis* showed little osmoregulation, and wilted at a leaf water potential of -1 MPa (see p. 132, Fig. 4.12, Fig. 5.10). It should also be noted that maintenance of turgor by osmoregulation may also permit stomata to remain open to lower leaf water potentials; although this could extend the range of environmental conditions over which CO_2 assimilation takes place, it would, presumably, be at the expense of water use efficiency (see below).

B. Conservation and Use of Water

1. Water Use Efficiency

As noted earlier, the assimilation of CO_2 is inevitably associated with loss of water to the atmosphere through open stomata. However, this is not a simple exchange of one CO_2 molecule for one molecule of water; since the diffusion pathway for water is shorter, and the concentration gradient driving water out of the leaf is steeper than that driving CO_2 inwards, the amount of water transpired is greatly in excess of the amount of CO_2 fixed. For example, Raschke (1976) has estimated, for an idealized leaf, that at least 20 water molecules are lost for each molecule of CO_2 taken up (20°C, 70% RH) rising to 430 at 50°C and 10% RH. On the other hand, Carlson (1980) calculates a theoretical minimum of 109 molecules of water per molecule of CO_2. Losses from real leaves will be many times higher than these values.

Since CO_2 uptake is essential for growth, some species, in particular xerophytes, but also many species from more mesic habitats, have evolved characteristics which reduce the transpirational loss of water during periods of active photosynthesis, albeit at the expense of growth rate in many cases. The water use efficiency data presented in Table 4.5 are derived from the pioneering experiments of Shantz and Piemeisal (1927) carried out in the dry rangelands of Colorado. Although such data should be treated with caution (note, for example, the substantial seasonal differences in efficiency between the driest season, 1916, and the wettest, 1915), they do demonstrate a broad relationship between water loss during photosynthesis and the water supply in the natural range of the species. For example, the C_4 species studied which originate from the seasonally-dry tropics (*Sorghum, Zea*) or are native to

TABLE 4.5. The water use efficiency of selected plant species grown in the open in pots at Akron, Colorado, expressed as g water transpired g^{-1} of dry matter produced (from Shantz and Piemeisal, 1927).

	1914	1915	1916	1917	Mean
Evapotranspiration (mm)	1063	848	1198	1085	1049
C_3 species					
Triticum aestivum	518	405	636	471	508
Hordeum vulgare	501	404	664	522	523
Gossypium hirsutum	574	443	612	522	538
Avena sativa	607	447	843	636	633
Linum usitatissimum	—	622	955	605	727
Medicago sativa	890	695	1047	822	864
C_4 species					
Sorghum bicolor	284	203	296	272	264
Amaranthus retroflexus	306	229	340	307	296
Sorghum vulgare					
var. *sudanense*	394	260	426	378	365
Zea mays	368	253	495	346	366

Colorado (*Amaranthus*) were more efficient at using water than the C_3 species, which are generally, but not exclusively (e.g. cotton), adapted to more mesic environments (see below and Chapter 2 for a discussion of C_3 and C_4 photosynthesis); on the other hand the choice of sorghum or millet rather than maize as the staple crop in the drier areas of Africa is at least partly explained by their more efficient use of water. These large differences in efficiency are not normally the result of a single adaptation, but of a variety of (commonly inter-related) anatomical, morphological and biochemical adaptations. A number of these adaptations and combinations of adaptations are examined in the remainder of this section.

2. Canopy and Leaf Characteristics

As shown in the preceding section, perennial species from drier habitats tend to develop deep and/or extensive root systems, giving high values of the ratio:

$$\frac{\text{length of absorbing root}}{\text{area of transpiring leaf}}$$

However, high values of this ratio can also be achieved by reduced canopy leaf area, either as a consequence of loss of turgor during leaf expansion or as a permanent morphological feature (Begg, 1980). This will normally be effective only in less competitive situations because reduction in canopy leaf area will also reduce radiation interception and potential dry matter production. Many xerophytes have smaller, thicker leaves than those of mesophytes, giving a higher ratio of photosynthetic mesophyll to transpiring leaf area. This ratio can be expressed by the specific leaf area (dm^2 of leaf area per g of leaf dry weight—see p. 48); thus, as we pass from the extremely arid Sahel zone to the more humid areas of central Europe, the mean specific leaf area of woody perennials rises from 0·36 through 0·70 (N. Sahara) to 1·10, which is very similar to the value of 1·03 measured in the Ivory Coast rainforests (Stocker, 1976). Similarly, Larcher (1975) tabulates values greater than 1·5 for species growing in mesic conditions (*Fagus sylvatica*, 1·4–1·6; *Oxalis acetosella*, 1·8; *Impatiens noli-tangere*, 2·2) but much lower values for xerophytes (*Sedum maximum*, 0·12; *Opuntia camanchica*, 0·026). However, these values must be treated with caution since they are, in part, a response to temperature and irradiance (p. 49, 192), and in some xerophytes, one consequence of extreme succulence, as discussed below.

Although reductions in leaf dimensions and in specific leaf area tend to reduce the leaf area involved in transpiration, these changes are also associated with the thinning of the leaf boundary layer (Fig. 4.10) and, in many cases, with increased stomatal frequency (number of stomata per unit of leaf area). Consequently, both the stomatal and boundary layer components of the leaf diffusive resistance will tend to be lower rather than higher, resulting in a potential increase in water loss by transpiration rather than a decrease. This effect may be slightly counteracted by increased mesophyll (or residual, p. 287) resistance, which is normally negligible for the diffusion of water vapour but more important in thick leaves. The effect of leaf blade dissection upon boundary layer resistance depends upon whether or not the boundary layers of the individual leaflets overlap and merge (p. 52). These findings suggest that changes in leaf dimensions, dissection and specific leaf area are more important in the thermal economy of the leaf than in the water economy (although, of course, the specific effects of drought and high temperature stress on the plant are difficult to distinguish in the field; Chapter 5).

Lewis's (1972) work on *Geranium sanguineum* from a variety of European habitats is an unusually well-documented account of the effects of the morphological characteristics of leaves on their temperature and water relations. As we move from wetter woodland sites to the more xeric steppe, and shallow, well-drained, soils over limestone (limestone pavement sites), the leaves of this species become smaller,

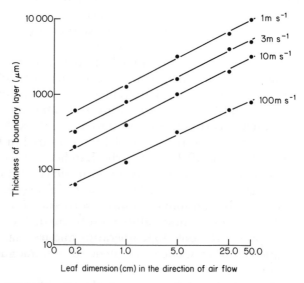

Fig. 4.10.   Thickness of the boundary layer over a leaf as a function of wind velocity and leaf size (from data of Nobel, 1974).

thicker and more dissected. In parallel with these changes, the stomatal frequency rises and the leaf boundary layer becomes thinner, giving large, but not statistically significant, reductions in leaf diffusive resistance when the stomata are open (Table 4.6).

This variation in leaf morphology does not, therefore, appear to be significant in terms of water loss, but because the lowering of the boundary layer resistance to the diffusion of water vapour can also increase the flow of heat from the leaf surface, especially when the stomata are closed and there is no transpirational cooling (p. 195), the leaf characteristics observed at the drier sites have presumably evolved to give some protection against the higher irradiances and temperatures experienced at these sites. For example, it can be calculated (Lewis, 1972) that a "woodland leaf" will be several degrees warmer than a "limestone pavement leaf" over a range of irradiance levels, when the stomata are open. This would tend to reduce the photosynthetic efficiency of the woodland leaf at high irradiance if the thermal optimum were exceeded. More important, when the stomata are fully closed and heat can be lost only by convection, the larger, less-dissected woodland leaf would be exposed to temperatures above the thermal death-zone for this species (47–50°C), if grown in the open. In direct contrast, the bulky shapes and daytime closure of the stomata of desert succulents appear to favour water conservation to the detriment of the thermal economy of the plant, although some protection is afforded by the high specific heat of water (see below, and p. 160). The many interacting factors influencing leaf size are reviewed by Parkhurst and Loucks (1972).

The damaging effects of water stress upon plant tissues which are actively growing and photosynthesizing can, in certain perennials, be avoided by bearing leaves only when the supply of water is adequate for normal function. At the onset of prolonged drought, water loss can be reduced dramatically by leaf abscission, with the result that the water stored in the soil is used much more economically. This response, commonly induced by seasonal changes in environmental factors other than water supply, or by endogenous rhythms, can be observed in a variety of dry environments from deserts to mediterranean zones, but when induced by drought it can cause premature and damaging defoliation of trees in late summer in normally more mesic areas (e.g. in the eastern United States; Parker, 1968). For example, in tropical savannas, trees and shrubs tend to show a regular seasonal pattern of abscission, shedding leaves soon after the beginning of the dry season (e.g. Fig. 4.3), but re-forming the canopy several weeks before the main rains begin. Since differences in day length are small in these

TABLE 4.6. Differences in leaf anatomy and diffusive resistances to water vapour movement, between eight populations of *Geranium sanguineum* from contrasting European habitats. Components of the leaf diffusive resistance (R) are indicated by r_a (boundary layer at 0.15 m s^{-1}), r_c (cuticle) and r_s (stomata) (from Lewis, 1972).

| Habitat type | Leaf anatomy | | Diffusive resistances ($s\ cm^{-1}$) | | | | |
	Thickness (μm)	Stomatal frequency[a] (per mm²)	r_a	r_c	r_s	R open stomata	R closed stomata
Limestone pavement 1	313	281	0.19	12	0.8	1.1	12.7
Limestone pavement 2	305	277	0.21	14	1.0	1.2	14.4
Steppe	276	308	0.19	58	1.4	1.7	58.0
Woodsteppe 1	240	275	0.29	23	3.4	3.5	23.8
Woodsteppe 2	264	235	0.32	31	1.3	1.8	31.0
Coastal	240	318	0.37	10	3.3	3.9	10.1
Woodland 1	230	179	0.48	16	2.5	2.9	16.0
Woodland 2	243	195	0.42	12	3.5	3.7	12.7
Probability[b]	†	*	‡	‡	‡	n.s.	n.s.

[a]Compare with the range of stomatal frequencies for mesophytes in Table 4.1.
[b]Probability of significant difference between habitat types estimated by analysis of variance (* = 5%, † = 1%, ‡ = 0.1%, n.s. = differences not statistically significant).

areas, it is thought that these woody species leaf out in response to increasing temperature, and are thereby able to make full use of the entire wet season for completion of the annual life cycle. On the other hand, the seasonal dimorphism (large leaves during the wetter winter months, replaced by smaller leaves in the dry summer) which is a feature of dominant shrubs in the plant communities of the Middle East (Orshan, 1963), permits photosynthetic activity throughout the year. In other cases, CO_2 assimilation can continue after leaf abscission by means of photosynthetic stem tissues; this can be particularly effective in those plants whose leaves make use of the C_3 pathway of photosynthesis but whose stems and other tissues can exploit the C_4 or CAM pathways (Osmond *et al.*, 1982; see below).

3. Stomatal Responses

Without doubt, the closure of stomata is the most important process in the protection of plants from exposure to severe water stress. It is possible to consider the rapid responses of stomatal aperture to changes in the humidity of the air to be a "first line of defence", protecting the leaf from tissue desiccation even before low leaf water potentials have developed (Mansfield and Davies, 1981). Recent work has shown that such responses to humidity are not restricted to species from dry environments, but are common to plants from habitats ranging from deserts through mesic temperate zones to the Arctic (Schulze and Hall, 1982).

Where necessary, the "second line of defence", i.e. closure in response to lowered leaf water potential, can come into play. In general, stomata remain fully open until a critical or threshold leaf water potential is reached (Ludlow, 1980); from this value, the aperture begins to narrow as a result of further water loss, and closure can be complete, causing the cessation of photosynthetic CO_2 uptake as well as stomatal transpiration, within 0·5 MPa of the threshold (Hsiao, 1973). There is considerable variation in both threshold and full closure values amongst species and growing conditions, and it is particularly important to note that the leaf water potential for complete closure is generally much lower for field-grown plants than for those raised under controlled conditions. This has caused serious difficulties in the interpretation of field water relations from growth cabinet or glasshouse studies. The threshold and closure water potentials, and the magnitude of the difference between these values, are dependent upon the water relations in the natural range of the species. As extreme examples, the stomata of *Larrea divaricata* close at leaf water potentials between −4·0 (controlled conditions) and −5·8 MPa (field), whereas the corresponding values for *Vicia faba* are -0.6 and -1.0 MPa (Ludlow, 1980). In spite of these difficulties, Hsiao *et al.* (1976) proposed generalized threshold values of -0.5 to -1.0 MPa for mesophytes (Fig. 4.2) and -1.0 to -2.0 MPa (and lower) for xerophytes. Thus mesophytes, which are generally exposed to water deficits for only short periods, use stomata to conserve water and avoid the effects of moderate to severe water stress; furthermore, since partial closure has a greater influence upon water loss than CO_2 uptake (Raschke, 1976, and see below), water use efficiency (but not dry matter production) is progressively improved between the threshold and closure leaf water potentials.

In contrast, xerophytes must continue to photosynthesize under drier conditions than those to which most mesophytes are exposed. The lower solute potentials of their cells and the process of osmoregulation make it possible for turgor to be maintained, and for stomata to remain open, down to much lower leaf water potentials; however, these characteristics must be associated with a greater ability to tolerate tissue desiccation (see below). There is no unique relationship between leaf water potential and leaf water content. For example, Bannister (1971,1976) found that the threshold water potential values in *Erica cinerea* (from dry heathland) and *Calluna vulgaris* (from wetter areas) were very similar (-1.8 to -2.0 MPa), but that at this water potential the leaves of the two species had very different water contents (88% for *Erica*, 75% for *Calluna*). Thus in *Erica*, as in other species from dry habitats, rapid lowering of leaf water potential is associated with relatively modest losses of water from the leaf tissues.

In a photosynthesizing leaf, CO_2 molecules must traverse the leaf boundary layer and the stomatal pore before diffusing through the cell walls and plasmalemma membranes of mesophyll cells to the site of fixation. Because the mesophyll resistance can also include a poorly-characterized biochemical component (see p. 287), the term residual resistance is more appropriate. In contrast, there is no mesophyll or residual component in the pathway of water vapour loss by transpiration. Consequently, the diffusive resistance offered by a leaf to CO_2 uptake is greater than that offered to water loss, and any change in the resistance of the common part of the pathway (stomata and boundary layer) will have a greater influence upon transpirational loss of water than upon CO_2 uptake.

For example, we have already noted that partial closure of stomata, by increasing stomatal resistance, improves the efficiency of water use during photosynthesis. In addition to this universal, but temporary effect, many species possess permanent structural features which favour photosynthesis over transpiration by increasing the diffusive resistance of stomata. For example, stomata can be situated at the bottom of depressions in the epidermis (e.g. *Canna indica*), in a deep pore, below an "antechamber" (*Allium cepa, Pinus sylvestris*) or surrounded by a "chimney" of cutin (*Euphorbia tirucalli*) (Meidner and Mansfield, 1968). In conifers, the stomatal resistance is further increased by the occlusion of the antechamber by loosely-packed plugs of wax (Jeffree *et al.*, 1971). However, these adaptations also reduce the uptake of CO_2, and any improvements in water use efficiency are gained at a cost in terms of net photosynthesis and dry matter production. Evaluation of the adaptive significance of stomatal frequency is complicated by the fact that it is sensitive to a number of environmental factors (water relations, irradiance, temperature) during leaf growth. Furthermore, if the cell size of any species (mesophyte or xerophyte) is reduced by loss of turgor during expansion, the resulting leaves will have a higher stomatal frequency than those grown in the absence of water stress, because the number of potential guard cells will be unaffected. However, there are some differences in stomatal frequency between groups of plants which appear to be adaptive in relation to water supply; in particular, CAM plants, from the most arid environments, have frequencies which are approximately one tenth of those for C_3 and C_4 plants (Osmond *et al.*, 1982).

Once stomatal closure is complete, CO_2 uptake and stomatal transpiration cease, but water loss continues, at a lower rate, through the cuticle. Consequently, for plants growing on limited water supplies, it is important for cuticular transpiration to be kept to a minimum. Thus many xerophytes have cuticular resistances in the range 60–400 s cm^{-1} compared with typical values of 20–60 s cm^{-1} for mesophytes (Cowan and Milthorpe, 1968); without such high resistances, which are achieved by the laying down of thick layers of cutin and additional coatings of wax (Martin and Juniper, 1970), most desert species could not survive prolonged droughts lasting from months to years.

4. Other Xeromorphic Characteristics

Features which are characteristic of certain groups of plants from hot, dry environments include:

Leaf pubescence. Depending upon their location, frequency, dimensions and colour, leaf hairs can affect the water and temperature relations of a leaf in different, and sometimes conflicting, ways. For example, as discussed in Chapter 5, the correlation between the density of pubescence and the severity of the environment (temperature, water deficit) in the natural range of perennial shrubs of the genus *Encelia* can be explained in terms of the fraction of incident solar radiation which is reflected by the leaf surface. However, increase in pubescence is also associated with an increase in the boundary layer resistance to loss of water and heat from the leaf surface, and to the uptake of CO_2 by the leaf. Protection of the leaf from lethal temperatures and severe water stress by increased reflection of intercepted radiation can, therefore, be at a cost in terms of net photosynthesis; in this context, it is interesting to note that the highly-pubescent white leaves of *E. farinosa* are replaced by green, less-pubescent leaves under wet conditions. On the other hand, the clustering of hairs round stomatal pores can increase stomatal resistance to water loss with less influence upon the overall temperature relations of a leaf. However, it should be noted that, in many cases, pubescence appears to have evolved as a protection against herbivory (Johnson, 1975; p. 319).

Adjustment of leaf canopy properties, for example, leaf rolling, which tends to reduce the water loss from one surface of the leaf, and adjustment of leaf angle, which can influence the quantity of solar radiation which is intercepted. As discussed in Chapter 5, this phenomenon can be exploited to increase or reduce the absorption of radiation, with implications for the water economy of the plant.

Storage of water in a wide variety of tissues — bulbs, tubers, swollen roots, the trunks of baobab trees, and the stems and leaves of desert succulents. For example, many herbaceous dicotyledons survive the prolonged dry season in tropical savanna regions as leafless under-ground tubers (e.g. the African arrowroot, *Tacca leontopetaloides*). However, storage of water is by no means a universal feature of the plants of arid zones; in a study of three perennial species growing successfully at the same desert site in California, the water released from storage could support maximum rates of transpiration of the CAM succulent *Agave deserti* for 16 h compared with only 7 min for the C_3 shrub *Encelia farinosa* and 4 min for the C_4 grass *Hilaria rigida*. However, the difference between *A. deserti* and the other species is at least partly the consequence of a much lower rate of transpiration (Nobel and Jordan, 1983).

Leaflessness either by seasonal, or less regular, leaf fall at the onset of severe water stress, or as a permanent feature, relying upon photosynthetic stem tissues (Lange *et al.*, 1976; Kozlowski, 1964; Parker, 1968).

5. *Alternative Pathways of CO_2 Assimilation*

As a consequence of their CO_2-concentrating mechanism, plants using the C_4 pathway of photosynthetic CO_2 fixation (see p. 58) can maintain the optimum intercellular concentration of CO_2 in the mesophyll at higher stomatal resistances (i.e. lower leaf conductances/smaller stomatal apertures) than can C_3 plants. The efficiency of use of water is, therefore, generally higher for C_4 than for C_3 plants. Furthermore, the temperature optima for C_4 photosynthesis also tend to be higher, and the photosynthesis of many C_4 species is not saturated at full sunlight (up to 1000 W m^{-2}), whereas many C_3 plants, especially crop species, are saturated by irradiances in the range 200–300 W m^{-2}. These characteristics, which became apparent during early comparisons of the relative productivities of stands of C_3 and C_4 plants, suggested that C_4 plants in general were well-adapted to hot, dry environments, with high levels of irradiance; this hypothesis appeared to be confirmed by the observation that the classic C_4 grasses were native to the dry tropics and sub-tropics.

However, the possession of a higher water use efficiency can serve only to extend the period of availability of water, and may not be an advantage in competition with more profligate users of water. Ultimately, unless a species has evolved to avoid water deficits, its success in a dry environment involves the survival of low tissue water potentials; possession of the C_4 pathway alone does not confer a distinct advantage on cells exposed to severe water stress, and it must be associated with appropriate life-cycles, leaf morphologies, etc. Furthermore, high optimal temperatures are of value only if the entire photosynthetic apparatus, and photosystem 2, in particular (Table 5.4), is adapted to high temperatures over prolonged periods. Recent studies have shown that the photosynthetic apparatus of C_4 plants is not necessarily more thermostable than that of C_3 plants. Furthermore, examination of a wider range of species has revealed C_4 species whose photosynthesis saturates at one tenth of full sunlight, and C_3 plants which do not saturate within the normal range of irradiance values (Osmond *et al.*, 1982; Pearcy and Ehleringer, 1984).

Reservations about the adaptive value of C_4 photosynthesis in dry environments have been reinforced by detailed studies of the relative distribution of C_3 and C_4 plants. The progressively-decreasing proportion of C_4 species in N. American grass floras from the hot, dry south-west deserts to cool, moist areas (Fig. 4.11a), or in E. Africa with increasing altitude, does suggest that possession of the C_4 pathway confers an advantage as the climate becomes hotter and dryer.

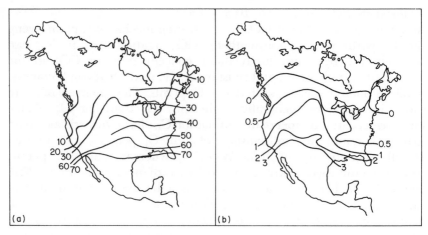

FIG. 4.11. "Contour" lines indicating (a) the percentage of grass taxa, and (b) the percentage of dicotyledon taxa, which are C_4 plants, in different parts of N. America. Constructed using regional flora data from Teeri, J. A. and Stowe, L. G. (1976). *Oecologia* **23**, 1–12 and Stowe, L. G. and Teeri, J. A. (1978). *Am. Nat.* **112**, 609–623.

However, multiple regression analysis has shown that the distribution shown in Fig. 4.11(a) is determined primarily by night temperature, since most C_4 plants are chilling-sensitive. Furthermore, although the dicotyledons show a similar overall trend (Fig. 4.11(b)), C_4 plants represent a very small proportion of all dicotyledons throughout N. America, and when the total number of all plant species is measured in many dry areas of the world, C_3 species tend to predominate. Finally, the C_4 pathway appears unexpectedly in some plants from warm, shaded and well-watered environments, such as the understorey of Hawaiian rainforests (Osmond *et al.*, 1982).

It can, therefore, be concluded that possession of the C_4 pathway is by no means indispensable for success in, or restricted to plants growing in, arid environments (compare the C_3 and C_4 species of the Sonoran Desert, p. 64; Fig. 4.12). By contrast, the CAM pathway (strictly, the stomatal CAM pathway; Cockburn, 1985) of photosynthetic CO_2 fixation (p. 60) is virtually restricted to plants native to hot, dry environments (desert succulents) or dry habitats within otherwise moist environments (tropical epiphytes, temperate crassulacean succulents growing on cliffs or walls with very restricted rooting volumes). It is also, with a few exceptions, associated with a range of other adaptations which facilitate the continuation of photosynthetic CO_2 fixation under severe drought—a degree of succulence (providing storage for both water and organic acids, as well as thermal buffering), nocturnal opening of stomata (admitting CO_2 at lower bulk air VPD than during the day, thereby improving water use efficiency), very high cuticular resistance to water loss, and relatively high critical leaf water potentials ($\simeq -1.0$ MPa) for the closure of stomata. Furthermore, in many cases a high proportion of the above ground tissues are photosynthetic, thereby counteracting, to a certain extent, their low daily rates of net photosynthesis (typically 5% of C_4 plants per unit leaf area under the same conditions). On the other hand, because their root:shoot ratios tend to be low, desert CAM plants are generally unable to exploit groundwater, and must rely on current precipitation or water stored within their tissues. This, and their low rates of convective cooling, is presumably the reason for their absence from the most arid habitats within deserts (Osmond et al., 1982).

There are few obligate CAM species; most are facultative, fixing CO_2 by the C_3 pathway under favourable conditions but switching to CAM within a few days of the onset of water stress (caused by drought or salinity) or in response to a critical photoperiod or temperature, which signals the start of dry conditions in an area of regular wet and dry seasons. Other species rely more heavily on CAM as development proceeds (Cockburn, 1985). The life-cycles of plants which rely upon the CAM pathway alone for net photosynthesis are essentially conservative, making use of the water available for survival, protecting their photosynthetic apparatus, and growing very slowly; successful reproduction by seed may occur only after a period of unusually heavy rainfall. On the other hand, the switch from C_3 to CAM at the end of a wet season can, in some cases, extend the growing season sufficiently to permit the completion of the annual life-cycle. However, the induction of CAM can be accompanied by no net fixation of CO_2, although the system appears to be recycling CO_2 released by respiration; this process, known as idling, has been interpreted as a means of conserving carbon and water, and, at the same time, protecting the photosynthetic apparatus from photoinhibition (Cockburn, 1985; Daniel et al., 1985). Overall, the possession of alternative pathways confers a flexibility which can be invaluable in irregularly-fluctuating environments.

6. Ontogeny

In general, the ability of plants to use water efficiently and avoid the damaging effects of water stress varies with ontogeny. In particular, most plants are most sensitive to drought at the beginning of the reproductive phase of development, but relatively insensitive during vegetative growth (Kaufmann, 1972). The reasons for this phenomenon include the large leaf areas carried by plants at the end of vegetative development, the diversion of assimilate from roots to developing fruit at the onset of flowering and, in the Gramineae, the temporary but severe disruption of the stem vascular system during rapid internode elongation (Hay, 1978). It is difficult to recognize plant adaptations which have evolved specifically in response to water stress at flowering, but the very rapid growth and maturation of desert ephemerals in arid zones could be interpreted as an adaptation favouring the completion of reproduction before the water supply has been exhausted. Passioura (1976) considers that the distinct seminal and nodal root systems of cereals tend to lead to improved water relations during grain filling, thereby maximizing grain yield.

C. Tolerance of Desiccation

In some environments, adaptations such as those outlined in the preceding two sections can be sufficient to prevent the exposure of the native species to more than moderate stress, and a relatively modest loss of tissue water (Table 4.2; Fig. 4.1). In contrast, many of the species adapted to arid zones experience severe water stress as a normal feature of their life-cycle, as a result of prolonged loss of water by cuticular transpiration, or of rapid water loss through stomata remaining open by osmoregulation. The survival of these plants depends upon the ability of their cells to tolerate dehydration. The most celebrated examples are the so-called resurrection plants (at least 83 species from a number of genera, predominantly from extremely arid regions of South Africa) whose mature leaves can tolerate many months of severe desiccation and recover to function normally when rehydrated. In the laboratory, resurrection plants can survive prolonged equilibration with dry air (0% RH; Gaff, 1980).

The desiccation of plants has been less thoroughly studied than other aspects of plant/water relations, primarily because it is not of great significance in agriculture; few crops are grown in environments where a serious risk of crop desiccation and total loss of yield is a consistent feature of the growing season. However, it is also an intrinsically difficult subject for research since it should involve the simultaneous study of whole plant physiology, cell ultrastructure and several metabolic pathways. Furthermore, it is difficult to establish the cause of cell death when so many potentially lethal processes are occurring in the same dehydrated cell (disruption of all major metabolic pathways, the denaturation of macromolecules, the failure of membrane function, as well as mechanical damage to the protoplast as a whole (Parker, 1968; Gaff, 1980)).

Although it is clear that there are substantial differences amongst plant species in their resistance to injury by dehydration, it is difficult to quantify such differences because of uncertainties in establishing appropriate indices of dehydration "stress" and the resulting "strain" or injury (see p. 128 for a discussion of the problems involved in defining and measuring water stress) (Parker, 1972). Barrs (1968) proposes that the level of dehydration should be expressed by a combination of Ψ_{leaf} and relative water content of the leaves, whereas Parker (1972) suggests that injury should be monitored by assessing cell survival using tetrazolium dyes. However, because detailed studies of both stress and injury are rare, it becomes necessary to fall back on the minimum recorded leaf water potential (Ψ_{min}) or the lowest relative humidity of the air which the plant can tolerate, in order to establish differences in desiccation resistance between groups of species. For example, Table 4.7 shows that species from xeric sites are exposed to, and can survive, much lower values of Ψ_{min} than those from more mesic sites. A comparable range of minimum leaf water potentials in terms of air relative humidity is given by Gaff (1980).

Because of these inherent difficulties, the mechanism of desiccation tolerance is poorly understood. Different experiments have suggested that tolerance involves changes in the viscosity of the cytoplasm during drought hardening, the protection of membrane properties by the release of organic solutes, or reduction in the number and reactivity of thiol groups carried by macromolecules (Fig. 5.9). In general, younger tissues appear to be more tolerant than more mature, possibly because

TABLE 4.7.   Minimum recorded leaf water potential values for plants from a range of habitats (from Richter, 1976; Gaff, 1980).

	$\Psi min\ (MPa)$
Resurrection plants	− 16·0 (and lower)
Other desert plants	− 1·8 to − 16·3
	(mainly − 6 to − 10)
Plants from zones with periods of pronounced drought	− 3·2 to − 7·0
Woody plants from mesic sites	− 1·5 to − 2·6
Mesophytic herbs	− 1·4 to − 4·3

they have a higher ratio of non-vacuolate to vacuolate cells (Parker 1968, 1972). Bewley and Krochko (1982) confirm that little real progress has been made, but suggest that species such as the resurrection plants which can survive very severe dehydration (in spite of serious damage, including the total loss of chlorophyll) show three important abilities:

    (i) to limit damage during dehydration;
   (ii) to maintain physiological integrity in the dry state (as do the seeds of most species); and
  (iii) to initiate repair processes (to membranes, etc.) on rewetting.

D. Contrasting Life-histories in Arid Environments

It is clear that survival and reproduction in a given dry environment do not depend upon any one specific adaptation or association of adaptations for the acquisition, conservation and optimum use of water. Some idea of the variety of possible life-histories, morphologies and physiological responses to the environment can be given by the range of species native to the Sonoran Desert which includes arid regions of Mexico, California and Arizona (Shreve and Wiggins, 1964). Species which have been studied in detail (mainly from Creosote Bush (i.e. *Larrea divaricata*) Scrub; Fig. 4.12) include:

Atriplex hymenelytra (C_4 photosynthesis). A long-lived, small-leaved evergreen perennial shrub (height 20 cm to 1 m), whose life-history is conservative, making use of episodic precipitation for the survival of the individual plant rather than annual reproduction. The adaptation of the photosynthetic apparatus to high temperatures and to the seasonal variation in temperature (adjustment of optimum by up to 10°C, Fig. 5.3) means that assimilation is possible throughout the year if sufficient water is available. However, dry matter production is generally negligible during the hot, dry autumn and winter months (Fig. 5.10) because of long-term closure of stomata and the secretion of (white) salt on the surface of the leaves, which increases the reflection

---

FIG. 4.12.   Morphology of species native to the Sonoran Desert, California. (a) *Larrea divaricata* (leaf length 0·5–1 cm); (b) *Agave deserti* (leaf length 15–40 cm, flowering stem up to 12 m); (c) *Encelia farinosa* (leaf length 3–8 cm); (d) *Tidestromia oblongifolia* (leaf length 1–3 cm); (e) *Atriplex hymenelytra* (leaf length 1·5–3·5 cm). (From Jaeger, 1940; Munz, 1974.)

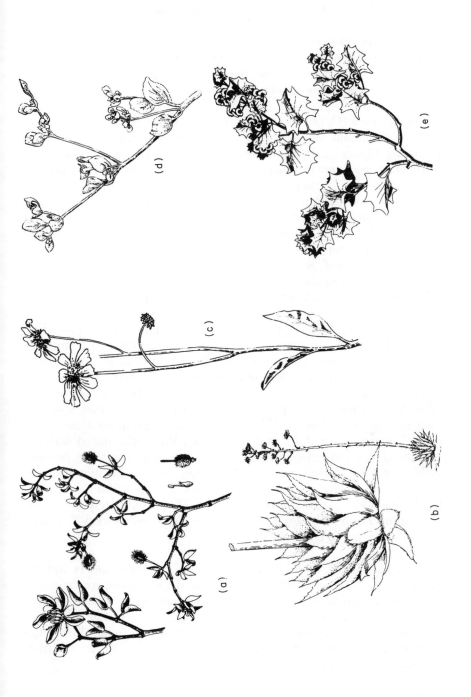

of intercepted radiation in much the same way as the thick, white pubescence of *Encelia farinosa* (see above). In addition, the interception of solar radiation can be reduced by adjustment of leaf angle (Gulmon and Mooney, 1977).

Larrea divaricata (C_3 photosynthesis, evergreen perennial shrub of up to 3 m), has a similar life-history to *A. hymenelytra* but is less conservative, achieving higher rates of net assimilation throughout the year (Fig. 5.10) as a consequence of biochemical and physiological adaptations to the environment rather than the morphological adaptations of *A. hymenelytra* (and in spite of using the C_3 pathway). For example, leaf turgor and stomatal opening can be maintained down to leaf water potentials of -5.8 MPa, and Mabry *et al.* (1977) quote a Ψ_{min} value of -11.5 MPa from New Mexico, with positive net photosynthesis recorded at a leaf water potential of -8 MPa. Under extreme drought, *Larrea* can shed leaves, twigs and branches, retaining young, dormant leaves in buds which are extremely tolerant of desiccation.

Tidestromia oblongifolia (C_4 photosynthesis). Unlike the two preceding perennials, which can remain active under water stress at a range of temperatures, plants of *T. oblongifolia* (15–30 cm in height) are highly adapted to life at high temperature (Fig. 5.3) but are relatively drought-sensitive. Their activity is restricted to the hot summer months (Fig. 5.10) when the deep root system exploits the stored water in moister habitats within the desert (water courses and depressions where run-off water accumulates and infiltrates). When the stored water is exhausted, the individual plants die but the species survives because viable seed is released annually from the end of the first growing season. Gulmon and Mooney (1977) class *T. oblongifolia* as a "long-lived ephemeral", with a normal life-span of approximately five years. The genus *Tidestromia* also includes a true desert ephemeral *T. lanuginosa* (Shreve and Wiggins, 1964).

Cammissonia claviformis (C_3 photosynthesis), a classic desert annual (ephemeral), completes its life-cycle within the wet spring and summer season by means of very high rates of net photosynthesis and growth (e.g. achieving a height of up to 50 cm) (Fig. 5.10). As a general rule, the ephemeral plants which flourish after episodic rainfall during the (hotter) autumn and winter use the C_4 pathway of CO_2 assimilation (Pearcy and Ehleringer, 1984).

Agave spp., *Opuntia* spp. and a variety of other bulky succulents whose conservative life-cycles, based on CAM photosynthesis, are made possible by their tolerance of high temperatures (p. 224).

V. SOME SPECIAL PROBLEMS IN TREE/WATER RELATIONS

Because of their great height and the structure of their leaf canopies, tall trees face unique difficulties in maintaining favourable leaf water relations; as a result, tall forest (e.g. tropical and temperate rainforest, northern deciduous and coniferous forest) is the characteristic vegetation of only the most humid areas of the world.

For example, as a consequence of the normal increase in wind velocity with height, the leaves of trees tend to transpire more rapidly than do the leaves of shorter plants, especially in discontinuous stands. A more serious problem is that the "sun" leaves at the top of the tree canopy are exposed to greater water and temperature stresses than leaves lower down, or shaded within the canopy. Thus, the leaves which require the most water are furthest from the point of supply, the soil. (Note that although most mature forest trees do not exceed 30–60 m in height, mechanisms must exist to supply water to the leaves of the worlds tallest trees, which in the recent past have exceeded 110 m; Sutcliffe, 1981.) A third feature of tree physiology is that although the leaf area index of forests or plantations can be similar to those of agricultural crops (values between 2·5 and 8·5; Rutter, 1968), in many cases a very large leaf area is supported and supplied by a single trunk rather than by a number of smaller stems; thus the water relations of an enormous number of leaves or needles can depend upon the health and normal function of the vascular system of a single trunk. In addition to these physiological difficulties, there is also the fundamental physical problem that work must be performed to raise water against the force of gravity. In short vegetation, this effect can be neglected, but when considering water potential in tall trees, a gravitational component must be included. Thus equation 4.3 must be expanded to give:

$$\Psi = \psi_s + \psi_m + \psi_p + \psi_g \qquad (4.15)$$

where ψ_g, the gravitational potential, increases by 0·01 MPa for each metre increase in height. Consequently, for water to move from the root to a transpiring leaf at a height of 50 m, a lowering of leaf water potential of 0·5 MPa is required in addition to the lowering required to overcome the hydraulic resistance of the plant (see p. 144). Therefore, as long as they are taking part in transpiration, the water potential of these leaves can never be higher, and is normally considerably lower, than -0.5 MPa, a value at which leaf turgor, leaf expansion and several other processes are significantly reduced in many species (Fig. 4.2) (Zimmermann and Brown, 1971).

Tall trees, therefore, require:

(i) A vascular system of high capacity which can deliver water rapidly and preferentially to those parts of the canopy which are most active in transpiration. The system must also be highly resistant to environmental and biological stresses, especially frost, wind and pathogenic infection;

(ii) leaves which use water efficiently but also continue to grow and assimilate at low water potentials.

A. Vascular System

As in all higher plants, the movement of water in the trunks and branches of trees takes place in the lumina of dead xylem elements; in trees, of course, the lignified xylem is also important in providing the mechanical strength necessary to support the leaf canopy. The structure of the vascular system differs markedly amongst groups of trees (Esau, 1965). In ring-porous trees (including many north-temperate deciduous species), water moves predominantly in an outer ring of wide vessels (60–400 μm, Table 4.8) laid down at the start of the current growing season, which are arranged to give long, continuous pipes without cross walls. Many of the older vessels will be gas-filled. In contrast, the functional xylem in diffuse-porous deciduous trees and in conifers is not restricted to an outer ring of new xylem but is composed of finer, shorter tracheids, interconnected through pores (Table 4.8).

According to Poiseuille's Law, the flow of water along a cylindrical pipe (down a constant gradient of hydrostatic pressure between its ends, and at a constant temperature and water viscosity) is proportional to the square of the radius of the pipe. Consequently, the conducting elements of ring-porous trees are able to transport water at much greater velocities (normally between 15 and 45 m h^{-1}; Table 4.8) than can the elements of diffuse-porous (1–6 m h^{-1}) or coniferous trees (up to 2 m h^{-1}). However, this does not necessarily mean that ring-porous trees can support higher rates of transpiration, because the rate of flow of water to the canopy also depends upon the total cross-sectional area of functional xylem. For example, Jordan and Kline

TABLE 4.8. Diameters of xylem elements (vessels, tracheids) of selected tree species and the midday peak velocities of water movement which have been observed through such elements (data measured at breast height, compiled by Zimmermann and Brown, 1971).

Tree species	Element diameter (μm)	Velocity ($m\,h^{-1}$)
Ring-porous		
Quercus robur	200–300	43·6
Robinia pseudacacia	160–400	28·8
Quercus rubra	250	27·7
Fraxinus excelsior	120–350	25·7
Castanea vesca	300–350	24·0
Ailanthus glandulosa	170–250	22·2
Carya alba	180–300	19·2
Rhus glabra	—	16·0
Ulmus effusa	130–340	6·0
Laburnum anagyroules	60–250	3·9
Diffuse porous		
Populus balsamifera	80–120	6·25
Juglans regia	120–160	4·12
Juglans cinerea	—	3·79
Tilia tomentosa	25–90	3·43
Salix viridis	80–120	3·00
Liriodendron tulipifera	50–120	2·62
Acer pseudoplatanus	30–110	2·40
Magnolia acuminata	—	2·06
Alnus glutinosa	20–90	2·00
Betula pendula	30–130	1·60
Carpinus betulus	16–80	1·25
Pyrus communis	50–80	1·11
Fagus sylvatica	16–80	1·07
Aesculus hippocastanum	30–60	0·96
Conifers		
Larix decidua	up to 55	2·1
Pinus strobus	up to 45	1·7
Picea excelsa	up to 45	1·2
Tsuga canadensis	up to 45	1·0

(1977) demonstrated that the rate of transpiration of tropical rainforest trees and of Douglas Fir is highly correlated with the sapwood area of the trunk, irrespective of species and soil type.

In temperate and boreal zones, the geometry of xylem elements is probably more important in ensuring xylem function in spring after severe winters. In ring-porous trees, new xylem elements are laid down just before the development of the leaf canopy; during the following growing season, a few wide and long vessels can provide sufficient conduit volume to supply the water requirement of the transpiring leaves. However, in a severe winter, after leaf fall, the contents of the xylem will tend to undergo cycles of freezing and thawing, leading to the release of small volumes of previously-dissolved gases; in the long continuous vessels of ring-porous trees, the released gases merge together to give large bubbles of air which lodge at the constrictions between elements, thereby breaking the continuity of liquid water (embolism). Since the vessels can no longer contain water columns under tension, they cease to function as water conduits and must be replaced by new water-filled xylem (Zimmermann and Brown, 1971). Although they achieve an efficient functional relationship between xylem development and leaf production, ring-porous trees are faced with other serious problems. For example, mechanical damage (e.g. the effect of strong winds) can also cause embolism and, as the recent spread of Dutch Elm disease has shown, their xylem is peculiarly susceptible to pathogenic organisms (Zimmermann and McDonough, 1978).

In contrast, conifers are much more resistant to environmental stress. In particular, gas released in the xylem tends to be trapped in small quantities within the fine tracheids where it can be redissolved in the xylem water; consequently, only a small proportion of the conducting volume will be lost during cycles of freezing and thawing. Similarly, since the functional xylem is not restricted to the periphery of the trunk, and since successive tracheids are arranged in a complex, commonly spiral pattern, the vascular system is less susceptible to mechanical damage and pathogens (Zimmermann and Brown, 1971; Zimmermann and McDonough, 1978). These features presumably contribute to the prevalence of coniferous species in more extreme boreal zones.

The Huber value ($mm^2 g^{-1}$):

$$\frac{\text{transverse section xylem area } (mm^2) \text{ of a stem or branch}}{\text{fresh weight of leaves (g) supplied with water by that stem or branch}}$$

has been used to demonstrate that the upper branches of a tree enjoy a better supply of water than do the lower branches. In the original study, Huber (1928) (reviewed in Zimmermann, 1978) measured values of 0·5 in the lower stem and lateral branches, and 4·26 in the terminal shoot of *Abies concolor*. Huber values have been subject to criticism because the measurement does not discriminate between functional and non-functional elements. However, by measuring leaf specific conductance rather than the Huber value, Zimmermann (1978) has confirmed the phenomenon in three diffuse-porous deciduous tree species; as a consequence of constrictions in the xylem elements leading to lateral branches, the hydraulic resistance of the pathway supplying water to the "sun" leaves at the top of the canopy is lower than that supplying lower and more shaded leaves.

B. Leaves

Because the resistance to water flow in the stem and branches of a tree is relatively high, considerable gradients in water potential (of the order of $0·01$ MPa m^{-1}, irrespective of xylem type) must be set up within the tree to ensure a supply of water to the canopy sufficient for maximum rates of transpiration. Consequently, the total gradient, including the gravity component, will be up to $0·02$ MPa m^{-1}, giving leaf water potentials of the order of $-1·0$ MPa at 50 m. It is, therefore, imperative that tree leaves use water efficiently so as to minimize further lowering of leaf water potential. Many of the relevant characteristics of tree leaves (sunken stomata with waxy plugs, thick cuticles, leaf abscission, etc.) have already been discussed in detail above, and are reviewed in Kozlowski (1976) and Kramer and Kozlowski (1979).

Under certain circumstances, measured gradients in water potential in tall trees are not consistent with these theoretical gradients, and to explain this it has been suggested that tree leaves are capable of absorbing significant quantities of liquid water from their surfaces, thereby raising leaf water potential (Connor *et al.*, 1977). In this connection it should be noted that tree canopies can intercept up to 80% of the rain falling on a forested area (Penman, 1963), and there are several reports of reversed movement of water from wet leaves towards the trunk (e.g. Daum, 1967). Nevertheless, the leaves of tall trees must be able to grow and function at leaf water potentials substantially lower than $-1·0$ MPa; under these conditions, expansion must take place largely at night, with turgor maintained by osmoregulation (Morgan, 1984; see above).

In conclusion, it is likely that in cold regions trees are most at risk from drought during the early spring when transpiration (in evergreens) is beginning to increase in response to seasonal increases in irradiance and air temperature. At the same time, the roots will be unable to extract water if the soil is frozen or root permeability is low as a consequence of low temperatures (Kramer, 1969). (The importance of water deficit in determining the altitude of the tree-line is discussed on p. 203.) However, the leaves (needles) of most of the coniferous species native to such areas are equipped with appropriate xeromorphic features, whereas *Larix* species avoid the hazards of winter and early spring by leaf abscission. (The advantages and disadvantages of an evergreen life-cycle are discussed on p. 208.) The water relations of woody plants are treated in much greater detail in Zimmermann and Brown (1971) and Kramer and Kozlowski, (1979).

Part III

Responses to Environmental Stress

5. Temperature

I. THE TEMPERATURE RELATIONS OF PLANTS

Unlike homeothermic animals, higher plants are unable to maintain their cells and tissues at a constant optimum temperature. Their growth and metabolism are therefore profoundly affected by changes in environmental temperature. However, it is difficult to establish precise relationships between plant processes and environmental temperature because of the extreme variability of soil and air temperatures. For example, the temperature of a leaf depends upon:

   (i) time of day (regular diurnal variation);
  (ii) month of the year (regular seasonal variation);
 (iii) cloudiness and wind speed (irregular, short-term variation);
 (iv) position in the canopy (e.g. "sun" or "shade" leaves);
  (v) height above the soil surface;
 (vi) leaf shape and dimensions;

whereas root temperature depends mainly upon (i) and (ii), but also upon:

 (vii) depth below the soil surface; and
(viii) soil properties controlling the energy balance at the soil surface and the transfer of heat through the soil (for example, soil moisture content, bulk density, colour and the vegetative or litter cover).

Consequently, the leaf canopy (and the soil profile) is a complex mosaic of rapidly fluctuating thermal regimes such that each group of leaves (or roots) is responding to a unique pattern of temperature fluctuation. This variability makes it very difficult to carry out field studies on the effects of temperature on processes such as photosynthesis.

It causes even greater difficulty in long-term investigations where plant growth rate (and, ultimately, plant distribution) can depend upon one or more of a number of thermal parameters; these include mean, minimum and maximum temperature and the amount of accumulated temperature (degree hours, degree days) above a threshold during the whole year, or a shorter critical period (e.g. Pigott, 1975; Wassink, 1972).

In addition to these problems of environmental variability, it has been found that different stages of plant development, and different physiological processes, can have different temperature optima. For example, in *Tulipa* spp., the temperature optima for the various stages of flower development vary from 8°C to 23°C, these values being correlated with seasonal temperature changes in the native range of these species (Pisek *et al.*, 1973). Furthermore, the reproductive development of certain species is controlled by night, rather than day, temperatures (Went, 1953; Leopold and Kriedemann, 1975), and some processes, especially germination, can be enhanced by temperature fluctuation (Thompson and Grime, 1983). Even more important, it may be difficult to establish the relative importance of soil and air temperature for plant processes; this is particularly true during the vegetative development of grasses and sedges in which leaf extension takes place from stem apices near the soil surface (controlled by soil temperature; Peacock, 1975; Hay and Tunnicliffe Wilson, 1982), whereas the mature photosynthesizing leaves are subject to air temperature.

Because of these problems, most investigations of plant response to temperature have been carried out under controlled conditions with root and shoot at the same constant temperature or, more rarely, with root and shoot at different, but still constant, temperatures. Some experiments have involved different day and night temperatures but very few workers have attempted to simulate even the regular diurnal variation (Precht *et al.*, 1973; Cooper, 1973). It is clear that extrapolation of such data to the field must be carried out with caution.

The response of plant growth rate to a wide range of (constant) temperatures can commonly be represented by an asymmetric bell-shaped curve as shown in Fig. 5.1(a) (Pisek *et al.*, 1973; Cooper, 1973; Sutcliffe, 1977). From such a curve, it is possible to read off the three classic cardinal temperatures, i.e. the *minimum* and *maximum* temperatures at which growth ceases entirely, and the *optimum range* of temperature over which the highest growth rate can be maintained, assuming that temperature is the factor limiting growth. In practice, the optimum range normally covers the range of temperatures over which the rate of growth, or of a metabolic process, is within (say) 10% of the maximum rate (e.g. Fig. 5.2). The cardinal temperatures of

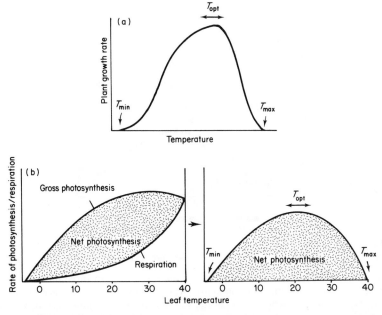

Fig. 5.1. Schematic representations of plant responses to temperature. (a) A generalized diagram of the response of plant growth rate to temperature, illustrating the three cardinal temperatures, i.e. the minimum (T_{min}) and maximum (T_{max}) temperatures, and the optimum temperature range (T_{opt}) for growth. (b) The influence of temperature on gross photosynthesis (net of photorespiration) and respiration in a typical plant (see text for full description) (adapted from Pisek et al., 1973).

higher plant processes vary widely within the range -10 to $60°C$ and are normally related to the temperature regime in the natural range of the species (e.g. Fig. 5.2).

This characteristic response of plant growth to temperature arises because increase in temperature affects biochemical processes in two mutually antagonistic ways. First, as the temperature of a plant cell rises, the velocity of movement (vibrational, rotational and translational) of the reacting molecules increases, leading to more frequent intermolecular collisions, and more rapid reaction rates; this effect is common to most chemical reactions. However, virtually all the reactions occurring in cells are catalysed by enzymes whose action depends upon the maintenance of precise tertiary structures (three-dimensional shapes) into which the reacting molecules must fit exactly in order for the reaction to proceed. As the temperature rises, increased molecular agitation tends to damage these tertiary structures, leading to reduced

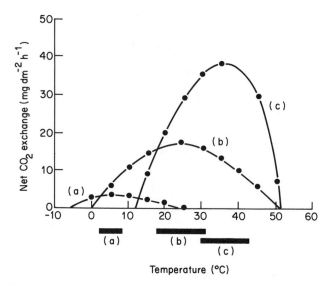

FIG. 5.2.   The cardinal temperatures of net photosynthesis in three members of the Gramineae from contrasting environments. (a) *Chionochloa* spp. tussock grasses (alpine, C_3 photosynthesis), (b) wheat (temperate, C_3) and (c) maize (sub-tropical, C_4). The horizontal bars indicate the optimum range (adapted from Wardlaw, 1979).

enzyme activity and reaction rates. Overall, at sub-optimal tempera-tures, thermal inactivation of enzymes is slight and the response to increases in temperature is therefore positive. In the optimal range, temperature change has little effect because the two antagonistic effects are balanced, i.e. the acceleration caused by increased agitation is matched by the reduction in rate caused by enzyme denaturation. However, above the optimum, thermal denaturation of the enzyme, or of the membranes to which it is bound, leads to a rapid cessation of the reaction. The asymmetry of the response curve is the result of a gradual, sigmoid pattern of increase in rate caused mainly by increased collision frequency, followed by an abrupt fall in rate caused by thermal denaturation of macromolecules (Sutcliffe, 1977).

The individual processes contributing to plant growth do not all respond to temperature in the same way. For example, in many temperate species, gross photosynthesis ceases at temperatures just below 0°C (minimum) and well above 40°C (maximum), with the highest rates achieved in the range 20–35°C (Fig. 5.1(b)). In contrast, respiration rates tend to be low below 20°C but, owing to the thermal disruption of metabolism (disruption of fine control of reaction rates, breakdown of compartmentation, etc.) at higher temperatures, they rise rapidly up to the compensation temperature, at which the rate of respiration equals the rate of gross photosynthesis, and there can be no net photosynthesis (Fig. 5.1(b)). In consequence, the response of net photosynthesis (gross photosynthesis less respiration) to temperature variation is broadly similar to that of growth (Fig. 5.1).

Since the exact pattern of response to temperature varies among species and processes, it is necessary to have a quantitative expression of these responses, especially in the sub-optimal temperature range. Most commonly, Q_{10} or Q_5 values are used, where

$$Q_{10} = \frac{\text{rate at temperature } T + 10°C}{\text{rate at temperature } T} \qquad \text{(similarly for } Q_5)$$

Because it has been found that Q_{10} values for chemical reactions *in vitro* usually fall around 2, it is normally assumed that measured Q_{10} values in excess of 2 indicate that the plant process is under metabolic control, whereas values below 2 are taken to indicate that the rate of the process under study is limited by a purely physical step such as diffusion or by a photochemical reaction (Sutcliffe, 1977; Berry and Raison, 1981). However, although these assumptions are generally useful, Q values for plant processes should be treated with caution because they are themselves temperature-dependent since plant response to temperature is not strictly exponential (Fig. 5.1); for example, the Q_5 value for the respiration of maize root systems fell from 1.85 to 1.35 over the temperature range 10–30°C (Wassink, 1972). Some typical Q_{10} values for plant processes are given in Table 5.1.

TABLE 5.1. Temperature coefficients (Q_{10}) for selected plant processes measured at varying intervals within the range 0–30°C. (Compiled from a range of sources.)

Process	Q_{10}
Diffusion of small molecules in water	1·2–1·5
Water flow through the seed coat of *Arachis hypogea*	1·3–1·6
Water movements into germinating seeds, various species	1·5–1·8
Hydrolysis reactions catalysed by enzymes	1·5–2·3
Respiration	2·1–2·6
Photosynthesis (light reactions)	$\simeq 1$
(dark reactions)	2–3
Phosphate uptake into beetroot discs	0·8[a]–3[b]
Potassium uptake into maize seedlings	2–5

[a] At high external concentrations (50 mM) where uptake is largely by passive diffusion.
[b] Active uptake at low external concentrations (0·1 mM) (see p.83).

A number of more recent field investigations have shown that, particularly for developmental processes (leaf initiation and appearance, leaf extension, spikelet initiation, germination etc.), the temperature response curve shown in Fig. 5.1(a) can be simplified to a pair of straight lines meeting at the optimum temperature. This is an extremely useful simplification because it allows the prediction of rates of developmental processes during periods of varying temperature using accumulated temperature (Monteith, 1977; Gallagher and Biscoe, 1979; Hay and Tunnicliffe Wilson, 1982).

As well as influencing the *rate* of plant growth, development and metabolism, environmental temperature can also play a part in controlling the *pattern* and *timing* of plant development. For example, many species (e.g. those forming bulbs or rosettes at the end of the first growing season, and many perennial grasses) are unable to flower until they have been vernalized by a period of low temperature; this adaptation improves the chances of successful reproduction by ensuring that reproductive development and seed production do not begin until the beginning of the next growing season. Similarly, seed dormancy, which is possessed by a wide variety of temperate plants, prevents premature germination in autumn and can be broken only by exposure to low temperature stratification which simulates the low temperatures of winter. In a similar way, the annual development and abscission of the leaf canopy in autumn-deciduous trees and shrubs are largely under thermo- and photoperiodic control (Went, 1953; Addicott and Lyon, 1973).

However, few developmental processes are controlled by temperature alone and responses to temperature can in many cases be modified by other factors, particularly the solar radiation environment. This is clearly illustrated by the finding that tuber initiation in the potato (*Solanum tuberosum*) depends upon the interplay of temperature, photoperiod, radiant flux density and nutrient supply (Hay and Allen, 1978).

As a further complication, the thermal regime during growth and development can influence the morphology and dimensions of the resulting plant parts, as well as the partitioning of dry matter within the plant. For example, in different species, root diameter, root branching and the size and shape of leaves can be determined by growing temperature (Bensink, 1971; Cooper, 1973; Abbas Al-Ani and Hay, 1983). There are numerous reports of the influence of temperature on root:shoot ratio, and there is evidence that leaf area ratio and specific leaf area can also be altered by the thermal environment (e.g. in *Phleum* spp.; Woodward, 1983).

In summary, it appears that the complexity of the thermal regime of plants in their natural environment is matched by the complexity of plant responses to temperature. In this chapter, we shall concentrate upon those adaptations of plants which enable them to survive in areas where extremes of temperature are commonly experienced. Although, in conformity with the remainder of this book, the treatment is largely restricted to higher plants, it should be pointed out that mosses and lichens play a major role in tundra vegetation, especially at extreme altitudes and latitudes.

A. Responses of Plant Species to Changes in the Thermal Environment

The adaptation of plants to changed or new thermal environments can occur through the evolution of genotypes with more appropriate morphologies, life-histories or physiological and biochemical characteristics, or by plasticity. For example, Fig. 5.2 gives some measure of the potential for evolution of the photosynthetic apparatus; beginning from a common ancestor, graminoid species have evolved which are well-adapted to alpine, temperate and sub-tropical environments. Similar conclusions can be drawn from a series of reciprocal transplant experiments carried out in California between a cool, coastal site (annual range of mean air temperature 13–18°C), and the floor of Death Valley (18–46°C)(Björkman et al., 1973/74; Björkman, 1980). The experimental plants were fertilized and irrigated to ensure that the principal differences were in temperature (and radiant flux density). Each of the wide range of species investigated fell into one of three classes: (a) those which were incapable of surviving the hot desert summer but survived and were summer-active at the coast. This class included the species which were native to the coast, e.g. *Atriplex glabriuscula* (C_3 photosynthesis) and *A. sabulosa* (C_4). (b) Those which could survive the desert summer but were winter-active in the desert and summer-active at the coast. This class included desert evergreens of both the old and the new world, e.g. *Atriplex hymenelytra* (C_4), *Larrea divaricata* (C_3) and *Nerium oleander* (C_3) (Fig. 4.12). (c) A class of only one species, *Tidestromia oblongifolia* (C_4) (Fig. 4.12) which is both native to and summer-active in Death Valley, but is incapable of surviving at the coast.

These patterns of growth and survival are reflected in the temperature relations of photosynthesis for each species (Fig. 5.3, at the appropriate growing temperature, i.e."cool" for *A. glabriuscula* and *A.*

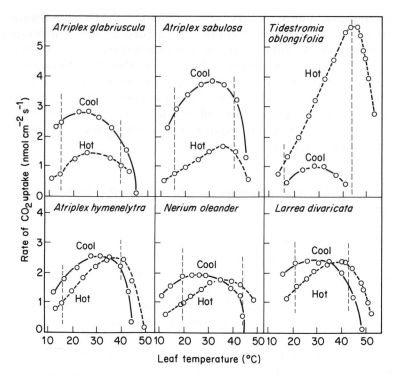

FIG. 5.3. The effect of growing temperature on the rate and temperature dependence of light-saturated photosynthesis for six species native to habitats with contrasting thermal regimes. The vertical broken lines indicate the daytime temperatures of the "cool" and "hot" growth regimes for each species (from Björkman, 1980).

sabulosa, and "hot" for the others). For example, the optimum temperature for *T. oblongifolia* was 42°C, compared with 25°C for *A. glabriuscula*. However, Fig. 5.3 also shows that alteration of the growing temperature can have a profound influence upon temperature responses. Thus, growing *T. oblongifolia* at 17°C rather than at 42°C lowered the temperature optimum of net photosynthesis to below 30°C, but since the rate of net photosynthesis was also severely depressed, presumably owing to damage to the photosynthetic apparatus (Berry and Björkman, 1980), this had serious implications for its survival at low temperatures. The opposite trend was observed for coastal species. Raising the growing temperature did raise the temperature optimum by a few degrees, but again there was a serious reduction in the rate of net photosynthesis. In contrast, the survival of the three evergreen species at both sites, albeit with lower growth rates than the more extreme types, can be attributed to the plasticity of their photosynthetic systems (change in optimum temperature by up to 15°C, associated with only modest penalties in terms of the rate of net photosynthesis; Fig. 5.3). Phenotypic plasticity of this kind is common in both hot (Berry and Björkman, 1980) and cold (Tranquillini, 1979) environments. For example, Billings *et al.* (1971) showed that the optimum temperature for net photosynthesis in alpine populations of *Oxyria digyna* fell by 1° C for each reduction of 3°C in growing temperature.

B. The Energy Budget of a Plant Leaf

When a leaf is illuminated, it absorbs between 20 and 95% of the incident radiation, depending upon the wavelength (Fig. 2.2). However, only a small fraction of this absorbed energy is used in photosynthesis; the remainder is transformed into heat and unless the leaf can lose this excess heat, its temperature will rise, leading ultimately to death by thermal stress. In a well-watered plant, there are three major processes which can act to remove heat from leaves, i.e. re-radiation (long wave), convection of heat, and transpiration.

If a leaf is to remain at a constant temperature, then its energy budget must balance, i.e.:

$$Q_{abs} = Q_{rad} + Q_{conv} + Q_{trans} \tag{5.1}$$

Q_{abs}	Q_{rad}	Q_{conv}	Q_{trans}
energy absorbed by leaf	energy lost by radiation	energy lost by convection of heat	energy lost by transpiration of water

Therefore, if:

$$Q_{rad} + Q_{conv} + Q_{trans} > Q_{abs} \tag{5.2}$$

the leaf will be cooled, whereas, if

$$Q_{rad} + Q_{conv} + Q_{trans} < Q_{abs} \tag{5.3}$$

leaf temperature will rise.

In a series of theoretical papers (reviewed in Gates and Papian, 1971; Gates, 1976), equation 5.1 is expanded to give the form:

$$Q_{abs} = \varepsilon \sigma T_1^4 + k_1 (V/D)^{\frac{1}{2}} (T_1 - T_a) + \frac{L \cdot d_1^s (T_1) - RH \cdot d_a^s (T_a)}{R_1} \tag{5.4}$$

$$\quad\quad (Q_{rad}) \quad\quad\quad (Q_{conv}) \quad\quad\quad\quad\quad (Q_{trans})$$

where

ε	is the emissivity of the leaf (long-wave radiation);
σ, k_1	are constants;
T_1 and T_a	are the temperatures of leaf and bulk air, respectively;
V	is wind velocity;
D	is leaf width;
L	is the latent heat of vaporization of water (dependent upon leaf temperature);
$d_1^s(T_1), d_a^s(T_a)$	are the saturation water vapour densities in the leaf and in the air, respectively;
RH	is the relative humidity of the bulk air;
R_1	is the leaf diffusive resistance (see Chapter 4).

Consequently the temperature of a leaf is determined not only by the interplay of a set of environmental variables (air temperature and humidity, radiant flux density, wind velocity), but also by a range of plant characteristics, including its radiative properties (reflection coefficient, emissivity, colour), dimensions, shape and angle, the responses of its stomata, and its height above the soil surface. For example, leaf temperature tends to fall under conditions favouring convection (high wind velocity, small leaves) and/or transpiration (low air humidity, low leaf diffusive resistance), whereas leaf warming will tend to occur with low wind velocity, large leaves, high humidity and high diffusive resistance (the latter following the closure of stomata).

As discussed in subsequent sections, modification of these plant characteristics leading to leaf warming (commonly associated with biochemical adaptations giving lower optimal temperatures; e.g. Fig. 5.3) has played an important role in the adaptation of plants to arctic and alpine environments; contrasting adaptations have led to leaf cooling in hot, dry regions. The remainder of this chapter, and parts of Chapter 4, concentrate upon these more extreme environments, but other important aspects of plant response to temperature, such as the comparative anatomy and physiology of species possessing C_3, C_4 and CAM photosynthetic pathways, are considered in greater detail in Chapters 2 and 4.

II. PLANT ADAPTATIONS AND RESISTANCE TO LOW TEMPERATURE

A. The Influence of Low Temperature on Plants

The primary effect of cooling plants below their optimum temperature range is the reduction of rates of growth and of metabolic processes (Fig. 5.1). Consequently, the length of time required for completion of the annual growth cycle increases as the climate becomes cooler, and there may be a critical mean temperature below which the plants of a given species cannot reproduce successfully (Tranquillini, 1979). Such an effect is responsible for the distribution of *Juncus squarrosus* in upland Britain. This perennial rush produces fertile seed and is widely distributed at altitudes up to 760–840 m; however, above 840 m, the mean temperature is normally too low to permit completion of reproductive development, and fertile fruit are produced only in unusually warm years. Consequently, plants of *J. squarrosus* are sparsely distributed above this altitude, occurring only where viable seed has been dropped by sheep or birds (Pearsall, 1968).

Cooling sub-tropical and tropical plants down to temperatures in the range 0–10°C tends to cause a very rapid fall in the activity of metabolic processes, especially respiration, and may result in severe damage and death within a few hours or days. The available evidence indicates that chilling of these species causes a phase change (liquid to solid) in membrane lipids, leading to the inactivation of membrane-bound enzymes, such as the respiratory enzymes attached to mitochondrial membranes, and disruption of the water and ion uptake activities of the root system (McWilliam *et al.*, 1982). Higher proportions of unsaturated fatty acids in the membrane lipids of temperate species appear to confer greater membrane stability and give protection against chilling injury (Lyons, 1973; Clarkson *et al.*, 1980).

In general, temperate plants are not susceptible to chilling injury at temperatures above 0°C and tend to show signs of serious damage only after ice has formed within their tissues. For example, it has been demonstrated for several tree species that photosynthesis does not cease completely until extracellular ice has formed (normally at -3 to -5°C; Neilson *et al.*, 1972), and even then the initial cessation of activity may be the result of a purely physical phenomenon—the blocking of CO_2 diffusion by ice (Larcher *et al.*, 1973). Under relatively low rates of cooling (< 1°C h^{-1}), ice tends to form preferentially in the apoplast of plant tissues (i.e. extracellular rather than intracellular ice) because of higher solute concentrations in cytoplasm and vacuoles. As long as such periods of freezing are not prolonged and the rate of thawing is not too rapid, the formation of extracellular ice may not cause significant tissue damage in hardened plants (see below).

However, if extracellular ice persists, the gradient of water vapour pressure between the apoplast and the cell causes water to migrate out of the cells and into the apoplast, where it freezes, thereby increasing the amount of ice in the tissue (Grieve and Povey, 1981; Ishikawa and Sakai, 1981). As well as causing mechanical damage, this process results in the progressive dehydration of the cell contents and an increase in the concentration of the cell sap (similar phenomena occur under water stress, p. 174). Consequently, the biochemistry of the cytoplasm is seriously disturbed; proteins, including enzymes, are denatured, various components are precipitated, compartmented substances such as hydrolytic enzymes are released into the cytoplasm, the buffer system becomes unable to control cell pH, and there may be a tendency for macromolecules to condense when forced together by the dehydration of the cytoplasm (Fig. 5.9). Under most circumstances, such effects lead inevitably to cell death (Burke *et al.*, 1976). Rapid thawing can also have a lethal effect upon frozen plants, owing to further disruption of cell metabolism and water relations.

As we shall see in a later section, hardened plants of many species can survive prolonged periods of very low temperature and high degrees of cell desiccation, owing to the resistant nature of their cytoplasm. However, it appears that the cells of even the most resistant plants cannot tolerate intracellular freezing, caused by very rapid cooling rates (rates of several degrees per hour). Overall, much of the earlier literature on frost resistance is highly confusing because (i) it has not yet been established that intracellular ice formation (as opposed to severe cell desiccation) does occur in nature—outside the laboratory (Levitt, 1978), and (ii) difficulties are experienced in discriminating between damage caused by intra- and extracellular freezing in the laboratory.

B. The Characteristic Features of Cold Climates—Arctic and Alpine Environments, Temperate Winters

In subsequent sections, we shall examine plant adaptations favouring survival in cold environments, especially arctic and alpine areas, but also in temperate regions during the winter months. However, it should be stressed from the outset that low temperature is only one of a number of unfavourable environmental factors in these areas, and that plant growth and distribution may not be determined by temperature alone, or even *directly* by low temperature (see below).

Although it cannot be defined precisely, the term arctic is generally used in ecology to describe regions stretching from the limit of tree growth (the treeline or timberline) into higher latitudes; it therefore includes tundra areas in both the Arctic and Antarctic. Many of the characteristic features of the arctic environment (with particular reference to plants) can be summarized as follows:

   (i) *Temperature* (air and soil): very low temperatures during the winter; low (and commonly sub-zero) temperatures in summer (Table 5.2).

  (ii) *Solar radiation*: very long photoperiod in summer (continuous for several weeks) although much of the radiation is received at relatively low radiant flux densities. The amount of radiation received by plants can be reduced by snow cover. Note that (i) and (ii) combine to give a very short growing season (typically 6–8 weeks).

 (iii) *Water relations*: in spite of the presence of very large quantities of snow and ice, arctic plants are exposed to drought in summer because of frozen soil moisture, low precipitation, high transpiration rates caused by high winds, and the direct sublimation of snow into dry air. In addition, since the processes of soil formation are so slow, mineral soils in the arctic tend to be unstructured sands and gravels, with very low water-holding capacities. Arctic peat soils have generally more favourable water relations.

 (iv) *Wind*: arctic regions are subject to persistent high winds throughout the year which can cause abrasion of plant tissues by sand particles and sharp snow and ice crystals, especially in winter, in addition to their effects on water relations.

  (v) *Inorganic nutrition*: as a result of the inhibition of soil microbial activity (mineralization of organic nitrogen, nitrogen fixation, etc.) by low soil temperatures and water stress (Marion and Miller, 1982) and the virtual absence of legumes in the tundra flora, the supply of inorganic nitrogen by arctic soils is poor. The supply of phosphate may also be low as a consequence of inadequate root development (Chapin, 1983).

 (vi) *Mechanical effects*: frost heaving of soils during cycles of freezing and thawing may cause the uprooting of plants, especially seedlings during the early stages of establishment.

(vii) *Reproductive problems*: because of the scarcity of insects, and the relatively sparse vegetative cover in many tundra areas, cross-pollination may be impossible. In addition, low population densities of mammals and birds rule out several important seed-dispersal mechanisms.

In summary, it can be seen that the arctic environment is extremely hostile and variable (Rosswall and Heal, 1975; Savile, 1972). Details of conditions at a typical arctic site are given in Table 5.2.

The alpine zone in mountainous regions is normally defined as those areas above the treeline. The altitude of the treeline boundary in this definition is extremely variable, depending upon latitude, exposure and distance from the sea; extreme examples include the Arctic, where the treeline is at sea level, and northern Chile, where forest is found up to altitudes of 4900 m (Wardle, 1974).

The fact that alpine conditions are generally similar to those in arctic environments is shown by the number of species which are common to both environments (e.g. the circumpolar species *Oxyria digyna* (Fig. 5.6), *Silene acaulis*, *Trisetum spicatum,* etc.; Billings and Mooney, 1968). However, they do differ in several important ways (Table 5.2):

(i) *Solar radiation*: irradiance and photoperiod depend upon latitude, season and snow-cover. For example, radiant flux density can be very high in the European Alps in clear summer weather, and can even exceed the solar constant because of reflection by snow, leading to very high soil temperatures (Tranquillini, 1964, 1979). Furthermore, the amount of ultraviolet radiation received increases with altitude (see p. 34).

(ii) *Composition of the atmosphere*: as a result of decreasing atmospheric pressure, the partial pressures of O_2 and CO_2 in the atmosphere fall with altitude.

(iii) *Variability*: the alpine environment is exceptionally variable. For example, there are large differences in radiant flux density and in temperature between north- and south-facing slopes at the same altitude. Tranquillini (1964) quotes a measured difference in soil surface temperature of 57°C, and such variation according to aspect is matched by variation with altitude in a number of other factors (see (i) and (ii) above; wind velocity, etc.).

Together, arctic and alpine regions account for a significant area of the

TABLE 5.2. Environmental conditions in three contrasting regions of North America (adapted from Billings and Mooney, 1968).

	Arctic tundra[a]	Alpine tundra[b]	Temperate forest[c]
Solar radiation			
Mean July flux density (W m^{-2})	209	391	405
Quality	Low in short wavelengths, particularly short UV	High in short UV	—
Maximum photoperiod	84 days	15 h	15 h
Air temperature (1 m, °C)			
Annual mean	−12·4	−3·3	8·3
January mean	−26·7	−12·8	−1·7
July mean	3·9	8·3	20·6
Absolute min.	−48·9	−36·6	−33·8
Soil temperature (15 cm, °C)			
Annual mean	−6·2	−1·7	8·3
Absolute min.	−15·5	−20·0	−10·0
Precipitation (mm)			
Annual mean	107	634	533
Wind (km h^{-1})			
Annual mean	19·3	29·6	10·3
Air composition			
CO$_2$ (mg l^{-1})	0·57	0·36	0·44
O$_2$ (partial pressure, mm)	160	100	122
Depth of soil thaw	20–100 cm	>30 cm	—

[a]Barrow, Alaska (altitude 7 m, latitude 71°20′ N).
[b]Niwot Ridge, Colorado (altitude 3749 m, latitude 40° N).
[c]Bummer's Gulch, Colorado (altitude 2195 m, latitude 40° N).

Earth's surface—23·6 million km^2 in the Northern Hemisphere and 1·3 million km^2 in the Southern (Good, 1964).

It is not possible to describe a typical temperate winter climate because conditions vary according to altitude and the warming influence of the sea. For example, in the lowlands of the British Isles, soil and air temperatures vary in frequent irregular cycles between -5 and 10°C, according to whether the prevailing air masses originated in the Arctic, the Atlantic or the sub-tropics (e.g. Hay, 1977). In contrast, the winter climate of continental North America tends to be much more severe, with sub-zero temperatures persisting through several months (i.e. approaching an arctic climate, Table 5.2).

The most important difference between arctic/alpine areas and temperate regions is, therefore, the length and quality of the growing season. In temperate climates, the growing season normally exceeds six months per year and native species are able to use this period of favourable conditions to grow vigorously and complete their annual cycle of growth and development; most species are able to avoid the following winter period in a state of dormancy (seed or hardened plant), although some "opportunist" species continue to grow whenever conditions permit (e.g. grasses, whose growing points near the soil soil surface are insulated from the extremes of air temperature). In contrast, the growing season in arctic/alpine regions is short, conditions are generally unfavourable for plant growth during the growing season, and severe weather can occur at any stage of plant development. In order to survive such conditions, plants must adapt to grow under unfavourable conditions, to survive and recover quickly from periods of stress during growth, and, in many cases, to complete their reproductive cycle over several years.

C. Adaptations Favouring Plant Growth and Development in Arctic and Alpine Regions

1. Limiting Factors at Arctic and Alpine Treelines

According to Savile (1972), the primary factor limiting the full development of trees beyond the *arctic* treeline is the level of leaf/needle abrasion caused by wind-blown snow crystals; the resulting loss of photosynthetic tissue, coupled with the short and unfavourable growing season, means that there is very little surplus assimilate to be invested in woody tissues (Elkington and Jones, 1974). The influence of snow abrasion on tree growth can be illustrated by the morphology of isolated spruce trees (*Picea glauca* and *P. mariana*) growing at the arctic treeline in Canada (Fig. 5.4). The lowest branches of these trees, up to about 30 cm from the soil surface, tend to be well-developed because they are protected from winter abrasion by a covering of early snow. Since most buds above this height are killed by abrasion, many trees are severely stunted and those that do continue to grow have bare trunks, stripped of branches. However, very little snow is carried by the wind at heights greater than 1–2 m, and therefore if trees do achieve sufficient height, abrasion is reduced and normal development can be resumed, leading to the "mop-head" effect shown in Fig. 5.4.

Because *alpine* treelines vary from sharp forest edges to extensive krummholz zones, in which trees show different degrees of damage and deformation, identification of the primary factor limiting tree distribution has proved to be more difficult. This is particularly true of the European Alps, where the activities of man from prehistoric times must be added to the range of factors common to all alpine regions. However, work over several decades in this area (reviewed by Tranquillini, 1979) has shown that tissue abrasion is less important than winter desiccation under the combination of high evaporative demand, low root permeability and frozen soil water. Although equally severe

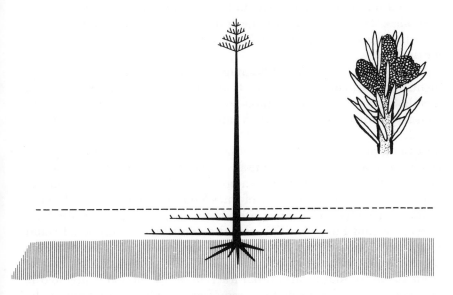

FIG. 5.4.   The form of an isolated spruce tree at the arctic treeline, where the lower branches are protected from abrasion in winter by a covering of snow. Above the level of intense abrasion, the terminal bud survives, protected by a "mop-head" of branches and a whorl of lateral buds (inset) (from Savile, 1972).

conditions can be experienced at lower altitudes, the evergreen coniferous trees at the treeline are particularly vulnerable because the cuticle
of the newest crop of needles has not fully developed by the end of the
(relatively short) growing season.

2. Dwarfing

In response to severe abrasion and drought in winter, most arctic and
alpine species have evolved a dwarfed habit, either as a permanent
feature of the genotype or as a plastic response. Consequently, tundra
plants are generally protected by layers of snow during winter, except
on exposed ridges. For example, arctic shrubs (e.g. *Betula nana, Salix*
spp.) are rarely taller than 15 cm, and tend to exceed this height only
when growing among large boulders. Similarly, tundra herbs tend to
occur as rosettes or cushions (dicots) or as short tufted grasses and
sedges whose apical meristems are insulated by soil and vegetation.
Other morphological features which protect young, growing tissues
from wind and abrasion include the clustering of twigs around the
buds of deciduous shrubs, the formation of dense grass tussocks, and
the accumulation of insulating layers of dead leaves and fruiting stems.
For example, dead material accounted for 94% of the above-ground
mass of a stand of *Luzula confusa* in the Canadian high arctic (Addison
and Bliss, 1984). Because of the low level of grazing and bacterial
activity (Baross and Morita, 1978), wind-breaks of this type can persist
for decades.

Dwarfing confers other advantages in arctic and alpine environments. In particular, it ensures that, in summer, plant growth and
metabolism occur in the warmest zone near the soil surface. This zone
is clearly illustrated by Bliss (1975) and Rosswall *et al.* (1975), and Bliss
(1962) has reviewed a series of earlier records which confirm that
maximum air temperature immediately above stands of arctic plants
can be several degrees higher than at standard meteorological screen
height (1·22 m). The thermal environment of plant tissues is further
improved by the development of thick boundary layers in and above
stands of closely-spaced prostrate plants, as in the cushion life-form
(see below and Fig. 5.5). This can also contribute to improved water
relations by increasing the resistance to the loss of water vapour from
the canopy (Grabherr and Cernusca, 1977). Finally, dwarfing also
tends to ensure that a substantial fraction of production is invested in
reproductive development, a common feature in hostile environments.

The giant rosette plants of tropical alpine areas in Africa and S.
America, which can grow to a height of 3 m, appear to be notable
exceptions to the rule of dwarfing above the treeline (Hedberg and

Hedberg, 1979). However, they grow in a unique environment where the daily range of temperature is as great as the seasonal range in most environments. Frequent night frosts are associated with high radiant flux densities during the day, and the combination of cold soils and high evaporative demand means that there is a high risk of water stress. The giant rosette life-form appears to be successful in this environment because of the thick insulating layer of dead leaves which protects the living tissues, and especially the central pith with its invaluable store of liquid water, against extremes of temperature (Goldstein and Meinzer, 1983).

3. Pigmentation
A characteristic feature of alpine species is the high concentration of anthocyanin pigments in their leaves (Billings and Mooney, 1968; Klein, 1978). The combined effect of these red pigments with the green of chlorophyll can give dark purple to black leaves which absorb a higher proportion of the intercepted solar radiation than correspond-

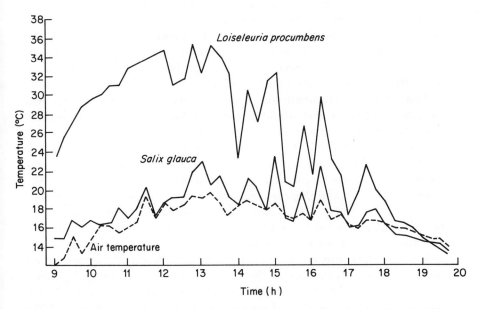

FIG. 5.5. Leaf temperatures in stands of *Loiseleuria procumbens* (cushion) and *Salix glauca* (single, erect) at Kongsvoll, Norway, 900 m a.s.l. on 17 July 1979. Air temperature was measured in a screen at 2 m above the ground surface (adapted from Gauslaa, 1984).

ing green leaves (i.e. reduction in transmission, commonly associated with lower reflection coefficient), leading to leaf temperatures which are higher than that of the surrounding air. Leaf heating has been confirmed experimentally by a number of authors in alpine (Tranquillini, 1964) and arctic regions (Bliss, 1975). Where heavy pigmentation is associated with a cushion life form in which pockets of warm air can be trapped, very high tissue temperatures can be achieved; for example, Salisbury and Spomer (1964) recorded leaf temperatures in cushions of *Arenaria obtusifolia* at 3800 m in Colorado which were up to 12°C higher than the surrounding air, and Gauslaa (1984) gives even more dramatic examples from the treeline in Norway (e.g. Fig. 5.5). The same effect can be demonstrated for lichens growing on bare rock surfaces in the Arctic; for example, Coxson and Kershaw (1983) recorded temperatures of 30–50°C in stands of *Rhizocarpon superficiale*. Because of these elevated temperatures, the growing season (expressed in day degrees) can be considerably extended, and growth and development accelerated. It may even prove possible for deeply-pigmented plants to absorb enough solar radiation to begin photosynthesis when still snow-covered at the beginning of the season (Billings and Bliss, 1959; Kimball *et al.*, 1973). In different species, the level of pigmentation appears to be genetically determined or induced by environmental conditions. However, it should be stressed that the synthesis of anthocyanin pigments is not a feature which is restricted to plants growing in cold environments; it is also a common response to other stresses such as drought and high temperature.

The degree of pigmentation of the corolla of flowering plants also varies among arctic and alpine species and, for example, on sunny days in Siberia, blue flowers can be 3–4°C warmer than the surrounding air, compared with up to 2° C for white flowers (Russian work quoted by Bliss, 1962). This warming of flowers, which should accelerate reproductive development, has reached an advanced stage in species like *Dryas integrifolia* and *Papaver radicatum* whose flowers track the sun across the sky (see p. 24). The resulting increases in temperature (up to 10° C at low wind velocity) are of direct benefit to the plant (for example, causing an increase in mean seed weight; Kjellberg *et al.*, 1982), but they also improve reproduction indirectly by encouraging pollinating insects to bask in the warm flowers for prolonged periods (Kevan, 1975). The remarkable increases in temperature (15–25°C) measured at the surfaces of catkins of arctic willows (Krog, 1955) appear to be the result of reduced emission of long-wave radiation by their very hairy surfaces.

Up to this point, it has been assumed that deep pigmentation is beneficial to alpine plants owing to the warming of illuminated tissues. However, in addition to improving the absorption of intercepted radiation, dark leaves are also more efficient at emitting long-wave radiation. Consequently, the accumulation of pigment molecules in leaves may lead to very low temperatures and the risk of freezing damage during long alpine nights in the growing season (Tranquillini, 1964). This has led to the suggestion (Klein, 1978) that high levels of anthocyanins, flavonoids and epidermal waxes may provide some protection against damage caused by ultra-violet radiation. This conclusion is the subject of some debate, but it has become clear from controlled experiments that not only are alpine plants more resistant to UV damage (possibly owing to efficient nucleic acid repair mechanisms), but also that the morphology, development and physiology of resistant plants can be partly controlled by the supply of UV radiation (Caldwell, 1968, 1981; see p. 34).

4. Life-histories and Resource Allocation

These adaptations which improve plant/temperature relations directly are commonly associated with characteristic life-histories and patterns of resource allocation (Johnson and Tieszen, 1976). For example, many perennials develop large storage organs for carbohydrates below the soil surface (swollen roots, rhizomes, corms, bulbs) containing a large proportion of the plant biomass (e.g. Bliss, 1975). In other species, lipids can be stored above ground in old leaves and stems. Investigation of arctic/alpine plants such as *Oxyria digyna* (Fig. 5.6) has shown that these reserves play a crucial role in enabling the plants to make the best use of short growing seasons. The translocation of stored carbohydrate from root to shoot at the beginning of the season supports the growth of new leaf and stem tissues at a time when the rate of net photosynthesis is low but respiration is less affected by low temperatures (see below). Thus, for a few weeks, the dry weight of the plant tends to fall; however, as conditions improve, net assimilation becomes positive and surplus carbohydrate can be partitioned between developing reproductive organs and recharging storage organs. This pattern of carbohydrate cycling, coupled with the widespread occurrence of preformed flower buds (developed one or more seasons earlier), increases the probability of successful reproduction and survival (Billings and Mooney, 1968; Billings, 1974). A similar phenology is shown by woodland species like the bluebell (*Hyacinthoides nonscriptus*), which flowers early in the year before the tree canopy has formed (see p. 45).

An alternative life-history, which ensures that the interception of

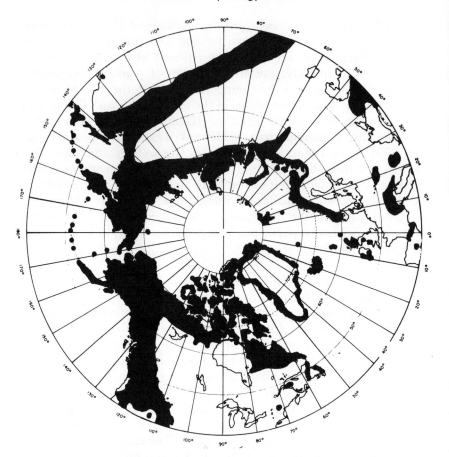

Fig. 5.6. Geographical distribution of the arctic/alpine species *Oxyria digyna* south-wards to 40°N (simplified from Hultén, 1962; and Billings, 1974).

solar radiation occurs from the beginning of the short growing season, is shown by evergreen tundra shrubs (e.g. *Ledum palustre, Vaccinium vitis-idaea*) and certain grasses with long-lived, overwintering leaves (Johnson and Tieszen, 1976; Robertson and Woolhouse, 1984). As is the case for evergreens in general (Schulze *et al.*, 1977), the photosynthetic potential of their leaves tends to be substantially lower than that of deciduous plants, but this must be set against increased leaf area duration and a lower annual investment of dry matter in new leaf tissue. Other responses which may contribute to improved interception of radiation by tundra grasses include the increases in leaf dimensions induced by long photoperiods (Heide *et al.*, 1985) and the decrease in specific leaf area with altitude shown by Woodward (1983). Responses of this kind are of particular value in scattered tundra vegetation, where interception of solar radiation is incomplete.

Overall, as the environment becomes more extreme, the life-histories of tundra species tend to become more conservative, geared to the survival of the individual plant rather than to regular annual reproduction. Thus, for example, high-arctic plants may produce only two new leaves, or a larger number of very small leaves, per stem each year, and it is not uncommon for the tillers of graminoid species to grow for four to seven years before flowering (Addison and Bliss, 1984; Miller *et al.*, 1984). Even in such cases, the season may not be sufficiently long for the production of mature viable seed and there may be difficulties in ensuring successful pollination (see above). Furthermore, seedlings attempting to establish in these hostile environments cannot fall back on substantial carbohydrate reserves, and they are at great risk from frost-heaving. Consequently, reproduction by seed is rarely successful and this accounts for the scarcity of tundra annuals, especially in the arctic. For some perennials, these difficulties have been overcome by vegetative reproduction (by rhizomes, layering of branches, vivipary, etc.) and, in some cases, this may confer a distinct advantage in dispersal since large propagules are more easily carried by the wind (Billings and Mooney, 1968; Savile, 1972).

5. Adaptation of the Photosynthetic Apparatus

However, the survival of plants growing at high altitudes and latitudes depends ultimately upon their ability to maintain a positive carbon balance at low temperatures around 0°C. This ability has been demonstrated for a variety of species and locations (reviewed by Pisek *et al.*, 1973; Billings, 1974); for example, with some notable exceptions, the optimum temperature range for net photosynthesis in species growing in the European Alps falls with the altitude of their origin (Fig. 5.7). The exceptions include valley evergreens (e.g. *Abies alba*), whose low optima permit photosynthesis to proceed during the greater part of the year (Neilson *et al.*, 1972), and high-altitude dwarf perennials (*Ranunculus glacialis* and *Geum reptans*), whose relatively high optima are presumably an adaptation to the high day temperatures of their natural habitat—sheltered rock crevices subject to intense insolation. It can be seen from Figs 5.2, 5.7 and Table 5.3 that the photosynthetic apparatus of tundra species is remarkably well adapted to the temperatures prevailing beyond the treeline. Not only are maximum rates of net photosynthesis attained at relatively low

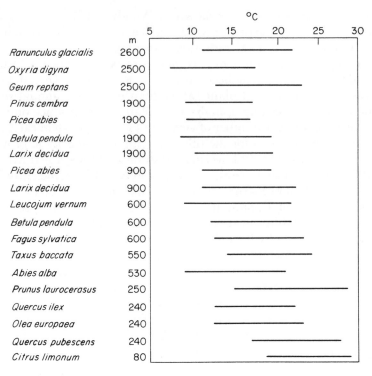

FIG. 5.7. Optimum temperature ranges for net photosynthesis at low radiant flux density (70 W m^{-2}), in species originating from warm temperature lowland (80–250 m), mountain valley (530–900 m), treeline (1900 m) and high mountain (2500–2600 m) regions of the Alps. At higher radiant flux densities, each optimum temperature range tends to move upwards by several degrees (redrawn from Pisek *et al.*, 1973).

temperatures (7–10°C), but also net photosynthesis can continue down to −6°C, with up to 75% of the maximum rate achieved at 0°C.

However, as shown by the Death Valley transplant studies (see above), cardinal temperatures for net photosynthesis can be subject to considerable phenotypic plasticity. Thus, although there may be pronounced genetic differences in photosynthetic characteristics between different populations of the same species adapted to different altitudes, the extent of the expression of such differences, in terms of cardinal temperatures, depends upon the growing conditions. Thus, by raising the growing temperature of plants of a high-altitude population, the temperature optimum for photosynthesis can be raised towards that of a lower-altitude population. Tranquillini (1979) reviewed the evidence for such plasticity in woody species from a wide range of experiments in N. America, Australia and the European Alps,

TABLE 5.3. Cardinal temperatures (°C) for net photosynthesis in selected arctic and alpine species. (Note that the optimum range will vary according to growing temperature and prevailing radiant flux density, without altering T_{min} significantly.)

Species	Origin	T_{min}	T_{opt}	% of max. rate at 0°C	Source
Chionochloa spp.	N.Z. Alps	−4	1–8	75	Fig. 5.2.
Oxyria digyna	Austrian Alps 2500 m	−6	7–17	50	Pisek et al., 1969
Geum reptans	Austrian Alps 2500 m	−3	10–25	25	Pisek et al., 1969
Loiseleuria procumbens	Austrian Alps 2000 m	−3	10–25	30–40	Grabherr and Cernusca, 1977
Alopecurus alpinus	Arctic Alaska	<0	15	c50	Chapin, 1983
Dupontia fisheri	Arctic Alaska	<0	15	c45	Chapin, 1983

and suggested that plasticity of the photosynthetic apparatus could even be found in a single population responding to the seasonal variation in temperature. Working with 17 ecotypes of *Oxyria digyna* from tundra areas throughout the Northern Hemisphere (Fig. 5.6), Billings *et al.* (1971) showed that alteration in mean growing temperature from 26·5 to 8°C reduced the optimum temperature for net photosynthesis by amounts varying from 0°C (Silver Gate, Montana) to 12·5°C (Tarmachan Crags, Scotland), but that this was associated with an increase in the maximum rate of net photosynthesis in only 9 out of the 17 ecotypes (all of the arctic and 5 out of 13 alpine ecotypes).

In spite of the considerable genotypic variation and phenotypic plasticity shown by a widespread species such as *Oxyria digyna*, several attempts have been made to detect overall metabolic differences between arctic and alpine ecotypes or populations, as well as between arctic and alpine species. This can prove to be a particularly difficult task because the term alpine encompasses such a wide range of environmental conditions; for example, the 13 alpine ecotypes studied by Billings *et al.* (1971) originated from areas as diverse as Arizona, sub-arctic Yukon, and very humid areas of the British Isles. However, most of the divergent characteristics which have been identified can be related to the higher day temperatures and radiant flux densities experienced in "classic" continental alpine zones. For example, Tieszen and Wieland (1975) showed that the optimum temperature for net photosynthesis tended to be 5–10°C higher for alpine than for arctic plants, whereas the photosynthesis of arctic populations of *Oxyria* and *Trisetum* is saturated at lower radiant flux densities than alpine populations (Mooney and Billings, 1961; Billings, 1974). However, there appears to be an important interaction between irradiance and temperature. Several workers have found that the radiant flux density required to saturate the photosynthesis of tundra plants rises with temperature (Billings, 1974). According to Wilson (1966), this effect is the consequence of lower demand for assimilate at lower temperatures, leading to the accumulation of sugars and the depression of photosynthesis by product inhibition. The ability of tundra species to maintain a positive carbon balance therefore depends upon a combination of factors, including the adaptation of the photosynthetic apparatus to low temperature (e.g. Figs 5.2, 5.7, Table 5.3), adaptations leading to tissue warming, characteristic canopy structures (e.g. prostrate canopies with high extinction coefficients), and the contribution of non-laminar tissues to photosynthesis. To these can be added the relatively high daytime temperatures and radiant flux densities in many alpine regions, and the possibility for arctic plants of assimilation throughout the entire 24 h period without respiratory losses in the dark (Tieszen and Wieland, 1975). At the molecular level, it appears that the thermal adaptation of photosynthesis is the consequence of improved membrane stability and increased levels of the rate-limiting enzyme RuBP-carboxylase, rather than any marked alteration of enzyme characteristics (Berry and Björkman, 1980; Chabot, 1979).

However, concentration upon net photosynthesis tends to divert attention from the fact that the rate of (dark) respiration at a given temperature tends to be higher for tundra plants than for those from warmer environments (Tieszen and Wieland, 1975; Berry and Raison, 1981). This, of course, means that rates of gross photosynthesis at low temperature must be proportionally even higher than those of net photosynthesis to make up for respiratory losses. Athough the precise response to temperature will be determined by the thermal history of the plant (phenotypic plasticity), low-temperature adaptation of respiration is a permanent feature of the genotype of high-altitude and latitude species and populations (Billings *et al.*, 1971; Tranquillini, 1979). The advantages of such adaptation are clear. For example, a small increase in temperature at near-freezing temperatures in spring could lead to a doubling of the respiration rate and a substantial growth response (e.g. when growing new leaves from stored reserves, see above). On the other hand, a doubling of the rate at higher temperatures could lead to a rapid exhaustion of carbohydrate reserves (Crawford and Palin, 1981). Thus alpine species may be restricted to high altitudes because of the adaptation of their respiratory apparatus to low temperature, rather than as a consequence of their susceptibility to thermal stress (Dahl, 1951). The sub-cellular and molecular details of this adaptation are not well characterized, although it has been shown that the low-temperature activities of isolated mitochondria and chloroplasts from a number of widely-distributed species are positively correlated with the altitude of origin of the populations studied (Klikoff, 1969; May and Villareal, 1974—but note that the low temperatures employed were 10 and 15°C respectively). There is also a need for more information on the effect of low temperature on the translocation of assimilate in tundra plants since this has been shown to be highly-sensitive to temperature in temperate plants (Berry and Raison, 1981).

The fact that the metabolism, morphology and life-histories of arctic and alpine plants appear to be so well adapted to their thermal environment has led Chapin (1983) to propose that their distribution is not limited directly by low temperature, but indirectly, as a result of the effects of temperature on other factors, principally water and nutrient supplies. Some assessment of water status has formed part of many studies of the environmental physiology of tundra species, and it has become clear that exposure to water stress is a characteristic and chronic feature of high-altitude and latitude areas (e.g. Addison and Bliss, 1984). However, field studies have not progressed much further than the measurement of leaf water potentials, and there remains a great deal of work to be done before the mechanisms of drought avoidance and resistance in these cold environments are fully understood. Much also remains to be discovered about plant/nutrient relations in arctic and alpine zones, although information is accumulating on the cycling of nutrients in tundra soils (e.g. Chapin, 1983) and on the response of vegetation to the addition of nutrients (for example, in the form of bird excreta; Cargill and Jefferies, 1984).

D. Adaptations Favouring Survival of Cold Winters—Dormancy

As noted earlier, temperate plant species face different problems from those of tundra species. Because climatic conditions are favourable for vegetative and reproductive development for at least six months in each year, temperate plants are normally able to complete their annual reproductive cycles successfully; however, in order to make use of the next growing season, they must be able to survive the intervening cold winter without sustaining substantial damage. This problem, which is particularly acute in north-temperate and boreal regions, is commonly overcome by the adoption of winter dormancy.

For example, the seeds of many species, both annual and perennial, are dormant when shed in the autumn and will germinate only after they have experienced a period of low temperature (stratification). In many cases, the temperature need not fall below 1–8°C and the cold period required for successful stratification can vary from a few days to a few months (Pisek et al., 1973). A particularly clear example of this adaptation is provided by Betula pubescens. When the seeds of this birch species are shed in autumn (short daylength) they are dormant and can be induced to germinate only when exposed to unseasonable long days. However after a period of chilling, dormancy is broken and the seeds will germinate whenever the temperature (and water supply) permits, even in continuous darkness (Black and Wareing, 1955). Thus under natural conditions, autumn germination is suppressed by daylength but, once the winter is over, birch seeds can germinate and establish early in the spring.

In the same way, the growing points of perennial species can cease growth and pass into a resting phase in autumn. For example, the shoot growth of most temperate woody plants ceases in autumn, and dormant terminal buds are formed under the influence of short days; as with seed dormancy, bud dormancy is broken naturally by winter chilling. In addition to developing dormant buds, deciduous woody plants avoid frost damage by autumn leaf abscission, which is stimulated by a combination of decreasing daylength and temperature. Winter-dormant buds are also found in the bulbs, corms, rhizomes and tubers of herbaceous perennials (Wareing, 1969).

Arctic and alpine plants also have to survive very severe winters. However, because of the shortness and irregularity of the growing season, it is essential that growth is resumed at the earliest possible date and continues to the end of the period of favourable conditions (compare with temperate perennials, which produce dormant buds in early autumn before the radiation and temperature regimes have deteriorated seriously). Consequently innate dormancy is less common, and growth is generally resumed whenever temperature permits. For example, of 60 alpine species examined by Amen (1966), only 19 possessed seed dormancy, and in eight of these, it was caused by the presence of a hard, impermeable, seed coat. The possession of a hard seed coat by several dominant alpine species presumably improves the probability of seedling establishment, by spreading germination over several years in soils where freeze/thaw cycles cause abrasion of seed surfaces (Amen, 1966; see p. 151).

In tundra perennials, the onset of winter, bringing reduced temperatures and daylength, causes a progressive slowing of metabolism, the hardening of plant tissues (see below) and the *enforcement* of dormancy in leaves and perennating buds. Since this dormancy is relieved in spring whenever temperatures rise to about 0°C, it is important that such dormant buds be able to respond rapidly to changes in temperature once the protective layers of snow have melted. Thus, unlike their temperate counterparts, the buds of arctic perennials tend not to be enclosed within heavy bud scales which would insulate their contents against fluctuation in air temperature (Savile, 1972).

Whether innate or enforced, dormancy is clearly a widely adopted way of avoiding severe winter conditions. However, the survival of plant tissues, even in the form of dormant buds, depends ultimately upon the ability of their cells to avoid freezing injuries.

E. Adaptations Favouring Survival of Cold Winters—Plant Resistance to Freezing Injury

Much of the existing literature on frost resistance deals with temperate species of economic importance, in particular field and horticultural crops and timber trees. However, where natural species have been studied, the degree of frost resistance of their tissues has been found to be broadly related to the temperature regime in their natural range. For example, Sakai and Weiser (1973) found that the freezing resistance of dormant twigs from a number of N. American tree species was dependent upon the winter minimum temperatures in their area of origin. Exceptions to this rule include the tolerance of very low subzero temperatures (-30 to -50°C, after hardening) in *Salix* species from the lowland tropics (Sakai, 1970).

Plant resistance to freezing injury, which is primarily resistance to the formation of intracellular ice (although tolerance of cell desiccation and resistance to the physical damage caused by the formation of extracellular ice are also very important), can be seen as a series of successive "lines of resistance". In general, as the winter climate becomes more severe, the possession of more of these lines becomes necessary for survival. The first line of defence is simply the depression of the freezing point of the water in the vacuole and cytoplasm as a consequence of its soluble solute content. Thus, even before hardening, the tissues of most temperate plants can be cooled to a few degrees below zero (typically, -1 to $-5°C$) before ice forms; similarly, halophytes can have freezing point depressions as great as $14°C$ below zero (Burke et al., 1976). This colligative property of the cell sap can give complete protection from frost damage in warm areas with a low incidence of frost, such as the Mediterranean and parts of California.

It has been suggested by several authors (reviewed by Larcher et al., 1973) that the accumulation of low molecular weight organic solutes (sugars, organic acids, amino acids as well as larger protein molecules) in the cell sap in autumn may give increased frost resistance by further depression of the freezing point. However, this additional depression cannot be more than a few degrees and does not explain the tolerance of extremely low temperatures discussed below. It is possible that the accumulation of solutes, resulting from the reduced demand for assimilate at low temperatures (Wilson, 1966; see p. 212), plays a more important role in protecting cytoplasmic macromolecules and membranes from the effects of severe desiccation (Li and Sakai, 1978).

The second line of defence, protecting critically important tissues of woody plants (dormant buds and xylem ray parenchyma) from freezing at temperatures down to about $-40°C$, can come into operation only after growth has ceased, dormancy has been established and the tissues have been hardened by exposure to temperatures below $5°C$ for several days (Langridge and McWilliam, 1967; Burke et al., 1976). This protection is, therefore, induced by autumn conditions, before the onset of winter. In spite of a great deal of biochemical work, the processes involved in such tissue hardening are poorly understood, although it is clear that membrane properties (physical strength, hydraulic conductivity, ion transport systems) are substantially modified (Li and Sakai, 1978). Tissues conditioned in this way tend to behave as if they contained ultra-pure water, lacking any nucleating sites where ice formation can begin; thus they can undergo deep supercooling down to about $-38°C$ (the spontaneous nucleation temperature of water) before ice forms. Although the causes of this deep supercooling are not clear, the effect has been observed in twigs of several tree species (intracellular freezing at -38 to $-47°C$; George et al., 1977). However, in other cases, supercooling gives protection only to -20 to $-30°C$. It is important to stress that whereas these critically important tissues are protected by deep supercooling, the remaining tissues survive by accommodating intercellular ice and tolerating cell dehydration (Ishikawa and Sakai, 1981).

It is clear that this line of defence will be of little value (and, in fact, positively harmful) to trees growing in areas where winter temperatures lower than $-40°C$ are common. However, there is a third line of defence, required only in the most severe climates, for example near the arctic treeline, by which hardened tissues can be cooled slowly to lower temperatures. For example, Sakai (1970) found that dormant twigs of *Salix* spp. from a number of provenances could be cooled to the temperature of liquid nitrogen ($-196°C$) without impairing subsequent growth; in these cases it may be that intracellular ice formation is prevented because all of the "freezable" water in the cell has been withdrawn into the apoplast, leaving thin protective layers of water molecules surrounding macromolecules, membranes and organelles (Burke *et al.*, 1976). There is no evidence that such deeply supercooled plant tissues can survive intracellular freezing.

However, this rather simple view of frost resistance as a series of lines of defence must be modified in several ways. In particular, as noted earlier, rapid cooling rates (several degrees per hour) tend to cause intracellular freezing and cell death even in hardened plants, at least in the laboratory. This presumably occurs because the hydraulic conductivity of the cell membranes is not sufficiently high to permit the rate of water efflux necessary for extracellular ice formation (Li and Sakai, 1978). Thus, for example, the rapid changes in temperature and the frequent cycles of freezing and thawing which are characteristic of the European Alps in spring, lead to tissue damage even in highly resistant species such as *Pinus cembra* (Tranquillini, 1964). Secondly, the tissues and organs of a given species are not uniformly resistant to frost injury; this is illustrated by Fig. 5.8 which stresses that below-ground parts, which are normally insulated from the extremes of temperature experienced by above-ground parts, are much more sensitive than stems, leaves or buds.

Plants which tolerate very low temperatures must be able to accommodate large volumes of ice in the apoplast without causing physical damage to cell walls and cell membranes; for example, in woody plants such as *Buxus* and *Camellia* ice is localized between readily-separated layers of leaf tissue (Sutcliffe, 1977). They must also be able to survive severe dehydration of the cytoplasm and the mechanical and biochemical stresses associated with cycles of dehydration and rehydration. The ability to tolerate these stresses improves with hardening, but the precise mechanisms involved are poorly understood, mainly because of the extreme difficulty of measuring appropriate cell characteristics. Speculation about the changes occurring during hardening have included the suggestion of an increase in

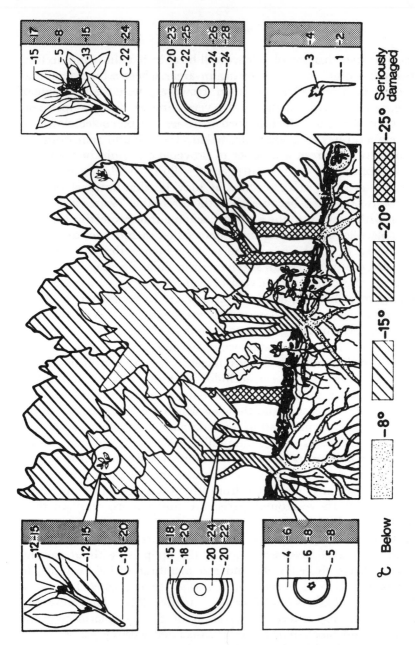

Fig. 5.8. Frost resistance in different zones of a community of *Quercus ilex* stressing the influence of age in increasing the resistance of above-ground parts. In the main diagram, the hatching indicates the critical temperature below which serious tissue damage is incurred. In the marginal diagrams, the left-hand column of figures indicates the temperature below which initial damage is caused, whereas the temperatures in the right-hand column (shaded) refer to 50% damage. C indicates cambium. Note that a whole year's crop of germinating acorns will be wiped out by soil temperatures below −4°C (from Larcher, 1975).

the amount of (protective) water bound tightly to macromolecules, membranes and organelles. Levitt (1972) considers that the denaturation of proteins, and cell death, which occur as a result of freezing and drought, are caused primarily by the formation of potentially damaging disulphide bonds between neighbouring protein molecules or amino acids (Fig. 5.9), and suggests that resistance to stress, in general (freezing, drought, high temperature), is related to a low incidence of thiol (SH) groups.

The relatively sparse amount of available information indicates that tundra plants also undergo a period of hardening at the end of the growing season and can then survive low temperatures by supercooling and extracellular ice formation. Because of the protection offered by layers of snow, the risk of frost damage in winter is probably lower than for trees in boreal zones, but unlike temperate plants, high-altitude and latitude species must be able to tolerate sub-zero temperatures at any time during the growing season, and recommence growth and metabolism without delay. The combination of frost resistance and active growth is unknown in temperate plants and its mechanism

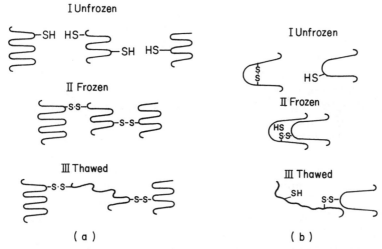

FIG. 5.9. Schematic diagrams of the disulphide bond theory of freezing injury to the tertiary structures of proteins. In scheme (a), the cytoplasmic dehydration associated with freezing forces three protein molecules together, with the result that they become linked together by disulphide bonds. On rehydration of the cytoplasm, the tertiary structure of the central molecule (determined largely by weak hydrogen bonds) is destroyed as the outer molecules regain their former positions. Alternatively, the shape of a protein molecule can be changed by the breaking and reforming of disulphide bonds (scheme (b)). Similar phenomena are thought to occur during periods of drought, high temperature and salinity stress (after Levitt, 1972).

remains to be explored. In contrast to the higher plants, tundra lichens appear to be able to tolerate extreme fluctuations in temperature, as well as very severe desiccation without sustaining serious injury (Savile, 1972).

III. PLANT ADAPTATIONS AND RESISTANCE TO HIGH TEMPERATURES

As noted earlier, plant tissues dissipate heat by three main processes (emission of long-wave radiation, convection of heat and transpiration), of which transpiration is the most effective in many natural situations (Gates, 1976). High plant temperatures ($> 40°C$) are therefore almost invariably associated with the cessation of transpirational cooling, following stomatal closure in response to drought. Consequently, the vegetation of drier areas in the tropics and sub-tropics commonly experiences a *combination* of water stress and thermal stress; this can also occur in xeric habitats (e.g. sand-dunes, shallow soils) in temperate zones and at high altitudes during periods of high irradiance.

Because of this close association between drought and high temperature, it can be very difficult to disentangle the effects of each stress on plants growing in the field. To do this, it is necessary to consider the stresses separately under controlled conditions, for example by studying the influence of high temperature on plants which are adequately supplied with water. The results of such experiments indicate that exposure to temperatures within a relatively narrow range (45–55°C) for as little as 30 min can cause severe damage to the leaves of plants from most climatic regions. The exceptions to this rule include shade-adapted and aquatic species which are injured by temperatures lower than 45°C and some desert plants which can tolerate slightly higher temperatures (Kappen, 1981; see below). This contrasts with the very wide range of temperatures required for low temperature damage ($+5$ to $-40°C$ or lower, see above).

The fact that chloroplast lipids *in vitro* have been shown to undergo phase transitions (abrupt reduction in viscosity) at temperatures within this narrow range (45–55°C) (Raison *et al.*, 1980) suggests that membranes may be the primary sites of high-temperature injury. This has been largely confirmed by detailed studies of the temperature responses of those species, adapted to different temperature environments, which were included in the Death Valley transplant experiment described earlier. Whole leaf studies established that photosynthesis was more temperature-sensitive than other processes (respiration, ion relations; see Table 5.4), although the precise temperature threshold for damage varied with species and growing temperature (Björkman, 1980). At the sub-cellular level, most membrane functions and enzyme activities were less sensitive than whole-leaf photosynthesis, with the exception of photosystem II which was consistently inhibited by temperatures similar to those affecting photosynthesis (Table 5.4; Björkman, 1980 for other species). It can, therefore, be concluded that the primary effect of damaging high temperatures involves phase

TABLE 5.4.   Minimum temperatures (°C, 10 min exposure) for the inhibition of whole leaf functions, membrane reactions and enzymes of *Atriplex sabulosa* (C$_4$ photosynthesis) and *Tidestromia oblongifolia* (C$_4$) (from Berry and Raison, 1981).

	A. sabulosa	T. oblongifolia
Whole leaf functions		
Net photosynthesis	43	51
Respiration	50	55
Retention of ions	52	56
Membrane reactions		
Photosystem I	> 55	> 55
Photosystem II	42	49
Soluble enzymes		
RuBP carboxylase	49	56
PEP carboxylase	48	54
NAD malate dehydrogenase	54	56
NAD glyceraldehyde-3-P		
dehydrogenase	51	56
NADP reductase	55	55
3-PGA kinase	51	51
Adenylate kinase	47	49
FBP aldolase	49	55
Phosphoglucomutase	51	53
Phosphohexose isomerase	52	55
Light-activated enzymes		
Ru5P kinase	44	52
NADP-glyceraldehyde-3-P		
dehydrogenase	42	51
NADP-malate dehydrogenase	—	51

changes in chloroplast membranes which disrupt photosystem II electron transport.

For practical reasons, these studies concentrated upon acute thermal stress imposed for only 10 min, and it should be stressed that somewhat lower temperatures (e.g. 35–40°C) would give similar degrees of inhibition and injury over longer periods of exposure. However, plants can also experience chronic high-temperature stresses which may not induce dramatic injuries, but which can seriously impair plant performance. In particular, plant reserves can be seriously depleted by high rates of respiration during long, warm nights (see p. 213, for alpine species).

The very wide range of morphologies and phenologies shown by desert plants (e.g. from small ephemerals to enormous, bulky cacti) indicates that survival in hot, dry environments can be achieved in a variety of ways, by combinations of different adaptations. Many species which are native to hot areas have evolved life-histories which

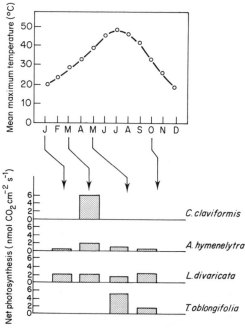

FIG. 5.10. Maximum photosynthetic capacities for four species native to Death Valley, California, and mean maximum air temperatures, during different periods of the year (Mooney, 1980). (*Camissonia claviformis*, winter annual, *Atriplex hymenelytra*, *Larrea divaricata*, evergreen perennials, *Tidestromia oblongifolia*, summer-active herbaceous perennial).

permit them to avoid the hottest period of the year. This can be achieved by leaf abscission, leaving hardened, heat-resistant buds (*Larrea divaricata*, Fig. 4.12), or, in desert annuals such as *Camissonia claviformis*, by completing the entire reproductive cycle during the cooler months (Fig. 5.10). Some of the latter species show apparently inappropriate adaptations, such as diaheliotropic sun tracking (maximizing the interception of solar radiation, by maintaining the leaves at right angles to the solar beam; Ehleringer and Forseth, 1980) and high rates of transpiration, which permit them to grow very rapidly within the short period of favourable conditions. Alternatively, individual leaves, or the canopy as a whole, may possess characteristics which reduce the interception of radiation or promote leaf cooling, with the result that lethal tissue temperatures are avoided, and the plant can continue to function throughout the year (Fig. 5.10). For example, control of leaf angle and leaf folding has reached a high degree of precision in those species whose leaves are held parallel to the direct rays of the sun (paraheliotropic sun tracking), thereby minimizing the interception of solar radiation (Ehleringer and Forseth, 1980). This phenomenon, which involves leaf-cupping as well as leaf movement, is common to a number of species with compound leaves (e.g. *Lotus, Lupinus, Stylosanthes* spp.). However, both dia- and para-heliotropic leaf movements will be most effective in areas with clear skies and clean air, where the contribution of diffuse radiation is small, and where the canopy is not unduly affected by wind.

If leaves do intercept excessive amounts of solar radiation under conditions of water shortage, survival will depend upon convection of excess heat from the leaf surface or reflection of incident radiation before it can be absorbed. As discussed in Chapter 4 (p. 164), the former process is enhanced by morphological features (e.g. small, non-overlapping leaves) which minimize the leaf boundary layer resistance. Thus, small leaflets of length less than 1·5 cm are a feature of the common evergreen shrubs (e.g. *Larrea divaricata*; Fig. 4.12) of N. American deserts (Smith, 1978). In general, as the climate becomes more hot and arid, the reflection coefficient of the leaves of native species increases. A particularly clear example of this trend is shown by the various perennial shrubs of the genus *Encelia* which are found in California and adjacent states (Ehleringer, 1980). The green, non-pubescent *E. californica* (absorbing 85% of intercepted *PAR*) grows at the coast, under warm moist conditions, but gives way successively, as the climate becomes more severe, to the lightly-pubescent *E. asperifolia* and *E. virginensis* (76%) and the highly-pubescent *E. palmeri* and *E. farinosa* (<60%)(Fig. 4.12). Furthermore, *E. farinosa* exhibits seasonal leaf dimorphism, with white pubescent leaves in summer (giving protection from lethal temperatures but restricting photosynthesis by high boundary layer resistances to CO_2 uptake) being succeeded under moister conditions by green, less hairy leaves which favour photosynthesis. Under the most severe stresses, these shrubs can be deciduous.

These morphological adaptations are commonly associated with biochemical adaptations favouring net photosynthesis at high temperatures (in particular C_4 and CAM photosynthetic pathways), although C_3 plants are by no means uncommon in desert floras. The comparative physiology of these pathways is considered in detail in Chapters 2 and 4.

However, there are many outstandingly successful species which rely on tolerance rather than avoidance of high temperatures. These include obligate thermophiles such as *Tidestromia oblongifolia* (Figs 4.12, 5.10), which is summer-active at temperatures in excess of 50°C in Death Valley but cannot survive the moderate climate at the coast (p. 193), and the various succulent species whose bulky shapes (low rates of convective heat loss) and CAM photosynthesis (no transpirational cooling during the day) give little scope for leaf cooling, although the high water content confers a degree of buffering against temperature change due to the high specific heat of water (p. 121). In fact, the highest recorded plant temperature (65°C) was measured in a species of *Opuntia,* and cacti in general can survive tissue temperatures above 60°C for prolonged periods (Smith, 1978). The thermal regime in deserts can be even more severe for seedlings which can be exposed to soil surface temperatures over 70°C (Nobel, 1984). In comparison, temperate, sub-tropical and tropical plants from mesic sites tend to be injured or killed by long-term exposure to temperatures as low as 35–40°C (compared with 45–55°C for short-term exposure, see above).

Previous work on thermal tolerance emphasized the importance of general protein stability but, as discussed earlier, more recent studies have identified membrane-bound systems, and photosystem II in particular, as the primary sites of heat injury. Table 5.4 shows that there is considerable potential for thermal adaptation of this system (difference between species in the temperature threshold for inhibition of 7°C), as well as for many soluble enzymes and whole-leaf functions, although the precise results obtained will depend upon the growing temperature. The limited information which is available (Raison *et al.*, 1980) indicates that adaptation to high temperatures involves the modification of the lipid composition of chloroplast membranes such that the phase changes, which disrupt electron transport systems, occur at higher temperatures.

6. Ionic Toxicity

I. THE NATURE OF TOXICITY

Among the many environmental factors that limit the growth of a plant are chemical toxicities: inhibitions of metabolism or physiological function brought about by excess concentrations of various molecules or ions. For almost all such factors there is a concentration range at which no injurious effects are suffered by any species, and one at which all species are susceptible; in between lies the area of interest to the ecologist, where only a few adapted or resistant species are able to survive. Classic examples are the halophytes of saline soils, calcifuges on acid soils, flood-tolerant plants on waterlogged soils, and, more topically, plants resistant to air pollution. The exception to this are radionuclides, where defined chemical substances (the radioactive elements) are involved, but the actual cause of the toxicity lies in the destructive power of the particles they emit. Consequently the toxic effect is simply statistical, depending on the number of cells killed by impact of α, β, or γ particles, and there is no definable lower threshold below which no toxicity occurs. Although plants can adapt to radiation stress—the grass *Andropogon filifolius* growing on uraniferous soil has more resistant seeds than other populations (Mewissen *et al.*, 1959)—we will not deal with this stress here. This chapter deals only with ionic toxicity; toxicity caused by gases whether in soil or air is covered in the next chapter. The ecological situations in which toxicities of 20 elements occur are shown in Table 6.1.

Ionic toxicity can be classified on the basis of two important distinctions:

(i) the concentration at which toxicity occurs;
(ii) whether or not the element is essential to plant growth—can a deficiency occur?

If an ion is toxic only at high concentrations, it may bring with it the problem of low soil water potential, because of the consequent increase

TABLE 6.1.   Ecological situations in which ions of various elements can cause toxicity.

Element	Essential (E) or not (N)	Situation
N	E	NH_4^+ toxic to calcicoles (Rorison, 1975).
P	E	Only under agricultural conditions of very high fertilizer applications (Green and Warder, 1973).
K	E	May compete with Na^+ in obligate halophytes (Austenfeld, 1974b).
Ca	E	Toxic to some calcifuges, particularly bog plants.
Mg	E	Toxic on serpentine soils with high Mg: Ca ratios (Proctor, 1971b).
S	E	Soils containing sulphide oxidize to produce sulphuric acid and give pH values below 3. Arid zone soils may accumulate Na_2SO_4 by evaporating.
B	E	Occurs at toxic levels in pulverized fuel ash (Hodgson and Buckley, 1975).
Cu, Zn	E	In natural ore deposits and mine spoil heaps; occasionally by aerial deposition.
Pb, Cd, Cr, Ni	N	
Fe	E	High concentrations of Fe^{3+} occur in soils of pH below 3·5. Fe^{2+} occurs in waterlogged soils.
Mo	E	As molybdate, in highly calcareous soils.
Na	(E)	In saline and sodic soils.
Cl	(E)	In saline soils.
Mn	E	Usually on acid soils.
Al	N	In soils below pH 4.
H	—	H_3O^+ ions are directly toxic at concentrations above 1 mm (pH below 3).

in the contribution of solutes to the lowering of Ψ (see p. 124). In this case the simplest resistance mechanism, straightforward exclusion of the toxic compound from the plant, would eliminate the necessary potential gradient between soil and root and so make it impossible for the plant to take up water, unless some other adjustment to plant Ψ could be made. Similarly if an ion is both an essential micronutrient and yet toxic at very low concentrations (Zn^{2+}, Cu^{2+}), exclusion may again be inappropriate, unless uptake can be precisely controlled. These two considerations greatly influence both the physiological and ecological responses of plants adapted to resist toxicity.

A. Salt-affected Soils

Soils with high dissolved ion concentrations are very common:

(i) salt-marshes arise from tidal inundation of low-lying areas and they are dominated by Na^+ and Cl^-. Ocean water is normally 460 mM Na^+, 50 mM Mg^{2+} and 540 mM Cl^-; other ions are relatively scarce. Its solute potential (ψ_s) is $-2 \cdot 7$ MPa. Soils affected in this way typically approach these values only if frequently inundated, as in the lower reaches of salt-marshes; upper regions, less frequently covered by the sea (often less than once a day, at or around spring tides), may have dissolved ion concentrations raised by evaporation or lowered by precipitation.

(ii) in inland areas, salt deserts may form where annual potential evaporation exceeds precipitation. Here soil water movement is predominantly upwards and leaching is minimal, so that solutes tend to accumulate at the soil surface. Such soils are often dominated by ions other than Na^+ and Cl^-; Ca^{2+} and SO_4^{2-} are often more abundant, and carbonate systems also occur.

Salt-marshes are normally coastal, although they may be found in inland areas where the surface water is saline. Coastal marshes occur in sheltered sites (frequently estuaries) where wave action is slight and deposition of silt allows higher plants to root. No angiosperms have managed to adapt to growing in the sea where wave action is severe, that niche being occupied by seaweeds, with holdfasts rather than roots. In sheltered conditions, the establishment of plants accelerates the deposition of silt and so salt-marshes are characteristically sites where successional processes are highly active, since raising of the shore surface reduces the frequency of flooding by salt water and so permits the invasion of differently adapted species, which experience greater fluctuations in soil NaCl concentrations.

Salt-deserts, on the other hand, tend to be more stable communities, though where salt input is prolonged there may be salination to such a degree as to eliminate all plants. Salt-deserts cover huge areas of the world where evaporation greatly exceeds precipitation, or where inland basins are served by rivers that never reach the sea. The same process is involved when irrigated land becomes useless through build-up of salt, as happened in ancient Sumer, after a thousand years of irrigation (Hyams, 1952; Jacobsen and Adams, 1958). Careful husbandry can avoid such damaging consequences, as in Mexico (Coe, 1964). The consequences for the development of civilization are immense; where irrigation is by the natural flooding of a river, as in the Nile valley, the annual rinsing of the soil by low-salt water prevents salination.

Plants that can grow on soils of high salt content are termed *halophytes*. To survive they have to overcome several problems:

  (i) osmotic effects—the solute potential (ψ_s, cf. Chapter 4) of sea water (*c.* 3% NaCl) is $-2\cdot7$ MPa and even one-tenth strength sea-water therefore has a potential of $-0\cdot27$ MPa. To take in water from such solutions a resistant plant must achieve an even lower intracellular potential.

  (ii) specific ion effects—high concentrations of Na^+ and Cl^- are toxic to most cells and Mg^{2+}, SO_4^{2-}, and many others can also be lethal.

  (iii) habitat effects—soils affected by salt tend to be extreme areas in other respects. Salt-deserts are of course dry, whereas salt-marshes tend to be waterlogged, and the high Na^+ content causes deflocculation of soil clay particles and loss of air-filled pore space. This occurs because the ionic interaction between the monovalent sodium ions and individual clay particles is quite different from that with the divalent calcium ions, the more usual dominant soil cation. The electrical double layer around Na-saturated clay particles is wider than in a Ca-system, and the repulsion between similarly-charged Na-particles acts at a much greater distance than between Ca-particles. Therefore aggregation fails to occur, pore space is low, and the soil acts like a cement. Anoxia is therefore a serious potential problem (cf. Chapter 7).

B. Calcareous and Acid Soils

One of the first things that any field botanist learns is the importance of limestone as a determinant of vegetation. Soils formed over parent materials containing a high proportion of $CaCO_3$ (limestones, chalks, and many other rocks and glacial deposits) tend to have pH values around or above 7, although during soil formation Ca^{2+} ions will be leached out, gradually reducing pH. Nevertheless, limestone soils are well-buffered by the reservoir of $CaCO_3$ and only in extreme conditions will the pH fall below 5. On the other hand soils formed on parent materials such as granite or base-poor sands are typically dominated by Al^{3+} as the major cation; such soils are strongly acid, their pH controlled by a complex hydrated aluminium ion buffer system, which, however, also sets a lower limit to pH. Al-based soils seldom have pH values below about 3·5 to 4·0. Lower pH values are encountered, but usually in soils with very little aluminium, such as ombrotrophic peat, or where soil sulphides are oxidized to sulphate, producing in effect sulphuric acid. This occurs in polders reclaimed from the sea (van Beers, 1962), mangrove swamps (Hart, 1963), and in pyrites-containing mine spoils, where pH values may rarely fall as low as 1·0. At such extremes no plants can survive.

Soil pH is a curious variable. In solution, pH defines the H^+ concentration and the measurement of soil pH normally involves making a soil suspension and determining the pH of that. In undisturbed soil pH will vary greatly from point to point, depending on local mineral types, CO_2 production and so forth, although these variations will occur around the average value referred to as soil pH.

Soil-forming processes greatly affect the distribution of acid and calcareous soils (see Chapter 3, pp. 71–74). Wide areas may be characterized by calcareous or acid soils, determined by the appropriate parent material and climatic conditions; local patches may differ due to topographic effects or the deposition of acid litter by vegetation, such as gorse (Grubb and Suter, 1971); and different vertical horizons within soil often vary greatly, due to movement of Ca^{2+} ions within the profile, caused by leaching or evaporation.

The differences between calcareous and acid soils are not just those of pH, nor only of Ca^{2+} and Al^{3+} ion concentrations. H^+ ions are toxic to most plants at pH values below 3, and below pH 4·0 to 4·5 mineral soils contain so much soluble Al^{3+} as to be severely toxic. But pH also controls the solubility of Mn^{2+}, Fe^{3+}, and many other cations. Both Mn and Fe are essential nutrient elements which may be present at toxic concentrations in acid soils and below deficiency levels in calcareous areas. Of the major nutrients, K^+ is displaced from exchange sites at low pH and lost by leaching, P availability varies in a most complex fashion with pH, but tends to be least at extreme values, and N availability is very low in acid soils because of impaired microbiological activity. Overall, therefore, plants of calcareous soils (*calcicoles*), must contend with deficiencies and plants of acidic soils (*calcifuges*) with toxicities, though there are important exceptions to this generalization. Table 6.2 summarizes these relationships.

TABLE 6.2. Ionic relationships of calcareous and acid soils.

Ion	Calcifuges on acid soils (pH < 5)		Calcicoles on calcareous soils (pH > 6·5)	
H^+	High:	may become toxic to non-adapted plants	Low:	not required
OH^-	Low:	not required	High:	may compete with anions for uptake
HCO_3^-	Low:	not required	High:	may compete with anions for uptake
Ca^{2+}	Low:	if very deficient may disrupt membrane function	High:	may cause phosphate precipitation at root surface, and compete with other cations for uptake
Al^{3+}	High:	not required and may cause precipitation of ions (e.g. $H_2PO_4^-$) at root surface, inhibition of Ca^{2+} uptake and transport, and interfere with DNA metabolism, *inter alia*	Low:	not required
Fe^{3+}	High:	acts similarly to Al^{3+} in phosphate precipitation	Low:	deficiency a major problem
Mn^{2+}	High:	toxicity relatively unimportant in relation to Al^{3+}	Low:	deficiency common in agricultural conditions
MoO_4^{n-}	Low:	deficiency may interfere with N fixation	High:	occasionally produces toxicity symptoms
NO_3^- / NH_4^+	Balance in favour of NH_4^+; nitrification inhibited		Balance in favour of NO_3^-; nitrifying bacteria active	
$H_2PO_4^-$ / HPO_4^{2-}	$H_2PO_4^-$ dominant species; adsorbed by Fe and Al		HPO_4^{2-} increases; adsorbed by Ca	
K^+	Displaced from exchange sites by H^+; leached		Ca^{2+} ions may interfere with uptake	

C. Metal-contaminated Soils

Few elements commonly reach toxic concentrations in soils. Some such as selenium and arsenic can naturally reach toxic levels, but the most important are a group of metals, including copper (Cu), zinc (Zn) and lead (Pb), and more rarely cadmium (Cd), chromium (Cr), cobalt (Co), and nickel (Ni). Ores of all these metals occur naturally, often in veins in rocks, and local high concentrations can occur. Such outcrops can be very small, or cover several hectares, as is often the case where ultrabasic rocks such as serpentine variously rich in Ni, Cr, or other metals, are involved.

Although such natural contamination provides certain fascinating problems, the type of metal-affected soil that has attracted most attention is where the soil is artificially contaminated—polluted by metals. Typically this occurs where waste from mining or smelting operations has been dumped. Such waste varies greatly in metal content, but older extraction methods were often very inefficient and it is not unusual to find mining wastes containing 2–3% by weight of metal. Alternatively soil can become contaminated by metals deposited from the air, typically in the neighbourhood of large smelters, but also from such improbable sources as galvanized fences!

Where mineral deposits are the source of contamination, release of the metal ions is usually facilitated by low pH. Copper, for example, typically occurs as the sulphide or carbonate ores, the latter being obviously more labile in acid conditions. Copper sulphide reacts with ferric sulphate, formed by the natural oxidation of ferric sulphide, to release soluble copper and sulphate ions (Peterson and Nielsen, 1978). The activity of oxidizing bacteria, particularly *Thiobacillus* and *Ferrobacillus* species, upon iron sulphide can indirectly release copper, nickel, zinc, arsenic, and even molybdenum from their native minerals (Firth, 1978). Almost the only important exception to the greater availability of toxic heavy metals in acid soils is molybdenum: in rather exceptional situations, molybdenum toxicity can occur on calcareous soils, as for example in parts of Derbyshire in northern England, where carboniferous limestone deposits have high native Mo levels.

II. EFFECTS OF TOXINS ON PLANTS

The range of substances capable of adversely affecting plant growth is enormous, and inevitably the specific effects of these toxins are too numerous to document. For instance Foy *et al.* (1978) suggest that aluminium alone may fix phosphate on root surfaces, and decrease root respiration, cell division, cell wall rigidity and the uptake and utilization of Ca, Mg, P, K, and H_2O. It is possible to classify these effects according to whether they exert their influence on the acquisition of resources by the plant or on the utilisation of those resources (Table 6.3). In the former case the toxins can be seen as in some sense acting by exacerbating unfavourability of a second factor, rather than by a direct toxicity. Thus a plant growing in an already phosphate-deficient soil may appear to suffer from toxicity through the precipitation of phosphate by aluminium at the root surface—what the plant experiences is in effect phosphate deficiency rather than aluminium toxicity.

A. The Acquisition of Resources

1. Water

Only three ions commonly reach solution concentrations sufficient to cause plants osmotic problems, without first having lethal specific toxicity. They are chloride (usually in wet saline soils) and sulphate (in dry sodic soils), both typically associated with sodium as cation, though magnesium and, rarely, calcium may also be involved.

TABLE 6.3.   A simple classification of the effects of toxins on plants.

A. *Effects on the ability to acquire resources*		
(i) acquisition of water:	(a)	osmotic effects arising from excess solute concentrations
	(b)	inhibition of cell division, reducing root growth
(ii) acquisition of nutrients:	(a)	competition between ions
	(b)	damage to membranes
	(c)	effects on symbionts
	(d)	inhibition of cell division
(iii) acquisition of CO_2 and light energy:	(a)	stomatal malfunction caused by toxic gases[a]
	(b)	chlorophyll bleaching[a]
B. *Effects on the ability to utilize resources*		
(i) inhibition of enzyme action		
(ii) inhibition of cell division		
(iii) loss of respiratory substrates; O_2 deficiency[a]		

[a]See Chapter 7.

If the solute concentration of the solution surrounding a plant's roots is suddenly raised, for example by flooding with saline water, the immediate effect is to reduce the water potential gradient between solution and root. Since water only moves into roots down such a gradient, this will inhibit water uptake. If the external water potential is lower (more negative) than the internal, water uptake will cease. In the short term this will reduce growth; in the long term it will cause wilting and ultimately death. Even plants which are highly sensitive to saline water, such as sunflower *Helianthus annuus*, react to this situation by osmotic adjustment (osmoregulation; cf. Chapter 4)—alteration of the internal osmotic pressure, which lowers the water potential (Fig. 6.1). Whichever solute is used to lower the external water potential (NaCl or PEG—polyethyleneglycol), the internal solute concentration increases. PEG is too large a molecule to be taken up by the roots, and increases in amino-acids (up 54%), reducing sugars (44%) and K^+ (104%) are responsible; where NaCl is used to lower water potential, a 50-fold increase in Na^+ concentration is the response.

Such adjustment can therefore allow non-halophytes to continue to grow at slightly raised external salt concentrations. Halophytes have the ability to withstand greater concentrations, because they can cope with higher internal concentrations of salt and other solutes and so maintain a positive water potential gradient. *Suaeda maritima*, for example, which is a common salt-marsh plant, could increase its root Na^+ concentration from 14 to 270 mM in response to an increase from 1 to 340 mM external NaCl in an experiment by Yeo (1974; reported by Munns *et al.*, 1983). Changes in Cl^- and K^+ are also important. As a result many halophytes have an optimum external NaCl concentration of around 100 mM, which would be damaging to all non-adapted species. There are costs associated with this response, however. It is remarkable that there are no halophytic trees taller than mangroves, which rarely exceed 10–15 m. This may be because the requirement for a water potential gradient from toots to leaves, coupled with the inevitable reduction in potential due to gravity, would mean that a tree of any size, growing in salt water, would have to withstand normal leaf water potentials of -4 MPa or less (cf. Chapter 4).

2. Nutrients

In soil, in contrast to physiological experiments, plant roots absorb ions from a complex medium, containing not only the dozen or so essential nutrient ions, but also a range of non-essential ions and organic compounds. If severe imbalances arise in this supply, the plant may not be able to take up nutrients efficiently, either because of direct

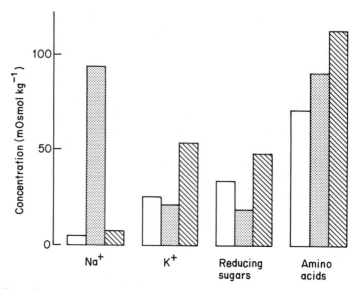

FIG. 6.1.   Solute concentrations in basal parts of sunflower hypocotyls in response to the addition of 100 mOsmol kg^{-1} NaCl (▦) or polyethyleneglycol (▨) to the control (open bars) medium. Data from Wyn-Jones, R. G. and Gorham, J. (1983). *Encyclopaedia of Plant Physiology*, Vol. 12C, pp. 35–58.

effects of the toxic ions on the root metabolism or function, or simply by competition or other interactions with nutrient ions. As a result even essential ions can become toxic (e.g. magnesium, see below), and plant species show great differences in the extent to which they can tolerate variation in ionic ratios.

The simplest effects are those where interactions occur outside the root. The process of transpiration causes accumulation of some ions at the root surface, if their rate of arrival exceeds their rate of uptake (see Chapter 3), and calcium can certainly accumulate in calcareous soils in this way. Since many calcium salts are insoluble, this may markedly inhibit diffusive supply of anions, such as sulphate and phosphate. In acid soils, aluminium may accumulate and aluminium phosphate deposits certainly occur on the rhizoplane (McCormick and Bowden, 1972). This Al-bound phosphate may, however, be exchangeable and so still in some measure available to the plant (Andrew and vanden Berg, 1973).

Damage to the plasmalemma, the selective boundary, may also occur. Calcium, as a plant nutrient, plays roles both as a messenger

(Hepler and Wayne, 1985) and in maintaining membrane function (Burström, 1968; van Steveninck, 1965), and one of the effects of aluminium toxicity is a reduction in Ca uptake (Clarkson and Sanderson, 1971). It is therefore striking that Lance and Pearson (1969) found that exposure of cotton roots to Al concentrations between 0·15 and 0·30 mg l^{-1} had identical effects in reducing uptake of Ca, Mg, K, P, nitrate, and water. The generality of the effect suggests that aluminium was causing some fundamental damage to the cell membranes. This is also reflected in increased leakage of K$^+$ ions across Al^{3+}-treated plasmalemmas, an effect also produced by Cu^{2+} (Wainwright and Woolhouse, 1975); Zn^{2+} on the other hand does not increase K$^+$ loss and does not appear to damage the plasmalemma .

More specific effects, however, occur without actual membrane damage, where ions interfere directly with each other's uptake. Generally such interactions may be competitive, where closely related ions compete for the same uptake sites, or non-competitive, where the toxic ion simply inactivates the uptake mechanism. The latter clearly grade into the general membrane disruption phenomena already discussed.

Competitive and non-competitive inhibition can be distinguished by their effect on uptake kinetics. In non-competitive cases, the actual uptake mechanism is incapacitated and so the maximum rate attainable (V_{max}) is reduced, whereas competitive inhibition manifests itself as a reduced K_m, or uptake affinity, and unchanged V_{max}, since the inhibition depends on the relative concentrations of the two ions. Competition occurs usually between chemically related ions, such as the alkali metals (Na$^+$, K$^+$, Rb$^+$) or alkali earths (Ca^{2+}, Mg^{2+}, Sr^{2+}). Thus saline soils may be toxic partly due to competitive inhibition of K$^+$ uptake, and Mg^{2+} certainly inhibits Ca^{2+} uptake on some serpentine soils with high Mg:Ca ratios (Proctor, 1971b).

Such interactions have received surprisingly little attention considering the complexity of natural soil solutions. These bear little relation to the artificial nutrient solutions in which ion uptake is normally studied, which have been designed to maximize plant growth or to simplify interpretation. Soil solutions are not only qualitatively and quantitatively different in their inorganic components, but also contain a range of organic compounds, some of which, such as phenolic acids, can profoundly alter ion uptake (Glass, 1973; Fig. 6.2).

Finally nutrient supply in soil may be heavily dependent on microbial symbionts, most obviously N-fixing bacteria and P-supplying mycorrhizas (see Chapter 3). Any adverse effects on these symbionts will severely reduce the nutrient supply. Thus most *Rhizobium* strains are more or less inactive below pH 5, though *Myrica gale* (bog myrtle), with an actinomycete symbiont, can continue to utilize atmospheric N at a pH as low as 3·3 (Bond, 1951).

B. Utilization of Resources

1. *Enzymes*

All metabolic behaviour ultimately involves enzymes, which typically operate optimally in well-defined ionic environments. Ions may be involved in enzyme functions as structural components of the enzyme molecule (Fe in cytochromes), chelated by an essential co-factor (Mo in nitrate reductase), or simply as an activator (K in pyruvate kinase). Ionic imbalances, therefore, are peculiarly capable of disrupting

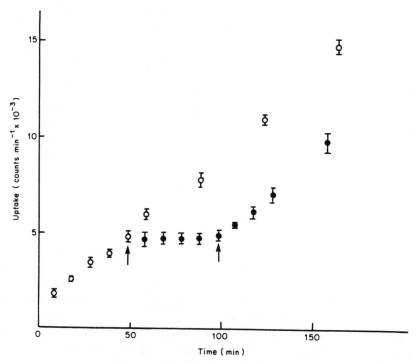

FIG. 6.2.   The effect of *p*-hydroxybenzoic acid on potassium uptake by excised barley roots. Arrows indicate application and removal respectively of the acid (from Glass, 1974). Control (no inhibitor) O — O; and 2·5 × 10^{-4} M *p*-hydroxybenzoic acid ● — ●. The first arrow indicates the time when the *p*-hydroxybenzoic acid was added to the root samples. At the time indicated by the second arrow roots were rinsed briefly and returned to inhibitor-free solutions. Each point (with standard error) is the mean of ten (prior to addition of inhibitor) or five (after addition of inhibitor) replicates.

enzyme action, but the results may be identified at three levels: the enzymic reaction itself, the metabolic process, or some measure of growth.

The direct effects of ions on enzymes depend on both the ion and the plant species. Some, for example Cl^-, tend to have rather general inhibitory effects at high concentrations (Porath and Poljakoff-Mayber, 1964), presumably the result of conformational changes. In contrast ions such as Cu^{2+}, Zn^{2+}, and Ni^{2+} may have specific reactions with particular enzymes at rather low concentration, often by complexing with chemical groups such as sulphydryl. Thus cadmium ions cause almost total inhibition of nitrate reductase extracted from leaves of *Silene vulgaris* at 0·01 mM but have no effect on iso-citrate dehydrogenase at concentrations as high as 1·0 mM (Fig. 6.3; Mathys, 1975).

2. Cell Division and Elongation

One of the characteristic symptoms of metal toxicity is the stunting of roots, often accompanied by browning and death of the meristem. This is particularly noticeable if roots of an aluminium-sensitive plant are exposed to low Al concentrations. Using a radioactive scandium isotope (^{46}Sc) as an analogue, Clarkson and Sanderson (1969) showed that primary cell walls in the root tip have a very low resistance to Al^{3+} movement, apparently owing to low exchange capacity, so that Al ions rapidly penetrate to the meristematic cells and inhibit DNA synthesis. Sainfoin (*Onobrychis viciifolia*) roots exposed to 20 p.p.m. Al^{3+} were able

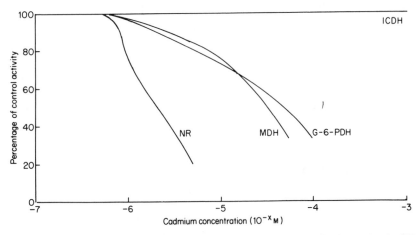

FIG. 6.3. Effect of cadmium on four enzymes from leaves of *Silene vulgaris*. NR, Nitrate reductase; ICDH, Isocitrate dehydrogenase; MDH, malate dehydrogenase; G-6-PDH, Glucose-6-phosphate dehydrogenase (after Mathys, 1975).

to incorporate only one third as much ^{32}P into nucleotides as were control roots (Rorison, 1965). In older root segments the primary effect of Al^{3+} and of many other di- and tri-valent cations appears to be on cell extension, the initial steps of which involve the orderly creation and breakage of cross-linkages within the pectic fraction of the cell wall. One of the major functions of Ca^{2+} ions is to regulate this process. Many polyvalent cations can interfere with this, including Al (Rorison, 1958) and Pb (Lane *et al.*, 1978). In the latter case, lead was found to be bound specifically to the pectinic acid fraction. It is not surprising, therefore, that levels of Ca^{2+} are critical in experiments on metal ion toxicity.

III. RESISTANCE TO TOXICITY

Some plants can grow in soils that contain levels of toxic ions lethal to other species. Four main mechanisms can achieve this.

(i) *Phenological escape*: where the stress is seasonal the active phase of the life-cycle may coincide with the most favourable season.

(ii) *Exclusion*: the plant may be able to recognize the toxic ion and prevent its uptake, and so not experience the toxicity.

(iii) *Amelioration*: the plant may absorb the ion but act upon it in such a way as to minimize its effects. Variously this may involve chelation, dilution, localization, or even excretion.

(iv) *Tolerance*: the plant may have evolved a metabolic system which can function at potentially toxic concentrations, possibly by means of distinct enzyme molecules.

Those species most able to resist toxic ions normally employ more than one such mechanism, but the adoption of any one or any combination imposes important physiological and ecological constraints.

A. Escape Mechanisms

1. Phenology
Stresses that vary over a sufficiently long-term and predictable time scale—typically seasonally—permit the survival of plants with life-cycles in which growth occurs at the most favourable time. This is particularly clear in the early spring growth of herbs in temperate deciduous forest

(Chapter 2, p. 44) and in the rain-triggered growth of desert ephemerals (Chapter 4, p. 151). There is little evidence as to the seasonality of toxic ion concentrations in soil, though where bacterial transformations of heavy metal ores are involved, one might expect spring levels to be low, since low winter temperatures should inhibit bacterial activity and high rainfall promote leaching. Saline soils, however, do fluctuate considerably. The upper reaches of salt-marshes are inundated by sea-water only rarely, at the extreme once a year, and at this time the soil water will be about 3% NaCl. Subsequently rainfall may reduce, or evaporation increase that concentration. As a habitat this zone is therefore subject to large, but mainly unpredictable fluctuations, and is typically characterized by few species. One such is *Juncus maritimus*, which strikingly, and in contrast to other *Juncus* species, shows little osmotic adjustment over a range of external NaCl concentrations from 0–300 mM, since its osmotic potential at zero NaCl concentration can be as low as − 1·9 MPa (Rozema, 1976).

Unlike the other species studied—*J. gerardi* (a lower salt-marsh plant), *J. bufonius* (a plant of wet, non-saline soils), and *J. alpino-articulatus* ssp. *atricapillus* (normally an alpine plant, exceptionally found near salt-marshes)—*J. maritimus* was unaffected by salinity in terms of growth rate, which was always low, and solute potential, which is maintained at a high level at low external salt concentrations by high internal potassium concentrations (Fig. 6.4). *J. maritimus*

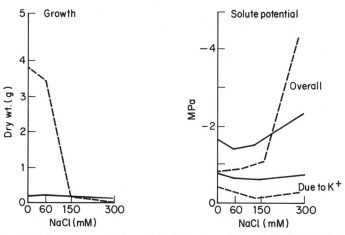

FIG. 6.4.   Growth and osmotic potential of *Juncus maritimus* (solid line) and *J. bufonius* (dotted line) over a range of NaCl concentrations. Note the association between the growth decrease in *J. bufonius* and the increase in solute potential, caused by ions other than K^+ (from Rozema, 1976).

apparently manages to survive by a constitutive adaptation to the most severe states of the environment, important because these are unpredictable. In contrast, inland saline areas tend to show a predictable rise in salinity due to evaporation as the summer progresses. Certainly salt concentrations in leaves of halophytes near the Neusiedlersee in Austria do rise during the season (Albert, 1975), but little critical work has been done on phenology in such habitats.

Escape may not, however, require cessation of growth but rather the timing of the most sensitive stage—typically seedling establishment—to coincide with the most favourable period. Dumbroff and Cooper (1974) showed that tomato plants were very sensitive to salt stress $(-0.6$ MPa) in the first 8 days of growth, and suffered irreversible damage. Thereafter both their resistance and their ability to recover from the stress increased. It is probably the benefit derived from protecting seedlings from salinities that explains the germination responses of *Limonium vulgare*, a common salt-marsh plant. Sea-water inhibits germination, which is low even in fresh-water; but if seeds are transferred from sea-water to fresh-water, there is a pronounced stimulation (Boorman, 1967; Fig. 6.5). For the seed, the transfer from sea-water to fresh-water might signal the onset of temporarily favourable conditions for seedling growth. Similarly soil pH fluctuates seasonally in many soils, particularly on waterlogged soils as redox potentials change. Gupta and Rorison (1975) found changes of the order of 1 pH unit in a podzol , and regular seasonal changes of 2·0–2·5 units may occur in waterlogged peat (Fitter and Smith, 1979). In mineral soils such changes will have dramatic effects on Al^{3+} concentrations, so that it may be significant that in some species, mature plants can grow in Al^{3+} concentrations that are lethal to seedlings (Rorison *et al.*, 1958).

2. Direct Environmental Modification

More rarely a plant may have the power to alter the environment and so reduce the toxicity . Where NO_3^- is the main N-source, anion uptake exceeds cation uptake and OH^- or HCO_3^- ions are extruded as counterions; in contrast, NH_4^+ uptake requires the loss of H^+ ions to maintain the electrochemical potential and this may cause acidification of the rhizosphere. Riley and Barber (1971) found that soybeans fertilized with NO_3^-–N raised rhizosphere pH by 1.4 units; NH_4^+–N fertilizer caused a reduction of 0·5 units. They were able to explain variation in P uptake on this basis. Rape *Brassica napus* roots also extrude H^+ ions and so typically have a very acid rhizosphere (Hedley *et al.*, 1982). Such effects can readily explain why low pH has a

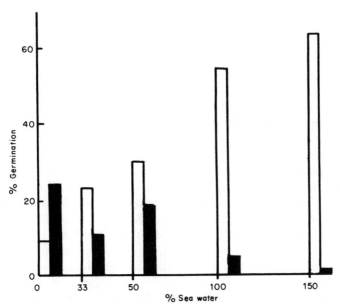

FıG. 6.5.   Effect of varying concentrations of sea water on germination of the seeds of the halophyte *Limonium vulgare*. Solid bars indicate germination in the stated medium; open bars germination in fresh water after 25 days at the stated salinity. Note the inhibitory effect of sea water and its potentiation of subsequent germination in fresh water (from Boorman, 1967).

deleterious effect on *Rumex acetosa* and *Scabiosa columbaria* (a neutrophile and calcicole respectively) in the presence of NH_4^+ but not if NO_3^- was supplied (Gigon and Rorison, 1972). In contrast the calcifuge *Deschampsia flexuosa* was inhibited by high pH in the presence of NO_3^- but insensitive to pH if NH_4^+ was the N-source. The ability of cereals to withstand Al toxicity appears to be closely related to their nitrogen nutrition: those that take up NH_4^+ most rapidly cause the greatest reduction in rhizosphere pH and are the most susceptible to Al (Taylor and Foy, 1985).

A further example of environmental modification can be seen in a wide range of wetland plants (e.g. rice; Tadano, 1975), which can oxidize Fe^{2+} to Fe^{3+} by O_2 excretion from the roots into the rhizosphere, so avoiding Fe^{2+}-toxicity (cf. Chapter 7).

B. Exclusion

Prima facie, exclusion of a toxic ion would seem to be the ideal resistance mechanism, but it presents a number of problems, both in respect of the ability of the plant to exclude the ion (recognition, damage to external structures), and of the consequences of exclusion (if the ion is either present at osmotically damaging concentrations, or required by the plant in small quantities).

1. Recognition

Plants typically have highly selective uptake systems, capable of distinguishing chemically similar ions, but a number of ion pairs cause problems, particularly K^+ and Na^+, and Ca^{2+} and Mg^{2+}. At low K^+ concentrations (around 0·5 mM), the K^+/Na^+ selectivity of barley roots is high, but at high concentrations of K^+ and Na^+, this selectivity is found only in halophytes (Epstein, 1969). Potassium is of significance in maintaining solute potentials in halophytes subject to fluctuating external salinity (Rozema, 1976, see p. 239), and a mechanism conferring some degree of stability of K^+ uptake over a wide range of NaCl concentrations is therefore necessary.

A similar situation occurs with Ca^{2+} and Mg^{2+}, between which most plants discriminate poorly, so that at very low Ca:Mg ratios in soil, excess Mg accumulates in plant tissue, with toxic effects (Proctor, 1971). Such conditions exist on many serpentine (ultrabasic) soils, and may account for the general paucity of vegetation there. Those plants that do grow on such soils, such as ecotypes of *Agrostis stolonifera* or *A. canina*, appear to take up the two ions to a greater extent in proportion to the external concentrations than do susceptible genotypes, so that their resistance mechanism must be internal. By contrast Olsen (1971) showed that the ability of rye (*Secale*) to withstand low Ca:Mg ratios lay in its ability to discriminate in favour of Ca, and the same is true of the serpentine endemic *Helianthus bolanderi*, when compared with sunflower *H. annuus* (Madhok and Walker, 1969, Fig. 6.6).

2. External Structures

Even if the plant can recognize the toxic ion and so exclude it, there are still metabolically active structures which cannot be protected in this way. Clearly all cell membranes in contact with the external solution are potentially at risk, and if the ion can move in the apoplast this may include all cells as far in as the endodermis. As already mentioned (p. 235), copper ions, in contrast to zinc ions, cause leakage of potassium from cells, presumably by membrane damage. Wainwright

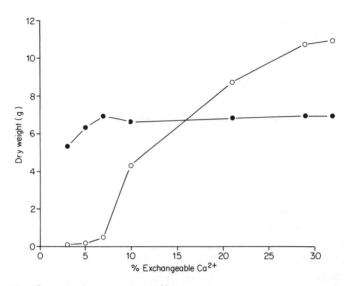

Fig. 6.6. Growth of two species of *Helianthus* in response to increasing Ca supply. *H. annus* (○) is the common garden sunflower; *H. bolanderi* (●) is an endemic Californian serpentine species, adapted to growth in soils of high Mg:Ca ratios. The serpentine species is almost insensitive to Ca (from Madhok and Walker, 1969).

and Woolhouse (1975) showed that a Cu-resistant race of *Agrostis tenuis* was only half as susceptible to such K^+ leakage as a Zn-resistant or a normal race.

In addition all plant roots have surface enzymes, of which the best known and studied are the acid phosphatases. The function of these enzymes is not clear; it is widely assumed that they are involved with the breakdown of organic phosphate in soil, but there is no evidence that plants are able to utilize directly soil organic P (Abeyakoon and Pigott, 1975). Nevertheless acid phosphatases are widespread and are inhibited by toxic ions in soil. Wainwright and Woolhouse (1975) found differential responses to Al^{3+} and Cu^{2+} of enzymes from different ecotypes of *A. tenuis*. In the case of Cu, kinetic analysis indicated a non-competitive inhibition, with the presumed adaptive differences resting in a difference in inhibitor constant (k_i) between the two ecotypes. The k_i for copper-resistant plants was 1·50 mM Cu^{2+}, for the susceptible 0·54 mM Cu^{2+}; the smaller value indicates that the susceptible enzyme has a greater affinity for copper, which presumably forms an ineffective complex with it.

Such results suggest a change in the molecular properties of the enzyme as a basis for resistance to toxicity. However, the preparation used was of cell-wall fragments rather than purified enzyme, which leaves open the possibility that the enzyme was in some way protected from free copper ions in the resistant ecotype. This is certainly more consistent with work on other systems exhibiting enzyme "tolerance" (see p. 257). Nevertheless cell walls, both inside and outside the plant, are liable to experience higher concentrations of toxic ions than other plant parts and there are many active molecules in or adjacent to these walls, most conspicuously the uptake systems for other ions. Whether or not these show a direct resistance to ionic toxicity is still not known.

3. Deficiencies: Ionic Imbalance

As a resistance mechanism, exclusion obviously leads to problems where the toxic ion is metabolically essential at low concentrations, such as Cu, Zn, or Fe. Is resistance to these ions, therefore, less often achieved by exclusion than is the case for Pb, Cd, Cr, and other non-essential toxins? The evidence is far from clear, and although it was possible for Antonovics et al. (1971) to state that "nowhere . . . is there any evidence for (tolerant) plants having an exclusion mechanism", Mathys (1973) clearly demonstrated exclusion of zinc in Agrostis tenuis (A. capillaris), a species which Wu and Antonovics (1975) found to accumulate Zn in the resistant ecotype.

At present it is not possible to make firm generalizations on this point, but it is likely that the stimulatory effects on growth and root elongation produced by low concentrations of Al^{3+} in calcifuge species (Clarkson, 1967; Fig. 6.7) are due to the presence of Al-binding sites on the root surface. When these are unoccupied by Al^{3+} ions they tend to bind the similar Fe^{3+} ions, and may thus cause Fe deficiency. Grime and Hodgson (1969) showed that growth responses to Al tend to be diphasic, with an initial decline at very low concentrations, due to the toxic effect of Al on cell division, and then a minor peak at slightly higher concentrations, apparently resulting from the liberation of Fe from binding sites on the cell wall. Finally, at high Al concentrations, growth falls off again. The position of the peak of growth varied according to the pH-resistance of the species or ecotype, falling at around 20–30 μM Al^{3+} in calcifuges such as Nardus stricta and Ulex europaeus, and nearer 5 μM in calcicoles (Bromus erectus, Scabiosa columbaria) (Fig. 6.8).

In those experiments the position of the initial growth depression also altered—from around 1 μM in calcicoles to as high as 10 μM in calcifuges—and there is evidence (Foy et al., 1978) that some Al-resistance may be due to an ability to prevent Al^{3+} migrating through

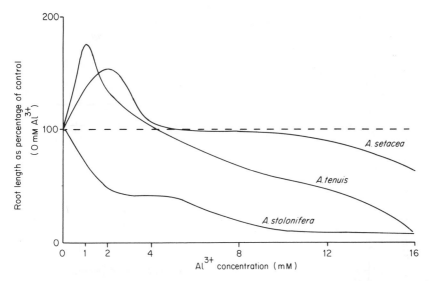

Fig. 6.7. Effect of Al^{3+} ions on root length growth in one week of three species of
Agrostis. *A. setacea*, calcifuge; *A. tenuis* (*A. capillaris*), mildly calcifuge; *A. stolonifera*,
mildly calcicole (from Clarkson, 1966).

the free space to the meristem, so protecting cell division. Hemming
(quoted by Foy *et al.*, 1978) found that a resistant wheat variety could
withstand a hundred-fold increase in external Al concentration before
Al entered the root meristem, as compared with a sensitive variety.
This again suggests a binding mechanism in the cell wall.

4. Osmotic Imbalance
There is little evidence that low water potentials are inherently
damaging, even to mesophytes, until extremely low values (less than
-1.5 to -2 MPa) are reached. As long as the xylem water potential is
lower than that of the soil water, uptake will continue. If such osmotic
adjustment does not take place, water deficits will occur. Tal (1971)
found that cultivated tomatoes (*Lycopersicon esculentum*) took up less Na
and Cl when subjected to high salinity than the wild *L. peruvianum*, with
adverse effects on both relative water content and growth rate. In
contrast, Greenway (1962) using barley and Gates *et al.* (1970) with
soybean varieties found precisely the opposite: it was the sensitive
varieties that took up most salt. Presumably osmotic adjustment was
achieved at the expense of a toxic effect of the salt itself.
   Generally, agricultural species achieve resistance by exclusion
(Greenway, 1973) and the same is true for a number of wild plants.

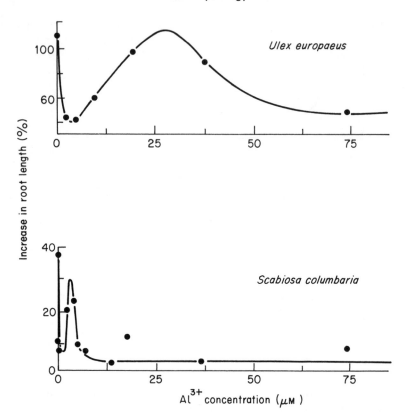

FIG. 6.8. Percentage increase in root growth shown by a calcifuge *Ulex europaeus* (gorse) and a calcicole *Scabiosa columbaria* (small scabious) in response to Al^{3+} concentration. Plants were Fe-deficient. Note the initial decline in elongation, the subsequent, but not sustained increase, and the very different positions of these two peaks. Redrawn from Grime and Hodgson (1969).

Tiku and Snaydon (1971) found that salt-marsh populations of *Agrostis stolonifera* had lower leaf Na concentrations than normal plants at the same salinity. It is dangerous to draw conclusions from leaf analyses (see below), but this conforms with the observations of Ahmad and Wainwright (1976) that leaves of salt-marsh *A. stolonifera* were less wettable and retained only 1/16th as much NaCl after immersion in sea-water as leaves of inland plants: apparently the exclusion mechanism here is in part physical.

    Cultivated plants, however, rarely experience severe salt stress, and appear to be able to make good a limited osmotic deficit created by salt exclusion by internal synthesis of solutes. Arid saline habitats, in contrast, present much lower soil water potentials, and halophytes there appear to carry out much of their osmotic adjustment by NaCl uptake. When this occurs, other problems arise, as salt is generally highly toxic at such concentrations. A fuller discussion of the significance of water potentials is given in Chapter 4.

C. Amelioration

For a variety of reasons it may not be possible for a plant to exclude a toxic ion. If high internal concentrations must be withstood, the ions will either have to be removed from circulation in some way, or tolerated within the cytoplasm. The former process is here termed amelioration. Four approaches are apparent.

  (i) *Localization*: either intra- or extracellularly, and usually in the roots;

 (ii) *Chemical inactivation*: so that the ion is present in a combined form of reduced toxicity;

(iii) *Dilution*: which is of primary significance in relation to salinity;

(iv) *Excretion*: either actively through glands on shoots or by the roots; or passively by accumulation in old leaves followed by abscission.

1. Localization

Analyses of both roots and shoots of a plant are rather rarely carried out, but data compiled by Tyler (1976) and summarized in Table 6.4 suggest that for *Anemone nemorosa* metals can be divided into three groups.

  (i) Those more or less equally distributed throughout the plant, including the essential major nutrients K, Ca, and Mg. Rb behaves as K in most plants.

 (ii) Those showing some accumulation in the root, including the three important micronutrients Cu, Zn, and Mn.

(iii) Non-essential, toxic ions such as Al and Cd, which are primarily stored in the root. Fe, a micronutrient, also falls into this class, presumably because of its high availability in soil, in contrast to Cu, Zn, and Mn.

Accumulation of toxic ions by roots is a widespread phenomenon. In all the examples quoted in Table 6.4, concentrations in the roots were higher than in the shoots. Even if roots have an inherently higher

TABLE 6.4.   Shoot concentrations and ratios of root concentration to shoot concentration for 12 elements in *Anemone nemorosa* (from Tyler, 1976).

		Shoot concentration (μg g^{-1})	root: shoot ratio
Widely distributed	Ca	7180	0·8
	Mg	2970	1·3
	K	11,400	1·3
	Rb	35	1·6
Slight root storage	Mn	405	2·3
	Cu	10.5	2·4
	Zn	113	3·6
Major root storage	Al	260	10·7
	Cd	1·24	11·7
	Fe	217	12·4
	Na	242	16·7
	Pb	1·04	62·6

tolerance to toxins than shoots, this clearly implies some form of localization.

Turner and Marshall (1972) found a linear relationship between the zinc resistance of a range of clones of *Agrostis tenuis* and the mass of Zn accumulated by their cell walls, and Peterson (1969) suggested that resistance might be related to the way in which Zn is retained by the wall. There is a very specific association between Pb^{2+} ions and the pectinic acid fraction of the cell wall (Lane *et al.*, 1978) and Al^{3+} may also accumulate in the pectic fraction (Matsumoto *et al.*, 1976). The pectic fractions of the cell wall have a considerable cation exchange capacity and the ability to form cross-linked structures—hence one of the major roles of Ca in plant metabolism—and so are well suited to the binding of polyvalent cations. However, several limitations must be borne in mind. First the toxicity due to NaCl will not be amenable to this solution, as both ions are monovalent; secondly cell walls are not inert, but contain active enzymes (cf. p. 243); and thirdly any such system is likely to be saturable, and so have a threshold effectiveness (Wainwright and Woolhouse, 1978).

There are two possible escapes from the last problem: binding sites could be continually synthesized within each cell wall, but this would certainly require enzyme action, and these enzymes would then be exposed to the toxic ion. Alternatively continued growth could provide new sites for chelation. In the latter case the maintenance of a healthy meristem is critical, and, particularly in the case of Al, therefore, it seems probable that the ability to protect the meristem from damage may be central to the resistance mechanism.

(a) *Intracellular.* If very much higher concentrations of an ion are found in the roots than in the shoots, it is strong evidence for extracellular localization, possibly by binding to the pectic fractions of the cell walls. Once ions have crossed the plasmalemma, there is no particular reason for their restriction to the roots. If the ion is to be kept out of general circulation, it must be accumulated in some particular compartment within the cell. Lead appears to accumulate in dictyosome vesicles in corn *Zea*, with mitochondria, plastids and nuclei remaining Pb-free (Malone *et al.*, 1974). Isolated mitochondria are capable of accumulating Pb^{2+} *in vitro*, so that clearly some specific affinity is exhibited by the dictyosomes.

The intracellular distribution of NaCl is more complex. The chloroplasts of the halophyte *Limonium vulgare* appear to accumulate Cl^- (Larkum, 1968), and those of *Suaeda maritima* grown at 340 mM NaCl contained about 200 mM Cl^- and 250 mM Na^+ (Harvey *et al.*, 1981). More generally, however, salt accumulated by plants to maintain osmotic integrity appears largely in the vacuole. The best evidence for this comes from X-ray microprobe techniques (Flowers and Läuchli, 1983) and from flux analysis, in which rates of efflux of ions from vacuole and cytoplasm are measured. Vacuolar Na^+/K^+ ratios are usually higher than that in the cytoplasm, but not all experimental data confirm this, and even so the total concentration of Na^+ and K^+ in the cytoplasm can approach 300 mM. Convincing indirect evidence comes from the total lack of specific tolerance to toxic ions exhibited by higher plant enzymes active within the cell (see below).

Enzymes from halophilic bacteria, which of course have extremely limited intracellular compartmentation, can operate *in vitro* at very high NaCl concentrations (Brown, 1976). In those eukaryotes that have been examined, from algae to higher plants, virtually no intracellular enzymes have been shown convincingly to exhibit any resistance to high concentrations of NaCl. Occasionally stimulations of enzyme activity are reported at low NaCl concentrations, usually less than 100 μM, but these can be found equally in glycophytes and halophytes. Even in *Dunaliella parva*, an alga found in highly saline lakes, most enzymes have no particular ability to function at high NaCl concentrations (Heimer, 1973); lactate dehydrogenase may be an exception.

The explanation appears to be that NaCl is isolated in the vacuole and so is not in the same compartment as the enzymes. However, this implies an osmotic imbalance between cytoplasm and vacuole: the concentration of Cl^- for example, in the vacuoles of the alga *Tolypella intricata* was five times that in the cytoplasm (Larkum, 1968). The cytoplasm must be at the same water potential as the vacuole, and this is achieved using "compatible" organic solutes (Borowitzka and Brown, 1974). Thus the halophilic alga *Dunaliella parva* contains large quantities of glycerol (Ben-Amotz and Avron, 1973); its nitrate reductase is unaffected *in vitro* by glycerol concentrations greater than 3 M, but although it can grow in 2 M NaCl (four times sea-water), nitrate reductase activity is reduced by 50% in 0·4 M NaCl. The alga *Ochromonas malhamensis* also uses glycerol for osmotic balancing (Kraus, 1969).

It appears to be a fairly general observation that the enzymes and organelles of halophytes are not tolerant of high NaCl concentrations, but can withstand the low water potentials of isoosmotic solutions of organic solutes. Von Willert (1974) found that malate dehydrogenase in several halophytes was in fact more inhibited by sucrose than by NaCl, but Stewart and Lee (1974) have found evidence that the amino acid proline is used by many halophytes for osmotic balance. In a wide range of plant species, they found proline to comprise on average 54·6% of the amino-acid pool in halophytes, as compared to 2·9, 2·4, and 4·0% respectively in calcicoles, calcifuges, and ruderals. Proline accumulation was directly related to the salinity of the medium and was found to have no effect on the activity of nine enzymes extracted from *Triglochin maritima*, at concentrations where NaCl was extremely inhibitory (Fig. 6.9). Since the concentrations used (up to 700 mM) were comparable to measured internal NaCl levels, and since Treichel (1975) found that the ratio of proline to Na^+ and Cl^- in the vacuole was maintained constant in three halophytes, the suggested role for proline is very attractive. Stewart and Lee were also able to show that inland races of *Armeria maritima* had a much lesser tendency to accumulate proline than maritime ones, when subjected to NaCl concentrations above 100 mM.

Of the eleven halophytes tested by Stewart and Lee, one, *Plantago maritima*, showed no proline accumulation; presumably it achieved osmoregulation by other means. Indeed Storey *et al.* (1977) have found that another amino-acid, glycine-betaine, is of greater significance than proline in many species, and CO_2 fixation and protein synthesis are unaffected by it, even at high concentrations (Wyn Jones and Storey, 1981). Different, though analogous, mechanisms, operate in relation to metals. Here there is no osmotic problem and Mathys

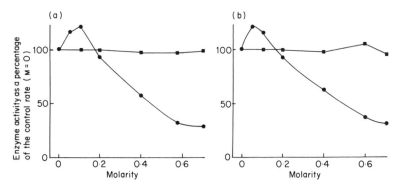

Fig. 6.9.   The effect of proline (■) and NaCl (●) on the activity of two enzymes extracted from *Triglochin maritima*. (a) Glutamate dehydrogenase; (b) nitrate reductase (from Stewart and Lee, 1974).

(1977) has suggested that Zn^{2+} is stored in vacuoles complexed with oxalate and transferred through the cytoplasm bound to malate. These ideas have not yet been tested critically.

As with halophytes, there is no evidence for actual tolerance by intracellular enzymes to high concentrations of ions, and the resistant ecotypes presumably maintain full enzyme function *in vivo* by means of compartmentation. Indeed zinc-resistant *Silene cucubalus* appears to have an increased zinc requirement, presumably because of an active chelating system: the activity of nitrate reductase in its leaves is increased by growing the plant in 0·4 mM Zn, whereas even 0·1 mM is inhibitory to the non-resistant ecotype (Mathys, 1975; Fig. 6.10).

2. Chemical Inactivation

Zinc resistance is a multiple phenomenon, partly involving exclusion in the root cell walls and partly localization in the cell vacuoles, where the ion is complexed with malate. Such complexes are probably widespread—copper, for example, is normally translocated chelated with polyamino-polycarboxylic acids (Tiffin, 1972)—and may even allow the plant to retain the toxic ion in the metabolic compartment, but in an inactive form.

When interest in the calcicole–calcifuge problem first appeared in experimental form much stress was laid on Ca levels. It is now generally accepted that the effects found on strongly acid or calcareous soils are better explained in terms of Al toxicity and Fe deficiency, amongst others. Nevertheless, Ca is one of the most variable elements

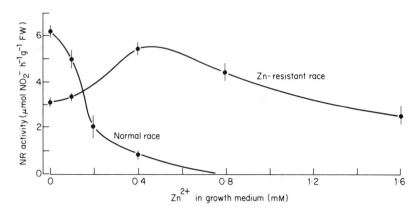

FIG. 6.10. *In vivo* activity of nitrate reductase in leaves of *Silene vulgaris* cultivated for four weeks at various levels of Zn^{2+} (from Schiller, 1974).

in different soils and cytoplasmic Ca^{2+} levels are normally very low (around 1 μM, Williamson and Ashley, 1982); free Ca^{2+} strongly inhibits many metabolic activities, including photosynthesis. As a result, Ca also plays an important role in less extreme conditions. Horak and Kinzel (1971) have classified plants in respect of their K:Ca ratios and the form in which Ca occurs in the plant, and Kinzel (1982) provides a recent account of this approach. Using standard diagrams (Fig. 6.11), they suggest the existence of three distinct forms of cation nutrition:

   (i) oxalate plants, which take up Ca but remove it from circulation as oxalate, such as Polygonaceae, Chenopodiaceae, Violaceae and most Caryophyllaceae (e.g. *Silene inflata*, Fig. 6.11);

  (ii) calciotrophes, which require high concentrations of free calcium, such as the Crassulaceae, and most of the Cruciferae and Fabaceae (Papilionaceae: e.g. *Coronilla vaginalis*, Fig. 6.11);

 (iii) potassium plants, which have large amounts of K and little free Ca, and are found in the Umbelliferae, Campanulaceae, and Compositae (e.g *Achillea clavenae*, Fig. 6.11).

The ecological significance of this metabolic diversity is unclear, since the ability to tolerate calcareous or acid soils follows no such clear taxonomic pattern. It may influence the ability of plants to colonize unusual habitats, such as serpentine soils with high Mg:Ca ratios. Serpentine-adapted plants do not generally exclude Mg (Proctor, 1971a), but may bind it as oxalate (e.g. *Tunica saxifraga*, Caryophylla-

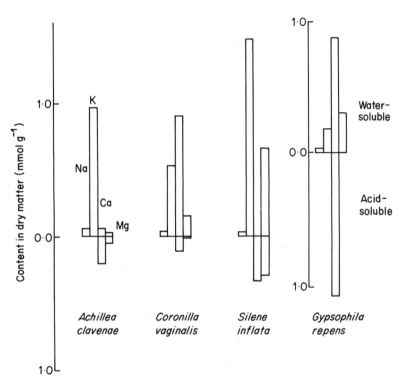

Fig. 6.11.   Diagrams to illustrate nutritional differences in response to soil pH. Each histogram shows the proportions of water- and acid-soluble Na, K, Ca, and Mg. The scale of Na is enlarged 10-fold relative to the others (from Kinzel, 1969).

ceae; Ritter-Studnička, 1971) or, in calciotrophe species, simply allow it to accumulate in the cell sap (*Sedum album, Biscutella laevigata*; Horak and Kinzel, 1971).

Possibly there is a link here with osmoregulation, as suggested by Kinzel (1982): calciotrophic species may use Ca^{2+} as an osmotic agent, while calciophobes use other ions, but the picture is still unclear. For example, Austenfeld (1974b) subjected *Salicornia europaea*, a characteristic salt-marsh plant, to a range of salinities, and found that the Na^+ content of the plant responded in a simple fashion to a range of NaCl concentrations from 0 to 250 mM. The bulk of the Na^+ was water-soluble, whereas the level of water-soluble Ca^{2+} was low and constant. The acid-soluble Ca^{2+} level, however, declined with increasing salinity, as did acid-soluble oxalate, whereas the water-soluble oxalate increased dramatically (Fig. 6.12). Possibly *Salicornia* uses oxalate to bind Ca^{2+} when grown at low salinity, and Na^+ at high salinities. If *Halimione portulacoides*, another salt-marsh plant, is grown at high NaCl levels, again water soluble oxalate increases, and at high $CaCl_2$ concentrations, it is the water-insoluble fraction that accumulates. The ecological significance of these ionic balancing operations is yet to be explained.

3. *Excretion*

Animals faced with excessive ion loads typically excrete them. Plants generally rely more on controls in uptake—simpler because their food is taken up in its component parts, not as a package—but are capable of excretion. A major disadvantage of excretion for plants, however, is that because they are stationary, the excreted substances will be returned to the root zone and may eventually lead to a build-up of toxin. This does not apply, of course, to plants growing in water.

The simplest form of excretion is the loss of an organ which has become saturated with the toxin. It is generally true that old leaves have much higher salt or heavy metal contents than young leaves and buds. Indeed, old leaves of tea may contain up to 30 mg g^{-1} (3%) Al, most of which is in the epidermis (Matsumoto *et al.*, 1976). Plants which accumulate Al in the shoots in this fashion are generally found

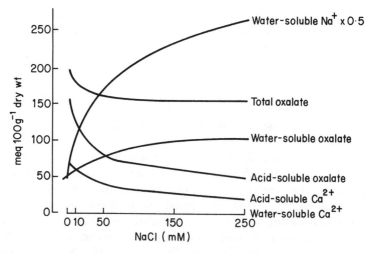

FIG. 6.12. Changes in amounts and extractability of Ca, Na, and oxalate in *Salicornia europaea* in response to salinity. Note the increase of water-soluble oxalate as internal Na rises and the correlation between acid-soluble Ca and acid-soluble oxalate (from Austenfeld, 1974).

in the more primitive, woody families (Chenery and Sporne, 1976), and nickel hyperaccumulators, which can contain similar or greater Ni levels, typically complexed with citrate (Lee *et al.*, 1978), are also taxonomically primitive (Jaffre *et al.*, 1976). Accumulation therefore seems to be an evolutionarily ancient phenomenon. It is noticeable that localization may also occur within a leaf: yellow margins of mustard leaves contained 2300 mg Mn g^{-1}, while the green parts had only 570 mg g^{-1} (Williams *et al.*, 1971). What appears at first sight to be a toxicity symptom, may be a part of the plant's resistance mechanism.

The relationship between toxin accumulation and abscission may be complex, for high salt levels may hasten senescence (Prisco and O'Leary, 1972). The same is almost certainly true of toxic heavy metals, as copper induces leaf chlorosis even in the very resistant *Becium homblei* (Reilly and Reilly, 1973). Interestingly, *B. homblei* appears to achieve Cu resistance by excluding Cu from the shoots, while *B. aureoviride* accumulates it (Duvigneaud and Denayer de Smet, 1963). In halophytes, abscission is most significant in rosette plants, which continually produce new leaves as the old ones senesce (Albert, 1975).

More active excretion also occurs, at least for salt. In non-halophytes Na$^+$ is actively withdrawn from the xylem back into the xylem parenchyma (Yeo *et al.*, 1977) and then possibly extruded from the roots back into the medium; certainly a potassium-stimulated Na$^+$ efflux can be shown to occur across root plasmalemmas (Jeschke, 1973). The most important route, however, is certainly through the glands found on leaves of mangroves, *Atriplex* species, and *Halimione portulacoides*, amongst others. In *Halimione* the glands are bicellular, the distal cell being highly vacuolated. When the plants are grown in high salt media the glands always contain much higher concentrations of Na$^+$ and Cl$^-$ than the sap from the leaves (Baumeister and Kloos, 1974). The glands are highly selective, Na$^+$, Cl$^-$ and HCO$_3^-$ being secreted against a concentration gradient and Ca^{2+}, NO$_3^-$, SO$_4^-$, and H$_2$PO$_4^-$ being retained against a concentration gradient. The movement of K$^+$ is less uniform. One estimate of the rate of excretion is 870 μmol g^{-1} dry weight of leaves d^{-1} for *Aeluropus litoralis* (Pollak and Waisel, 1970).

The activity of these glands, at least in *Limonium vulgare*, is not constitutive but is induced by growth in NaCl (Hill and Hill, 1973). A Cl$^-$-stimulated ATPase in these glands shows a 300% increase in activity if incubated *in vitro* with NaCl, as compared with water or even Na$_2$SO$_4$. Such a flexible system could be of great significance in environments showing wide fluctuations. A further advantage of inducibility is in the metabolic cost of an energy-consuming process such as salt excretion: in *Tamarix ramosissima* 65% of leaf Na^+, 82% of leaf Cl^- (and strikingly 90% of leaf Al^{3+} and 88% of leaf Si) could be removed by washing (Kleinkopf and Wallace, 1974); no other elemental losses exceeded 40%. Although net photosynthesis was actually stimulated by up to 200 mM NaCl, growth declined even at 10 mM and was reduced to 32% of the control at 200 mM. The growth decrease was apparently due to energy losses through increased respiration maintaining salt excretion.

4. Dilution

Toxic metals, such as zinc and copper, have rather specific affinities for particular biochemical groups. Thus copper reacts with sulphydryl groups, and nitrate reductase, an SH-enzyme, is particularly susceptible to Cu-poisoning; zinc reacts primarily with carboxylic acid groups. For these toxins, dilution by increasing cell water content is not a practicable resistance mechanism. Against salinity, with its osmotic and non-specific toxic effects (cf. p. 232) dilution is widespread and effective.

In many halophytes, but curiously not in monocotyledons, succulence, an increase in the water content per unit dry weight, is specifically stimulated by NaCl. Both Na^+ and Cl^- ions can produce succulence, which is typically manifested as an increase in cell volume. There is an intriguing link between succulence and CAM metabolism (cf. Chapters 2 and 4). In *Mesembryanthemum crystallinum* high salt levels induce both succulence and CAM, and the malate produced by CAM appears to have an additional role in balancing charge discrepancy between Na^+ and Cl^-. It is possible, though, that succulence has as much to do with temperature regulation and water conservation as with salt resistance. Many non-halophytes in arid regions are succulent: cacti are an obvious example. Typically they have CAM and do not open stomata in the daytime. Though they do not lose heat by evaporation, the high specific heat of water means that their internal water store can act as a heat sink, limiting the rise in temperature, but they do still have a very great resistance to high tissue temperatures.

The dominant effect of salinity is on water relations. If plants on saline soils have problems in obtaining water, they will respond by closing stomata and this entails a risk of overheating. Succulence may then be an alternative response in evolutionary terms to this dilemma.

D. Tolerance

Some resistant prokaryotes show true tolerance of their metabolic systems. Enzymes extracted from them operate *in vitro* in the presence of salt concentrations lethal to the enzymes of eukaryotes (typically KCl; Brown, 1976). Eukaryotes have much more complex intracellular compartmentation, and rely on this to ensure that their metabolic systems do not experience high concentrations of toxins. Nevertheless, such a stress may induce enzymic changes. Salt induces the synthesis of a new isozyme of malate dehydrogenase in peas (Hassan-Porath and Poljakoff-Mayber, 1969), which may be better equipped to operate at low water potentials.

If one examines plants under mild stress, it is apparent that enzyme adaptation can occur. Cultivated oat is a calcifuge and has a Mg^{2+}-activated ATPase system in the roots; under low-Ca conditions it operates adequately. By contrast the calcicolous wheat has a Ca^{2+}-activated ATPase inhibited by Mg^{2+}: if grown under low-salt conditions, however, ATPase activity is equally stimulated by Ca^{2+} and Mg^{2+} (Kylin and Kähr, 1973). Clearly some enzymic changes have occurred.

There are some enzymes, however, which cannot be protected by compartmentalization—those in the plasmalemmas and cell walls. In some cases cell wall enzymes, particularly acid phosphatases, have been shown to be tolerant of much higher levels of toxic ions (Cu^{2+}, Zn^{2+}) in resistant than in normal plants (Wainwright and Woolhouse, 1975). Unfortunately rather little is known of the function of these enzymes.

IV. THE ORIGIN OF RESISTANCE

Some forms of resistance to toxic ions are ancient. Saline habitats are certainly as old as the Angiosperms and the high proportion of halophytes in some families, in particular the Chenopodiaceae, including the genera *Chenopodium, Atriplex, Halimione, Salicornia, Suaeda* and *Salsola*, is strongly indicative of an ancient origin. Similarly Al accumulators, plants containing more than 1000 mg Al g^{-1} in their shoots, tend to be primitive on taxonomic grounds (Chenery and Sporne, 1976), suggesting that this physiologically crude method of excretion is also in evolutionary terms a first attempt.

It seems likely that the more sophisticated mechanisms have evolved more recently, and particularly the combinations of resistance mechanisms that characterize most specialist toxicity-resistant species.

In many cases resistance is found to be multi-layered—sometimes an exclusion mechanism at the root reducing the stress, and the ions that do penetrate being localized, inactivated, or excreted.

In the case of metal resistance at least, it has been clearly demonstrated by Bradshaw and others that individuals with a measure of resistance exist in normal populations not previously exposed to toxicity (Walley *et al.*, 1974). Table 6.5 shows that plants of *Agrostis tenuis* growing on Zn- or Cu-waste from old mines have high resistance to the appropriate metal, and to that metal only. It is possible, however, by growing commercial seed on contaminated soil to select for a few individuals (1–2%) which have Cu-resistance indices as high as the naturally-occurring populations. Zinc resistance appears to be more complex, since the most resistant individual selected was only about half as resistant as the mine population.

If such effects can be produced in a population in one generation, one would naturally expect to observe them occurring in the wild. Zinc-resistance has been observed under a galvanized fence after 30 years (Bradshaw *et al.*, 1965), and progressive increases in Cu-resistance in lawns established at different times around a copper refinery were found by Wu and Bradshaw (1972; Fig. 6.13).

The dramatic effects of metal toxicity on plants and the consequently massive selection pressures have highlighted the ease with

TABLE 6.5.   Selection for Cu and Zn resistance in *Agrostis tenuis* (Walley *et al.*, 1974). Plants were grown for four months in soil, copper-mine waste, or zinc-mine waste, and the resistance of survivors measured by rooting tillers in 0·5 g litre^{-1} $Ca(NO_3)_2$, containing the appropriate metal. Resistance is measured on a 0–100 scale, with 100 representing insensitivity to the toxin.

Population	Treatment	Mean and (maximum) tolerance of survivors to	
		Zn	Cu
Pasture	Direct measurement	0·6 (1·7)	2·0 (8·6)
Commercial	Grown on soil	N.D.	5·6 (8·5)
	Grown on Cu-waste	N.D.	48·0 (77·5)
	Grown on Zn-waste	31·8 (41·3)	0·3 (0·9)
Copper mine	Direct measurement	0·8 (0·8)	79·0 (87·5)
	Grown on soil	17·6 (20·2)	85·3 (93·5)
	Grown on Zn-waste	36·4 (47·8)	52·1 (66·9)
Zinc mine	Direct measurement	93·0 (93·0)	3·8 (3·8)

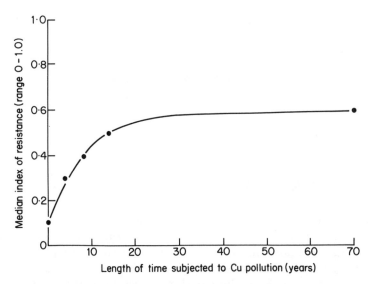

Fig. 6.13. Rate of increase of copper resistance in a population of *Agrostis tenuis* around a copper smelter. (Data from Wu and Bradshaw, 1972.)

which physiological attributes of plants can be altered. The adaptation of plants to many other environmental variables exhibits similar small-scale pattern. *Anthoxanthum odoratum* grows in many of the Park Grass plots at Rothamsted which have received different fertilizer treatments for over 120 years. Plants from closely adjacent plots are clearly distinct in their responses to Ca, Al, P, and Mg (see, for example, Davies and Snaydon, 1973).

Typically such adaptation involves ecotypic differentiation, that is the evolution of distinct genotypes. More rarely resistance is explained by phenotypic plasticity, as appears to be true for the resistance of *Typha latifolia* to zinc, although possibly *Typha* has constitutive Zn-resistance (McNaughton *et al.*, 1974). Nevertheless it seems likely that investigation of plastic responses to mild toxicity would be amply repaid.

7. Gaseous Toxicity

I. ANAEROBIOSIS IN SOILS

As we saw in Chapter 4 (p. 135), freely-draining soils cannot retain water in pores wider than 10–60 μm, and therefore even at field capacity they contain substantial air-filled pore spaces (normally in the range 10–30% of soil volume) which become even more extensive as water is withdrawn from the capillary pores by plant roots. Consequently, the oxygen content of the soil air is usually maintained at 15–20% by gaseous diffusion, giving adequate supplies of oxygen for root growth and metabolism.

However, in the many "wetlands" (marshes, mires, swamps) of the world, soil drainage is impeded by the low permeability of the underlying subsoil or rock, or by high groundwater levels, giving seasonal or permanent waterlogging when the soil pore space is almost entirely occupied by water. Under these conditions, the oxygen in trapped pockets of air is rapidly exhausted by microbial and root respiration, and further supplies of oxygen from the free atmosphere are effectively cut off by the very low rate of diffusion of oxygen through water (10^{-4} times the rate in air). Thus waterlogged soils rapidly become anaerobic and aerobic respiration rates fall to a very low level.

However, in the absence of oxygen, facultative and obligate anaerobic organisms (but not roots) can maintain soil respiration by transferring electrons from the respiratory chain to a range of electron acceptors other than oxygen. Thus, in contrast to the normal aerobic reaction which yields water:

$$O_2 + 4H^+ + 4 \text{ electrons} \rightarrow 2H_2O$$

a series of anaerobic reactions:

$$NO_3^- + 2H^+ + 2 \text{ electrons} \rightarrow H_2O + NO_2^-$$

$$\text{(further reduced to } N_2O \text{ or } N_2)$$

$$MnO_2 + 4H^+ + 2 \text{ electrons} \rightarrow Mn^{2+} + 2H_2O$$

$$Fe(OH)_3 + 3H^+ + 1 \text{ electron} \rightarrow Fe^{2+} + 3H_2O$$

$$SO_4^{2-} + 10H^+ + 8 \text{ electrons} \rightarrow H_2S + 4H_2O$$

$$CO_2 + 8H^+ + 8 \text{ electrons} \rightarrow CH_4 + 2H_2O$$

results in the loss of nitrate from the soil as gaseous N_2O or N_2 (denitrification) and the accumulation of phytotoxic substances such as H_2S and high concentrations of soluble Fe^{2+} and Mn^{2+} ions. Because these reactions proceed in the strict sequence as written above (i.e. sulphate will not be reduced until all the available ferric iron has been reduced to ferrous), the soil redox potential (E_h which is an index of the degree of reduction, or oxygen depletion of the soil; White, 1979) is also a useful guide to the chemical reactions which are taking place. For example, the reduction of ferric iron will be complete when E_h falls below $+120$ mV at pH 7. The anaerobic decomposition of soil organic matter also releases a variety of organic toxins (especially acetic acid and higher carboxylic acids) and a series of hydrocarbons (Drew and Lynch, 1980), but increases in the concentration of the gaseous plant growth regulator, ethylene, are normally a feature of the early stages of anaerobiosis. On the other hand, anaerobic conditions generally lead to increased nitrogen fixation by cyanobacteria in the surface layers and the accumulation of ammonium ions, to increased availability of soil phosphorus, and to increased concentrations of potassium ions in the soil solution (Stolzy et al., 1981; Ponnamperuma, 1984).

A. The Effects of Anoxia on Plant Roots

In plant cells receiving an adequate supply of oxygen, the oxidation of carbohydrate to yield energy for growth and metabolism normally takes place in three stages. In the first stage (the Embden–Meyerhof–Parnas Pathway or Glycolysis; Fig. 7.1(a)), the conversion of 1 mol of glucose to 2 mol of pyruvate is associated with the net synthesis of 2 mol of the high energy intermediate ATP (which is the main form in which energy is transported and utilized in plants) and 2 mol of $NADH_2$. In the second stage, the complete oxidation of each mol of pyruvate to carbon dioxide by the Krebs or Tricarboxylic Acid Cycle (Fig. 7.1(b)) is accompanied by the synthesis of 1 mol of ATP and a further 5 mol of $NADH_2$ (or reduced flavoprotein). Finally, the energy stored in each mol of $NADH_2$ is used for the synthesis of 3 mol ATP (2 for FPH_2), by

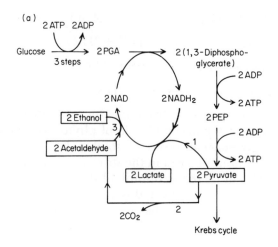

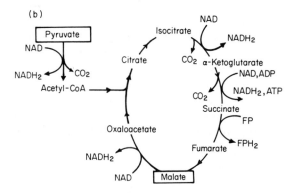

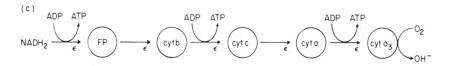

Fig. 7.1. The oxidation of glucose by plant cells. (a) The Embden-Meyerhof-Parnas Pathway. Under anaerobic conditions, reaction 1 is catalysed by lactic dehydrogenase, reaction 2 by pyruvic decarboxylase and reaction 3 by alcohol dehydrogenase. (b) The Krebs Cycle, and (c) The Respiratory (Cytochrome) Chain, showing the transport of electrons (ε) from $NADH_2$ to molecular oxygen. Boxed compounds are discussed at length in the text. (Where ATP and ADP are adenosine tri- and diphosphate respectively; NAD and $NADH_2$ are the oxidized and reduced forms of nicotinamide adenine dinucleotide; PGA is phosphoglyceraldehyde; PEP, phosphoenol pyruvate; Acetyl-CoA, acetyl coenzyme A; FP, flavoprotein; and cyt, cytochrome.)

means of the mitochondrial respiratory chain (Fig. 7.1(c)); in this process, electrons and protons are transferred, via the cytochrome chain, from $NADH_2$ to oxygen, giving water. Overall, the complete aerobic respiration of 1 mol of glucose yields 38 mol of ATP.

When a soil in which plants are growing becomes waterlogged, the cytochrome chain in the root cells ceases to function in the absence of molecular oxygen, and this in turn leads to the accumulation of $NADH_2$ and the suppression of the Krebs cycle. The resulting build-up of acetaldehyde, the first end-product of anaerobic glycolysis (fermentation, Fig. 7.1), induces the synthesis of the enzyme alcohol dehydrogenase (ADH) which catalyses the transformation of acetaldehyde to ethanol. Because this step uses the $NADH_2$ generated by fermentation, the reactions involved in fermentation can continue to generate ATP and pyruvate under anaerobic conditions as long as carbohydrate reserves last.

However, dependence on this pathway for energy supplies has serious consequences. In particular, the yield of useful energy, at 2 mol of ATP per mol of glucose, is much less than the 38 mol released by aerobic respiration. Consequently, when anaerobic conditions are imposed, the rate of fermentation must increase sharply if the cell is to maintain ATP supplies near to the aerobic level. This "Pasteur effect" can lead to the rapid and inefficient exhaustion of plant reserves and, eventually, to the death of both root and shoot, if anoxia is prolonged. The success of a plant in maintaining the energy supplies of its cells can be assessed by the adenylate energy charge (AEC, but commonly referred to simply as the energy charge):

$$\frac{[ATP] + 0.5[ADP]}{[ATP] + [ADP] + [AMP]}$$

i.e. the proportion of the total adenine nucleotides involved in energy metabolism which are at the higher energy states. Thus if all the nucleotides are in the form of ATP, the energy charge will be 1, but unstressed plant tissues normally give values of 0.8 to 0.95. With the imposition of anaerobic conditions, the energy charge will tend to fall to a low level (≥ 0.2), although the decline can be delayed by rapid ATP cycling under accelerated fermentation, or by reduction in the total quantity of nucleotides (Tripepi and Mitchell, 1984). Even before the exhaustion of reserves, such low levels of energy charge will result in the disruption of the organization and function of membranes and organelles (Jackson and Drew, 1984).

Such a wasteful use of energy resources would in itself be a very serious limitation to the survival of plants growing in flooded soils, but it is accompanied by a further hazard—the accumulation of potentially toxic end-products of fermentation (generally ethanol, but also lactic acid, particularly in germinating seeds, Fig. 7.1). Much of the evidence for the toxicity of ethanol, principally by the modification of cell and organelle membrane properties, came from earlier studies of bacteria and single cells, whereas the consensus now appears to be that ethanol will only rarely reach sufficiently high concentrations (of the order of 1 M) to cause damage even in the most flood-sensitive species such as pea (Jackson *et al.*, 1982). However, because this consensus is based on the results of experiments in which ethanol was applied exogenously, rather than generated endogenously, the toxicity of ethanol as an end-product of fermentation remains uncertain. This controversy is central to the study of plant adaptation to flooded soils because several of the hypotheses involved are based upon the assumption that ethanol is highly toxic and must be removed rapidly from the fermenting tissue, or transformed to a less harmful metabolite (see below). In addition to these effects, the imposition of anaerobic soil conditions leads to changes in the relative amounts of at least four growth substances, in shoots as well as below ground (ethylene, generated by both root and soil, see below; cytokinins and gibberellins, whose synthesis in root apices is disrupted or suppressed; abscisic acid; Reid and Bradford, 1984). Thus many of the classic symptoms of hypoxia (inhibition of internode extension, petiole epinasty, stomatal closure, wilting, premature senescence, adventitious roots, aerenchyma) can be induced or alleviated by the application of growth substances. Finally, there is evidence from arable soils that seedling mortality is high in the presence of acetic acid produced during the anaerobic microbial breakdown of organic matter (e.g. wheat straw; Lynch and Elliot, 1983).

In summary, it is clear that wetlands are very hostile environments for higher plants. Their roots are exposed not only to very low oxygen and very high carbon dioxide levels, but also to a wide range of inorganic and organic toxins, and to varying levels of salinity in coastal areas. Furthermore, the duration of flooding can vary considerably between locations and seasons. However, in spite of this formidable array of unfavourable factors, wetlands are amongst the most productive ecosystems, and rice, grown principally in flooded soils, is the second most important food crop in the world. Part of this success can be attributed to the absence of water and temperature stresses, and to the high nutrient status of many wetlands; however, as shown in the following sections, wetland plants could not benefit from these favourable aspects without specific adaptations which enable them to overcome or avoid the physical and chemical hazards of anaerobic soils.

B. Plant Adaptations Favouring Survival and Growth in Waterlogged Soils

1. Oxygen Transport

Of the many physical and chemical characteristics of waterlogged soils which can limit plant growth, lack of oxygen is the primary, but not necessarily the most important, problem. This is because, in most wetland species, the diffusion of oxygen from the free atmosphere through the extracellular spaces of their shoots and roots to the flooded organs can be sufficiently rapid to reduce or remove the risk of tissue anaerobiosis (e.g. in British bog plants—Armstrong, 1964; in *Spartina alterniflora*—Teal and Kanwisher, 1966; in the N. American hardwood tree, *Nyssa sylvatica*—Hook et al., 1971; Keeley and Franz, 1979). These rapid rates of diffusion are made possible by the presence of large, continuous air spaces in the root cortex (aerenchyma) which can be a permanent feature of the roots (e.g. in rice, John, 1977) or induced in developing roots by flooding or the application of ethylene (e.g. in maize, Fig. 7.2); this movement of oxygen through stomata, stem lenticels and root aerenchyma can be sufficient to oxidize the rhizosphere soil round the flooded root (Philipson and Coutts, 1978; Hook and Scholtens, 1978), as is clearly shown by the deposits of red ferric hydroxide on the root systems and rhizomes of many wetland plants (e.g. Armstrong, 1967). In addition, the development of aerenchyma may improve root aeration by reducing the volume of respiring tissue (i.e. root demand for oxygen). However, there is not necessarily a simple relationship between the success of a species in colonizing waterlogged soils and the degree of development of aerenchyma in its roots; for example, Smirnoff and Crawford (1983) found that although flood-tolerant species generally had higher root porosities (20–50% of root volume occupied by air spaces) than species from well-drained sites (< 10%), there were examples of common wetland species with very low porosities (e.g. *Myosotis scorpioides*, 8%). Clearly, although the occurrence of root aerenchyma is widespread amongst flood-tolerant plants, it is not universal.

Other morphological features favouring oxygen supply to roots include superficial mats of adventitious roots (as in rice; Alberda, 1954) which, in trees, can be associated with more vertical "sinker roots" (e.g. in *Pinus contorta*; Boggie, 1972), and the pneumatophores of mangroves and swamp cypresses (*Taxodium* spp.) whose lenticels above the water level give oxygen direct access to the aerenchyma of submerged portions (e.g. Scholander et al., 1955). In general, trees growing on waterlogged soils are shallow-rooted and prone to "wind-throw" (Armstrong et al., 1976).

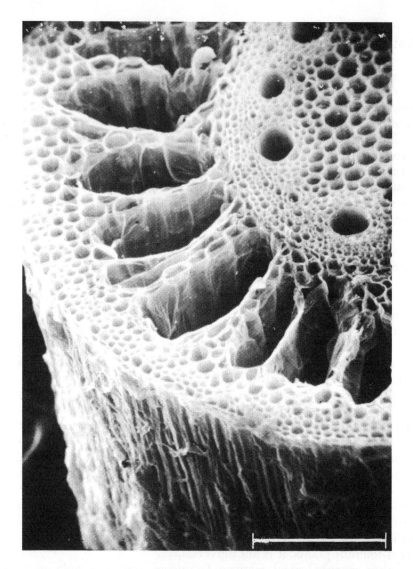

FIG. 7.2. Cortical air spaces induced by oxygen deficiency in a nodal root of maize. The parent plant was grown in aerated culture solution before being transferred to unaerated culture solution, which rapidly became oxygen deficient. The original root system died but, within 9 days, was replaced by new nodal roots with well-developed aerenchyma (cortical tissue containing large air spaces) formed by the collapse of cortical cells. The development of aerenchyma under these conditions can be inhibited by the presence of Ag^+ ions (which suppress the activity of ethylene in plant tissues). (Scanning electron micrograph of a transverse section 10 cm from the root tip, bar = 200 µm; Drew, 1979; Drew et al., 1981.)

Much of the quantitative information on oxygen transport from shoot to root and for the leakage of oxygen from the apical portions of roots has come from laboratory experiments on seedlings and young plants. Consequently, when Greenwood (1967) demonstrated that oxygen transport is a feature of the seedlings of several cultivated species which were considered to be flood-sensitive, Crawford (1972) proposed that this ability might be a characteristic of seedlings in general rather than of flood-tolerant species in particular. Whether this is generally true of seedlings or not, it has been clearly shown in the field that older, more established, bog plants can oxidize the rhizo-sphere (e.g. Armstrong and Boatman, 1967) and, therefore, if we are to understand the significance of this process for plant survival and distribution, it is essential to have quantitative information about the capacity of root aerenchyma to supply oxygen to roots growing in anaerobic soils, i.e. how quickly and how far can oxygen move to supply root respiration?

Because of the considerable technical difficulties involved (Jackson and Drew, 1984), experimental data on this subject are scarce. Armstrong and Gaynard (1976) showed that for rice and *Eriophorum angustifolium* plants whose roots were immersed in anoxic media, maximum rates of whole root respiration (aerobic) could be maintained at oxygen partial pressures in the cortical air spaces down to 2–4 kPa (2–4% O_2). In other words, the critical oxygen pressure for aerobic respiration in the roots of these species is 2–4 kPa. More recent work (Armstrong and Webb, 1985) has suggested even lower critical oxygen pressures (0·8 kPa) for rice root extension. Since aerenchyma oxygen pressures of this magnitude, or greater, have commonly been found in the roots of wetland plants, it would appear that aerobic respiration and root growth in rice will not normally be restricted by anaerobic soil conditions. However, it is not possible to discover from this, and other investigations, how *far* oxygen can move through the root aerenchyma to maintain aerobic functions, but it seems likely that the maximum distance will be of the order of 20–30 cm in the roots of herbaceous plants, and up to 2·5 m in large trees, the actual distance depending upon root diameter, root porosity, the sites and rate of leakage of oxygen to the rhizosphere and the respiratory demand for oxygen (as illustrated by Fig. 7.3). Some wetland species appear to be particularly well-adapted, by the development of an impermeable suberized hypodermis, to restrict the leakage of oxygen to the apical regions of the root system (e.g. in *Carex arenaria*, Robards *et al.*, 1979). In other cases, the problems involved in moving oxygen down long, fine roots have been avoided by having large underground rhizomes

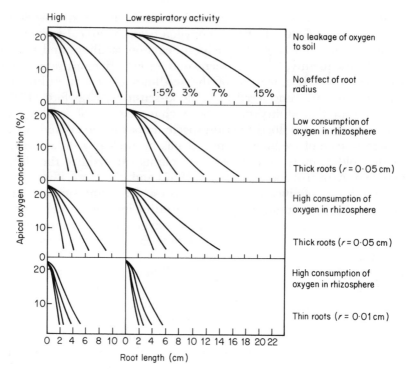

FIG. 7.3. The effects of plant and soil factors on the internal oxygen concentration at the apex of a root growing in soil, evaluated using an electrical analogue system. The assumptions used include:

(a) uniformity of respiratory activity in roots; high $= 120$ ng cm^{-3} s^{-1}, low $= 30$ ng cm^{-3} s^{-1};
(b) varying levels of root porosity (air spaces occupying 1·5%, 3%, 7% and 15% of root volume, from left to right in each figure);
(c) uniform oxygen consumption in aerated rhizosphere; high $= 5·27 \times 10^{-8}$ g cm^{-3} s^{-1}, low $= 5·27 \times 10^{-9}$ g cm^{-3} s^{-1};
(d) root wall (epidermis) permeability to oxygen declining from 100% at the apex to a minimum of 60% at 6 cm and beyond;
(e) cessation of root growth at an internal oxygen concentration of 2%.

The distance to which oxygen can be transported to maintain growth declines (i) with increase in respiratory activity, (ii) with increase in oxygen consumption in the rhizosphere, (iii) with reduction in shoot porosity and (iv) with reduction in the radius of the root (redrawn from Armstrong, 1979).

which supply a network of shorter fine roots. For example, an aerenchyma oxygen content of 8% was maintained at the distal end of a 75 cm rhizome of *Menyanthes trifoliata* immersed in an anoxic medium (Coult and Vallance, 1958).

In summary, it is now clear that the internal ventilation of roots by means of aerenchyma is a major factor, and possibly the most important single factor in the adaptation of plants to waterlogged soils. However, even in species whose oxygen transport system appears to be effective in maintaining a high energy charge ($\geqslant 0.8$) over prolonged periods, there is evidence of some ethanol excretion from their flooded parts, indicating that some anaerobic fermentation is taking place (e.g. in the bulrush, *Schoenoplectus* (*Scirpus*); *lacustris*; Duss and Brändle, 1982; Steinmann and Brändle, 1981; and in rice, as discussed below). Furthermore, deep-rooting species cannot depend solely upon oxygen movement through the roots to enable them to avoid the effects of anoxia, and internal ventilation will not be particularly effective in plants growing in normally well-aerated soils (and therefore lacking well-developed aerenchyma) which are subjected to seasonal or sporadic periods of waterlogging (for example on lake margins and river banks). Plants which are successful in these environments and wetland species without well-developed aerenchyma (e.g. *M. scorpioides*, see above) must therefore possess metabolic adaptations which overcome the problems of relying upon anaerobic fermentation.

2. Control of the Rate of Fermentation

According to Crawford (1982), there are three possible metabolic responses to the development of tissue anaerobiosis, whether partial or total:

(a) acceleration of glycolysis/fermentation (Pasteur effect), leading to the depletion of carbohydrate reserves, the accumulation of potentially toxic metabolites, and rapid death (flood-sensitive species).

(b) acceleration of glycolysis/fermentation associated with the efficient removal of ethanol by leakage, excretion into the root medium or in the gas phase via the aerenchyma/lenticel/stomata diffusion pathway. Because reliance on this response alone would result in serious depletion of reserves, it is likely to be restricted to well-ventilated species which normally experience only partial or short-term anaerobiosis.

(c) no acceleration of glycolysis/fermentation, with the result that reserves are protected, there is no risk of ethanol toxicity, but under long-term anaerobiosis or during periods of high demand for ATP, the energy charge will fall sharply, with serious consequences for cell integrity.

In spite of the obvious problems involved, earlier work by Crawford and his co-workers indicated that response (c) might be the more important for successful wetland species. The lack of Pasteur effect was attributed to:

(i) non-induction and, in some cases, inhibition of alcohol dehydrogenase (ADH) (Fig. 7.1) by anaerobiosis (e.g. in *Iris pseudacorus*, *Phalaris arundinacea*, Crawford and McManmon, 1968; Fig. 7.4)

(ii) the higher K_m values (and, therefore, lower affinities for acetaldehyde) of the ADH isoenzymes of flood-tolerant than of flooding-sensitive species (McManmon and Crawford, 1971).

Consequently acetaldehyde and pyruvate will tend to build up, eventually causing the rate of fermentation to slow (Fig. 7.1). These ideas were given support by the work of Marshall *et al.* (1973) and Brown *et al.* (1976), who demonstrated that the level of flood tolerance in cultivars of maize and in populations of *Bromus mollis* was related to the properties of the ADH isoenzyme(s) present in the roots of these generally sensitive species. However, there are a number of serious difficulties associated with the use of ADH activity in the roots as an index of flooding resistance. For example, at least one flooding-sensitive species, maize, has shown a peak of ADH activity at intermediate oxygen levels (8–13%, Wignarajah and Greenway, 1976) with lower levels at 20% and under anoxic conditions, and there are indications that ADH levels can respond to a number of other factors including rooting density. The results of these and other experiments using rice, barley and maize, led John and Greenway (1976) to propose that pyruvate decarboxylase activity (catalysing the transformation of pyruvate to acetaldehyde) might give a better indication of flood tolerance than ADH.

In contrast, the claim of accelerated fermentation with removal of ethanol (response (b)) to be the more important metabolic response of wetland plants has been steadily enhanced by the accumulating evidence of the extent and effectiveness of root ventilation, and the efficiency of the processes transporting ethanol away from the site of fermentation (e.g. in rice; Chirkova, 1978). Indeed, it has been suggested that an *increase* in ADH activity in response to flooding could be positively useful as a means of regenerating acetaldehyde and pyruvate during recovery from a period of waterlogging or in aerobic tissues of a flooded plant (Jackson and Drew, 1984). However, it

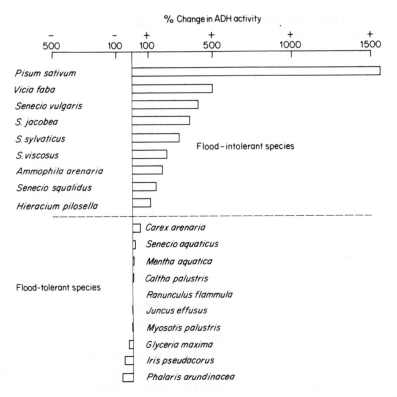

FIG. 7.4.   Changes in alcohol dehydrogenase activity (ADH, expressed on a soluble protein basis) in the roots of 19 species after flooding in sand culture for one month, as compared with unflooded controls (redrawn from Crawford, 1978).

should be stressed that neither response (b) or (c) is appropriate for those plants (possibly a minority amongst wetland species) whose active metabolism and growth appear to continue for prolonged periods under unventilated, anoxic conditions (Barclay and Crawford, 1982). Perennial species, of course, can survive months of seasonal waterlogging in a state of dormancy (Kozlowski, 1984), but it is clear that many questions about energy metabolism under anaerobiosis remain unanswered (Lambers, 1979)

3. Alternative End-products of Fermentation

It has been known for some time that the concentrations of a number of metabolic intermediates (carboxylic acids, alcohols, amino acids) increase in the tissues of tolerant species exposed to anoxia. This led to the suggestion that some wetland species might be able to divert fermentation to produce end-products which are less harmful than ethanol and which could, therefore, be retained in the tissue as biosynthetic intermediates, translocated to the shoot to take part in aerobic metabolism, or stored in the root until the re-establishment of oxygen supplies. Such a mechanism, which would overcome the considerable losses of energy associated with leakage or excretion of ethanol, would be analogous to the metabolic adaptations of diving mammals (Crawford, 1972, 1982).

Investigations of this hypothesis have tended to concentrate upon malate as an alternative end-product to ethanol (or lactic acid, see below) on theoretical grounds (Mazelis and Vennesland, 1957), and because it can be produced in large quantities under certain circumstances (e.g. in the xylem sap of flooded birch trees; Crawford, 1972). For example, as a result of a series of experiments comparing a range of flooding-tolerant herbaceous species (e.g. *Glyceria maxima, Mentha aquatica, Myosotis scorpioides, Senecio aquaticus*) with species from well-drained sites, McManmon and Crawford (1971) proposed that malate accumulates in the flooded tissues of the former group because of the absence, or inhibition by flooding, of malic enzyme (catalysing the conversion of malate to pyruvate which can, in turn, react to give acetaldehyde and ethanol). However, it has proved difficult to assess the importance of these findings because of conflicting evidence from other workers (Davies, Nascimento and Patil 1974; Smith and ap Rees, 1979). Other less common metabolites which can also accumulate under anaerobiosis include glycerol, shikimic acid, alanine, aspartic acid, glutamic acid, proline, serine and γ-aminobutyric acid (Crawford, 1982). Overall, there is increasing evidence that the production of these metabolites is a symptom of, rather than an adaptation to, flooding stress (ap Rees and Wilson, 1984; Smith *et al.*, 1984), particularly in view of the fact that several of the biochemical pathways involved generate less ATP than does ethanol production.

4. Seed Germination

As in other hostile environments, the greatest risk to the survival of a species, however well-adapted, probably occurs during germination and seedling growth, and this may be reflected in the fact that many herbaceous wetland species overwinter and spread by means of rhizomes containing substantial reserves; this is also a common feature in cold and dry environments (see pp. 158, 209). In certain permanently-flooded ecosystems in N. America, the conditions required for seed germination are broadly related to the depth and pattern of flooding (Hook, 1984). Where the depth of inundation is normally deep, the seeds of several wetland trees can survive prolonged submergence without germinating or deteriorating, but germinate freely when deposited on exposed soils during periods of drought (possibly in response to ethylene evolved by the drying soil; Etherington, 1983). The probability that seedlings establishing in this way will develop into mature trees is very small, but it is improved by rapid rates of stem elongation during the first few years of growth. Where flood waters are generally shallower, the seeds of a range of other wetland trees can germinate under water in the absence of oxygen and establish successfully if the shoot can extend into the air before the seed reserves are exhausted. In the case of rice, the seeds will germinate under water but root growth will not begin if the depth of water is too great for the coleoptile to reach the free atmosphere (Hook, 1984). However, it should be stressed that the seeds of many species show germination responses to anoxia which have no adaptive significance in their natural environment.

Under natural conditions in many environments, the energy required for the initial stages of seed germination is supplied by fermentation (Mayer and Poljakoff-Mayber, 1975). This is particularly true of those species whose seed coat has a low permeability to gases, and in wet soils where oxygen diffusion is restricted (e.g. Sherwin and Simon, 1969). Once the radicle has ruptured the seed coat and an adequate supply of gaseous oxygen has been established, the products of fermentation (lactate initially but ethanol later; Davies, Grego and Kenworthy, 1974), will be oxidized by aerobic respiration. However, the seeds of those wetland species for which germination must be completed under anaerobic conditions must rely upon fermentation and therefore face two serious problems.

(i) *Energy conservation:* the carbohydrate or lipid reserves in seeds are finite and cannot be replenished before germination has been completed and photosynthesis has been established. Consequently, since fermentation releases only 5% of the potentially available energy of carbohydrates, its rate must be strictly controlled to avoid wasteful consumption of these resources. Thus flooding-tolerant seeds (rice, lettuce) showed only a small Pasteur effect after 72 h of soaking, possibly owing to a lack of ADH induction (Crawford, 1977). Of course, this will lead to low rates of germination under anaerobic conditions, and reliance upon fermentation may result in natural selection for larger seeds, although only in species whose germination is restricted to anaerobic soils.

(ii) *Disposal of potentially-toxic metabolites:* unlike whole plants, seeds
have very limited vacuolar volumes in which to store the
products of fermentation, nor are they able to transport them to
the shoot. In spite of this, the seeds of a substantial number of
species can germinate very successfully under waterlogged conditions (e.g. Lazenby, 1955; Mayer and Poljakoff-Mayber,
1975; and see above) and it is still not clear how they dispose of
lactic acid (which can disturb the pH buffering of cells) and
ethanol. It may be that, as for whole plants, disposal is primarily
by leakage of ethanol, although Crawford (1977) suggests that
the more tolerant species tend to produce lactic acid rather than
ethanol for as long as possible.

5. The Importance of Metabolic Adaptation

Overall, the significance of metabolic adaptations for flooding-toler-
ance in plants and seeds is the subject of some controversy. Although
there can be no doubt that populations and species do vary in their
ability to control the rate and nature of the products of fermentation, it
is difficult to assess how widespread a given adaptation may be.
Furthermore, since most of the published work on flooding tolerance
has involved relatively short periods of hypoxia or anoxia (typically
from a few days up to a month), the findings and conclusions may not
apply to permanent wetlands. For example, Keeley (1978) has shown
that in roots of the hardwood tree *Nyssa sylvatica* the fermentation rate
rose during the first month of flooding, but thereafter it fell, over a
year, to a value below that of (year-old) control seedlings growing in
well-drained soil. These changes in fermentation rate did not correlate
with root malate content, making it difficult to reconcile this flooding
response with the ideas of Crawford outlined above. Instead, Keeley
(1978) and Keeley and Franz (1979) propose that the acceleration of
fermentation in the early weeks of flooding serves as an "emergency"
measure to support the metabolism of the existing root system, while a
fresh system of roots, rich in aerenchyma and able to support a
predominantly aerobic respiration, is developed.

6. Other Toxins

It is important to stress that metabolic adaptations leading to the
tolerance of prolonged fermentation will be of little use to wetland
plants which are not able to deal with the high concentrations of toxins
(Fe^{2+}, H_2S, etc.) generated by waterlogged soils. Thus it has been
found that in the British Isles, the plant species occupying the wettest
sites in woodlands (Martin, 1968), dune and slack areas (Jones, 1972)

and moorlands (Jones and Etherington, 1970) are the most efficient at excluding Fe^{2+} ions from the shoot, although there can also be pronounced differences in sensitivity to high ferrous iron concentrations among true wetland plants (Wheeler *et al.*, 1985). The mechanism of this exclusion has not been formally established, but the high iron content of resistant roots and the clearly-visible deposits of ferric hydroxide on the root surfaces of wetland plants (e.g. Armstrong and Boatman, 1967) indicate that soluble ferrous (Fe^{2+}) ions are oxidized to insoluble ferric (Fe^{3+}) ions in the rhizosphere and in the intercellular spaces of the cortex. The relative ability to exclude iron from the shoot will therefore depend upon the degree of oxygen movement from shoot to root (Green and Etherington, 1977). For the species, including some wetland plants, which do accumulate substantial quantities of ferrous iron within the root tissues, it appears that survival is determined by the activity of enzyme systems which can prevent cell damage by free radicals and hydrogen peroxide (Hendry and Brocklebank, 1985). In the case of *Erica tetralix*, there is the additional suggestion (Jones, 1971) that its xeromorphic growth habit, apparently ill-adapted to its wet moorland habitat, gives reduced transpiration rates which, in turn, reduce the mass flow of soluble Fe^{2+} ions to the root surface and their accumulation in the rhizosphere (see p. 234).

Plant responses to the two other important inorganic toxins in anaerobic soils, Mn^{2+} and H_2S, are not clearly defined. The total manganese contents of most soils are much lower than those of iron, and plant adaptations which exclude Fe^{2+} ions will also prevent uptake of Mn^{2+} ions. Sulphate ions are reduced to H_2S only under very reducing conditions when all of the labile iron in a soil has been converted to Fe^{2+}, and even then, according to Ponnamperuma (1972), the sulphide concentration in the soil solution should be very low owing to precipitation as FeS. Nevertheless, Armstrong and Boatman (1967) measured H_2S concentrations as high as 8 mg l^{-1} round the roots of *Menyanthes trifoliata* growing in a Yorkshire bog, whereas *Molinia caerulea* was absent wherever H_2S was detectable in the soil. This difference may reflect differences in the ability to oxidize H_2S in the rhizosphere. Havill *et al.* (1985) were also able to explain the distribution of *Salicornia europaea* in British salt-marshes in terms of its tolerance to sulphide in the root medium (Table 7.1), but most work on plant response to H_2S has been directed towards the solution of problems of rice cultivation. For example, there is the case of the relationship between the bacterium *Beggiatoa* and rice roots in Louisiana paddy fields. The bacteria are able to oxidize H_2S to sulphur intracellularly, thus removing the toxin from the rhizosphere, but the reaction also produces hydrogen peroxide, which is potentially harmful to both micro-organism and roots. However, the rice roots release the enzyme catalase which catalyses the breakdown of peroxide extracellularly. This is an excellent example of symbiosis, since neither partner could survive without the other (Pitts, 1969 and Hollis, 1967—cited by Armstrong, 1975).

Among other unfavourable characteristics of waterlogged soils, high carbon dioxide concentrations are normally considered to be of minor importance compared with low oxygen levels (Armstrong, 1975). Although the growth regulator ethylene does occur in anaerobic soils at physiologically active levels, especially during the early stages of waterlogging, its occurrence is probably more characteristic of imperfectly-drained soils and, therefore, its effects on plants are discussed in the subsequent section.

In conclusion, it seems clear that plants cannot colonize wetlands unless they possess a *range* of anatomical, biochemical and physiological adaptations leading to the *avoidance* or *amelioration* of the many unfavourable and toxic features of anaerobic soils. In many cases, this range will include:

(i)   anatomical and morphological features which improve oxygen transport to respiring root tissues; *and*
(ii) biochemical features permitting prolonged anaerobic fermentation in roots; *and*
(iii) mechanisms for the exclusion of phytotoxic substances in the rhizosphere.

Furthermore, if (iii) is not fully effective, it will be essential for such plants to be *tolerant* of toxins within their tissues, for example by compartmentation in vacuoles (see p. 249).

C. Imperfectly-drained Soils and Soils with Anaerobic Microsites

So far in this chapter, we have considered only the extreme cases of soil aeration. On the one hand, we have seen that in freely draining soils with wide air-filled pore spaces, the oxygen content of the soil air in the rooting zone can be maintained at a high level (e.g. 15–20%) by gaseous diffusion; on the other hand, waterlogged soils rapidly become anoxic. However, the aeration status of many undisturbed and agricultural soils is intermediate between these extremes. For example, throughout western parts of the British Isles, where soil drainage tends to be imperfect as a result of high precipitation coupled with fine soil textures, temporary anaerobiosis is common, especially during the winter.

The overall influence of poor soil drainage on vegetation depends upon the extent, duration and seasonal distribution of anaerobic soil conditions, but at any given time, plant growth, function and survival will be controlled by the same factors as in permanently waterlogged soils, i.e. oxygen supply, levels of organic and inorganic toxins, disturbance of the balance of growth regulators. There exists a wealth of information on the preference of certain species for poorly-drained, rather than either drained soils or wetlands (e.g. Harper and Sagar, 1953), but there is a serious lack of information on the physiological and biochemical adaptations involved. For example, it would be useful to have values of critical oxygen pressure for respiration in the roots of a range of species adapted to sites of different drainage status; however, few measurements have been made and even these, according to Armstrong and Gaynard (1976), are generally unreliable because excised roots rather than roots attached to whole plants were used. In concentrating upon the oxygen status of soils, we have seriously underestimated the extent of anaerobiosis in better-drained soils. For example, in moderate to freely draining agricultural soils, ethylene has been detected at levels varying from 0·1 to 10 p.p.m. (vol/vol) in soil air samples containing 5–20% oxygen (Smith and Dowdell, 1974). In these soils a product of anaerobic soil respiration (see below) is accumulating in an apparently well-aerated environment. This apparent paradox can be resolved if soil structure is taken into account. In medium to fine textured soils at field capacity, most of the air-filled macropores (> 60 μm diameter) are *between* soil aggregates whereas the narrower, water filled, pores are within aggregates (see p. 135). Thus oxygen can move rapidly through the macropore space, but since its diffusion to the centre of a water-saturated aggregate will be very much slower, the oxygen content within an aggregate in which there is respiratory demand for oxygen will fall rapidly with distance from its surface. Using simple diffusion theory, it is possible to predict that the smallest spherical aggregate to have an anaerobic volume at its centre will have a (critical) radius given by:

$$r_c = \left(\frac{6D_a . C . S}{Q} \right)^{\frac{1}{2}} \qquad \text{(Smith, 1977)}$$

where D_a is the diffusion coefficient of oxygen within the aggregate; C is the concentration of oxygen just outside the aggregate; S is the solubility of oxygen in water; and Q is the rate of uptake of oxygen by soil respiration within the aggregate.

Using data from a number of sources, Russell (1973) estimated that this critical radius (r_c) for aggregates of an agricultural soil at Rothamsted will be approximately 10 mm in summer and 20 mm in winter; the conclusion that r_c is higher in winter underlines the fact that anaerobic conditions at the centre of an aggregate are the consequence of demand for oxygen by soil respiration, which is reduced by lower temperatures in winter. It seems likely that these values of r_c will prove to be overestimates when more precise determinations of D_a, C and Q become available.

Clearly, r_c will vary considerably with soil conditions, and the values quoted do no more than indicate the magnitude of aggregates in which anaerobic volumes can occur. However, this diffusion analysis is very valuable in providing a visual model of soil aeration, i.e. a mosaic of anaerobic microsites set in an aerobic matrix (Fig. 7.5). This alternation of aerobic and anaerobic soil appears to provide an optimum environment for the production of ethylene by (aerobic) saprophytic fungi. For example, studies using pure cultures of *Mucor hiemalis* (reviewed by Lynch, 1975a) have indicated that:

(i) evolution of ethylene (in aerobic and substrate-deficient anaerobic soils) is enhanced many-fold by the addition of glucose and methionine, the latter being considered to be a precursor in the biosynthesis of ethylene. These substrates appear to be supplied naturally in the soil during periods of anaerobiosis.

(ii) evolution of ethylene is most effective when fungal growth is restricted by unfavourable soil conditions. In particular, ethylene production is maximal at a soil atmosphere oxygen content of 1%.

(iii) evolution of ethylene can be stimulated by ferrous ions which are generated under anaerobic conditions.

However, subsequent work has shown that the biosynthesis of ethylene by soil organisms under field conditions may be more complex than these observations suggest (Drew and Lynch, 1980). In general, the production of ethylene by anaerobic soils is dependent upon temperature, organic matter content, pH and the wetting/drying history of the soil, but it is virtually independent of the nitrate status (Goodlass and Smith, 1978).

In conclusion, when a soil atmosphere sampling probe (e.g. Dowdell *et al.*, 1972) is inserted into a soil, it collects gas predominantly from the aerobic macropore space, enriched with ethylene (and, in some cases,

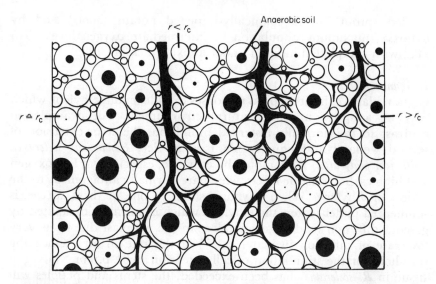

FIG. 7.5. Schematic profile of a well-drained soil whose macropore space between aggregates is aerobic but whose water-saturated aggregates (drawn as circles of varying radius, as if they are all exactly cut in half) contain anaerobic microsites if their radius is greater than the critical value, r_c (see text). Note that the roots are restricted to aerobic zones.

the gaseous products of denitrification, N_2, N_2O) diffusing out of anaerobic microsites and from the poorly-aerated volumes surrounding these microsites. Since roots are, by virtue of their dimensions, largely confined to the macropores, soil atmosphere compositions obtained in this way are reasonable estimates of the gaseous environment of the root system.

D. Plant Responses to Exogenous Ethylene in the Soil Atmosphere

Ethylene is produced normally by most plant tissues and it appears to be involved, in association with other growth substances, in the regulation of a wide range of developmental processes, including renewal of growth after dormancy (seeds, bulbs, etc.), root growth, stem extension, leaf expansion, flowering, fruit ripening and tissue senescence (Abeles, 1973; Roberts and Tucker, 1985). Endogenous ethylene production by plants can also be stimulated by damage or disease (for example, giving rise to morphological distortions such as "coiled sprout" in mechanically-impeded potato shoots) and by hypoxia (maximum stimulation at intermediate oxygen levels, 3–5 kPa, with complete inhibition under anoxia, Jackson, 1985).

1. Aquatic plants

Certain plants exhibit atypical responses to endogenous ethylene which may be adaptive in their natural environment. For example, in contrast to the normally observed inhibition, the stem extension of several wetland and aquatic plants (e.g. rice, *Ranunculus sceleratus*, *Callitriche platycarpa*) is promoted by ethylene, especially at low oxygen and high carbon dioxide levels, which are normally antagonistic to the action of ethylene. When a young shoot of one of these species is submerged under water, the ethylene which is normally dissipated by gaseous diffusion tends to accumulate in the apoplast due to its very low rate of diffusion in water (10^{-4} of the rate in air). Once the threshold for the stimulation of cell extension (0·01 p.p.m. vol/vol of tissue in *R. sceleratus*) has been exceeded, the stems and petioles will extend until the leaf laminae have been returned to the aerial environment, where normal rates of photosynthesis and dissipation of ethylene can be resumed. Exogenous ethylene from the underlying soil may also contribute to this effect (Ku *et al.*, 1970; Musgrave *et al.*, 1972; Musgrave and Walters, 1973). Detailed investigation of the mechanism of this phenomenon has shown that the extension response can involve increases in cell number and length, or length alone, depending upon the age of the leaf at the time of submergence, and that the cell extension effects are primarily the consequence of increased cell wall extensibility, although in some cases (e.g. *Nymphoides peltata*) turgor pressure can be supplemented by the tension exerted by the buoyant petiole and lamina (see p. 130) (Osborne, 1984; Ridge and Amarasingh, 1984).

2. Root Growth and Development

As a result of the great heterogeneity of the soil environment and the practical difficulties involved in studying roots *in situ*, the effects of exogenous ethylene in the soil atmosphere upon root growth and development in undisturbed soils are not clear. (Note that the development of root aerenchyma described earlier is a response to the accumulation of *endogenous* ethylene in the root apoplast under flooded conditions; Fig. 7.2.) In fact, much of the available information comes from experiments in which young root axes, growing down filter papers moistened with culture solution, are exposed to different gas mixtures (e.g. Smith and Robertson, 1971). These experiments, on a variety of crop species, have shown that levels of ethylene above 1 p.p.m. cause the inhibition of root apex extension, coupled with increases in axis diameter and the proliferation of lateral roots. In contrast to many other effects of ethylene, the inhibition of root extension is not affected by carbon dioxide concentrations up to 20%, and is essentially reversible if exposure to ethylene is not prolonged. The sensitivity of the species examined:

Tomato > Tobacco > Barley > Rye > Rice

correlates well with their sensitivity to waterlogging (Smith and Jackson, 1974; Fig. 7.6). At lower concentrations (< 1 p.p.m. for rice, 0·15 p.p.m. for pea and 0·02 p.p.m. for tomato, the value depending upon the rate of endogenous production of ethylene) root extension is stimulated (Jackson and Drew, 1984).

The ecological implications of these findings remain largely

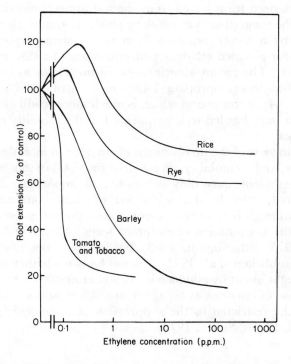

FIG. 7.6.   The influence of ethylene on the extension of seedling root axes during three days at 20–25°C in water culture. The ethylene concentrations used in the figure are the concentrations in the air which was in equilibrium with the culture solution. From Smith and Jackson, 1974.

unexplored (Jackson, 1985), partly because of lack of information about the composition of soil atmospheres in natural ecosystems. It has been suggested that shallow, highly-branched root systems, formed under the influence of ethylene in poorly-drained soils, could be well-placed to exploit the nutrient-rich surface layers of undisturbed soils (Smith and Jackson, 1974). However, such root systems would render plants susceptible to drought in seasonally-fluctuating environments, and trees susceptible to "wind-throw".

3. Growth Regulation
In addition to these direct effects on roots, several workers have observed that the shoot symptoms of waterlogging stress in sensitive species (epinasty; wilting, desiccation and senescence of leaves; adventitious rooting) can be induced by treating aerated root systems with ethylene (Smith and Jackson, 1974). However, although these effects are associated with increases in leaf ethylene content, not all of this ethylene has been transported from the soil atmosphere via the root, leading to the suggestion that ethylene production in the leaf could be stimulated by a factor exported from root systems exposed to low oxygen and/or elevated ethylene concentrations (Jackson and Kowalewska, 1983). The recent identification of this factor as the ethylene precursor 1-aminocyclopropane-1-carboxylic acid (ACC)(Reid and Bradford, 1984), a compound which is much more readily transported in the xylem sap, has led to a reappraisal of the mobility of ethylene within plants.

   Overall, in view of the involvement of ethylene in most developmental processes, and its mobility (as ethylene or ACC) in soils and plants, exogenous ethylene may play a number of important, and as yet unappreciated, roles in the regulation of plant communities. For example, although it has been demonstrated that exogenous ethylene stimulates the germination of dormant seeds (Mayer and Poljakoff-Mayber, 1975; Etherington, 1983) but inhibits root nodulation in legumes (Grobbelaar *et al.*, 1971), it is not known whether these effects can be brought about by ethylene in the soil atmosphere. However, the proposed role of ethylene as an agent in soil fungistasis (Smith, 1976) appears to be restricted to the suppression of pathogenic fungi only (e.g. Lynch, 1975b; Smith, 1978).

II. AERIAL POLLUTION

Up to this point, we have considered the effects on plants of naturally generated stresses associated with poor aeration. However, it has been clear for many years that pollutant gases originating from urban and industrial areas and from the internal combustion engine can have a profound influence upon the composition, biomass and distribution of natural vegetation.

A. Sources and Levels of Pollutant Gases

Gaseous pollutants originate from a variety of sources. For example, sulphur dioxide, SO_2, is a primary product of the combustion of sulphur-containing fossil fuels, whereas nitrogen oxides, commonly written as NO_x, are by-products of combustion, arising out of the thermally-induced combination of atmospheric nitrogen and oxygen. The initial product of this reaction, NO, is slowly oxidized to NO_2 in the atmosphere. The burning of fuels also gives rise to a series of hydrocarbon gases including ethylene, C_2H_4; thus, for example, motor vehicle exhaust contains varying proportions of CO_2, H_2O, CO, SO_2, NO_x and C_2H_4.

When NO_2 is released into the atmosphere, it can take part in a series of photochemical reactions leading to the synthesis of ozone, O_3, and peroxyacetyl nitrate (PAN). These reactions occur most rapidly in warm sunny regions and the resulting "photochemical smog", as in California, can be very harmful to plants, animals and humans. However, it has now been established that even in a cool area such as the United Kingdom, phytotoxic levels of ozone can occur frequently during clear weather in summer (Bell, 1984). Other important sources of pollutants include the chemical industry, which releases varying quantities and mixtures of gases (principally SO_2, H_2S, NO_2, NH_3 and HCl), and brickworks and metal smelters, which produce localized enhancement of SO_2, HF and particulate fluoride (as well as toxic metal) levels in the atmosphere.

It has been estimated that gaseous aerial pollution is responsible for global crop losses of the order of $\$10^8$ p.a., excluding the widespread damage caused to forestry enterprises and natural vegetation (Bell, 1984). In spite of the economic importance of these effects, there is still a scarcity of precise information on the levels of phytotoxic gases in urban and industrial areas, and with the development of methods for more efficient dispersal of pollutants (e.g. tall chimneys), there is an increasing need to monitor the composition of rural, apparently unpolluted, atmospheres. Plants growing in the United Kingdom will normally be exposed to levels of the most important pollutants within the ranges presented in Table 7.1, but, at a given site, the concentration, and the ratios of pollutant species ($NO:NO_2$, $NO_x:SO_2$, etc.) will depend upon a number of environmental factors including distance from the source, topography, altitude, precipitation, irradiance, and wind direction and velocity. There are also considerable seasonal variations, with O_3 levels rising in clear weather in summer, whereas SO_2 and NO_x concentrations tend to be highest in winter owing to increased use of fuel for heating.

Some confusion can be caused by the fact that a range of different concentration units are used in air pollution studies, including the vol/vol units (p.p.m., p.p.b., $\mu l\, l^{-1}$ and $nl\, l^{-1}$) and $\mu g\, m^{-3}$. However, if the molecular weight of the chemical species is known, then these units are interconvertible (Mansfield and Freer-Smith, 1981; Lenzian and Unsworth, 1982). For example, at STP, 1 p.p.m. SO_2 ($1\,\mu l\, l^{-1}$) = 2860 $\mu g\, m^{-3}$, 1 p.p.m. $NO_2 = 2054\,\mu g\, m^{-3}$ and 1 p.p.m. $O_3 = 2000\,\mu g\, m^{-3}$.

B. The Influence of Gaseous Aerial Pollution on Plants

Although the effects on plants of pollutant gases, especially SO_2, have been studied for nearly a century, much of the information collected is

TABLE 7.1. Representative concentration levels of selected gaseous pollutants in the air over the British Isles (5 min averaging period), all values p.p.m.

Pollutant	Rural areas	Urban areas	Industrial areas
SO_2	<0·001–0·05	0·02–0·5	<0·001–1
NO_x	0·005–0·05[a]	0·02–0·2[a]	up to 1[b]
O_3	0·02–0·06[c]	0·08–0·25[d]	
C_2H_4	<0·001–0·01	0·005–0·05	<0·001–0·5

[a]Primarily as NO_2.
[b]About 1:1 $NO:NO_2$.
[c]Natural background.
[d]Photochemical pollution.
Data compiled for 1979 by Dr R. Harrison, Department of Environmental Sciences, University of Lancaster. More complete data for the United Kingdom, Europe, North America and the Far East can be found in Fowler and Cape (1982) and Koziol and Whatley (1984).

of doubtful value because of deficiencies in experimental procedure and in the interpretation of results.

(i) *Experimental deficiencies.* In the majority of earlier experiments in which plants were fumigated with pollutants in controlled environments, the air was inadequately stirred. Consequently, the boundary layer over each leaf offered a high resistance to the entry of pollutant gas molecules into the leaf (see pp. 143, 196), with the result that the effects recorded at a given level of pollutant were much milder than would have been observed in the field (Ashenden and Mansfield, 1977). Confusion of this kind can be avoided if all reports include details of the pollutant concentration, the period of exposure and sufficient data (e.g. windspeed) to permit calculation of the boundary layer resistance (see the discussion of pollutant dose below).

(ii) *Seasonal effects.* Inadequate attention has been paid to variations in the sensitivity of plants at different stages of development and at different times of the year. For example, investigations since 1970 in the United Kingdom have indicated that SO_2 damage to pasture grasses is much more severe in winter, not only as a result of seasonal increases in the level of pollutant, but also because plant resistance and regrowth are limited by unfavourable environmental conditions (low temperature and irradiance). This has been confirmed by Davies (1980) who showed that the sensitivity of *Phleum pratense* plants depends upon the irradiance, but Davison and Bailey (1982) have also demonstrated that fumigation of *Lolium perenne* plants with SO_2 can reduce their ability to tolerate freezing.

(iii) *Pollutants as fertilizers.* It is likely that the rather mild effects of SO_2 fumigation on plants which are found in some experiments are a consequence of the low sulphur status of the soils used; under these conditions, atmospheric SO_2 can act as a fertilizer, giving a stimulation rather than a reduction in growth (e.g. Cowling *et al.*, 1973). A similar effect can be observed with NO_x, and SO_2 can also improve dry matter production by controlling plant diseases.

(iv) *Pollutant mixtures.* In spite of the fact that polluted air invariably contains a mixture of gaseous pollutants, most of the experimental work reported in the literature involves the exposure of plants to varying levels of a single pollutant. This accumulated information is of limited value in explaining or predicting damage in the field because the effects of different pollutants in

combination are only rarely additive; in fact, most investigations show synergistic depressions in dry matter production (e.g.
Table 7.2), even in cases where exposure to one of the consti-
tuents, singly, results in the stimulation of growth. Study of the
effects of pollutant mixtures on single plants and vegetation has
now become a major area of research (Ormrod, 1982; Kondo,
1984).

(v) *Invisible injury.* A major factor hindering progress in air pollution research has been the tenet, widely held until recently in N.
America, that gaseous pollutants, and SO_2 in particular, do not
have a significant influence upon plant growth and physiology
until tissue lesions are visible (Heath, 1980). This theory, which
arose partly out of the experimental deficiencies described
above, has now been generally discarded. More recent work,
starting with the pioneering investigation of Bleasdale (1973),
has established that "hidden" or "invisible" injuries (such as
reduction in the rates of net photosynthesis and dry matter

TABLE 7.2.   Percentage changes (relative to control) in dry weights and leaf
areas of grass plants exposed to atmospheres containing SO_2 or NO_2 alone, or
in combination for 20 weeks (from Mansfield and Freer-Smith, 1981).

	SO_2^a	NO_2^a	$SO_2 + NO_2^a$	Effect
Dactylis glomerata				
Leaf area	−5	+21	−72	Synergistic
Dry wt of green leaves	−28	−7	−83	Synergistic
Dry wt of roots	−37	−11	−85	Synergistic
Poa pratensis				
Leaf area	−28	−17	−84	Synergistic
Dry wt of green leaves	−39	−29	−88	Synergistic
Dry wt of roots	−54	−47	−91	Additive
Lolium multiflorum				
Leaf area	−22	+1	−43	Synergistic
Dry wt of green leaves	−28	−10	−65	Synergistic
Dry wt of roots	+7	+35	−58	Reverse
Phleum pratense				
Leaf area	−11	+30	−82	Synergistic
Dry wt of green leaves	−25	+14	−84	Synergistic
Dry wt of roots	−58	+1	−92	Synergistic

[a]All concentrations 0·068 p.p.m.

production, disruption of water and ionic relations and of cell biochemistry) occur commonly in plant tissues exposed to doses of pollutants which do not cause visible injury (e.g. Table 7.2). The most important consequence of this misconception is that the concentrations of pollutants used in most experiments before 1960 were unreasonably high.

Overall, there is no doubt that it is difficult to carry out realistic field investigations of the influence of gaseous air pollutants on natural vegetation and crops (Unsworth, 1982). However, improvements in experimental techniques and in the theoretical foundation of the subject have resulted in such an expansion in air pollution research (Lenzian and Unsworth, 1982) that only a few salient topics can be discussed here.

1. Gas Exchange between Leaves and Polluted Atmospheres

It is unfortunate for plants growing in polluted air that adaptations favouring CO_2 assimilation tend also to favour the uptake of other gases into the leaf mesophyll. Thus, for example, many plant species are more sensitive to SO_2 during the day, when the stomata are open, than during the night; an exception to this observation is the potato whose stomata remain open at night (Mansfield, 1976).

According to the conventional resistance analysis (p. 140), the diffusive flux of CO_2 or of pollutant molecules to the site of assimilation or damage in the leaf mesophyll is determined by the concentration gradient and the leaf diffusive resistance, the latter of which is made up of a set of component resistances:

r_a boundary layer resistance, outside the leaf surface;

r_s stomatal resistance ⎱
r_c cuticular resistance ⎰ arranged in parallel;

r_m mesophyll resistance (this component can be attributed to the pathway through the mesophyll apoplast and cells to the site of assimilation or damage, together with the "resistance" of the enzyme systems involved in assimilation or damage. However, because r_m is poorly characterized, diffusion is in the liquid phase, and the diffusing pollutant may have been transformed by chemical reactions (see below), it is more correct to use the term "residual resistance" to signify that the other resistances in the system are insufficient to account for observed rates of diffusion (Unsworth, 1981). It is generally difficult to evaluate the residual resistance for pollutant molecules and, in many cases, it is assumed to be much smaller than the other resistances);

where r_a and r_s are dependent upon the molecular weight of the diffusing molecules (Unsworth et al., 1976).

In still air or at low windspeeds, r_a will be the dominant resistance over a range of stomatal apertures (see p. 144). Consequently, as noted above, inadequate ventilation during experimental fumigation can lead to unexpectedly low rates of pollutant uptake and damage at doses which are clearly phytotoxic in the field (e.g. Ashenden and Mansfield, 1977). Many studies have shown that at higher windspeeds (lower r_a), the uptake of pollutant gases (especially SO_2), like CO_2, is controlled by stomatal aperture, since the cuticle offers a much higher resistance (Spedding, 1969; Mansfield and Freer-Smith, 1981).

Despite these similarities between CO_2 and SO_2 uptake, the processes involved in the exchange of pollutant molecules between air and leaf are much more complex. First, some phytotoxic gases are able to override the plant's control of stomatal aperture; in particular, SO_2 can cause stomatal opening, even when the leaf is subject to water stress, leading to increased SO_2 uptake and water loss (Mansfield, 1976). For example, Biscoe et al. (1973) found that stomatal resistance in leaves of Vicia faba fell by 20% at 140 µg SO_2 m^{-3} and above, although later work indicated that stomatal response is highly dependent upon the humidity of the air (Black and Unsworth, 1979). In contrast, O_3 tends to cause stomatal closure (Mansfield, 1976).

Secondly, the cuticle appears to be relatively permeable to HF and HCl and may represent the major pathway from air to mesophyll for these gases (Guderian, 1977). Furthermore, the cuticle can be damaged by gaseous pollutants, presumably leading to reductions in r_c, and several workers have suggested that long-term residence of acidic pollutants (resulting from dry or wet deposition, see below) in water films covering the leaf surface, especially in winter, will lead to substantial annual rates of uptake into the plant tissues, even if r_c is considerably higher than r_s (Garsed, 1984). In some cases, the quantity of SO_2 adsorbed in these external water films can represent a substantial fraction of the total amount of pollutant removed from the air by vegetation (Unsworth et al., 1976; Unsworth, 1981).

Study of the movement of gaseous pollutants to the site of damage or detoxification can be further complicated by the chemical transformations which occur when they dissolve in the water films bound to the mesophyll or epidermal apoplast, or to the outer surface of the leaf. For example, SO_2 dissolves readily to give hydrated SO_2 ($SO_2.H_2O$)

which acts as a strong acid, dissociating to give HSO_3^- and SO_3^{2-} ions in proportions determined by the pH of the solution (pK_a for $HSO_3^- \rightleftarrows SO_3^{2-} + H^+ = 7\cdot2$). However, because the movement of SO_2 species into cells appears to be passive, and the cell walls are negatively charged (i.e. excluding anions), only the uncharged species ($SO_2.H_2O$) can enter the cell. This, in turn, will dissociate within the cytoplasm to give phytotoxic SO_3^{2-} ions ($pH > 7$) which can then be oxidized to less toxic sulphate, primarily by the chloroplasts. (The cell buffer system must also deal with the accompanying hydrogen ions.) Consequently, the actual quantity of damaging pollutant (sulphite) to which the cell contents are exposed depends upon the relative rates of entry and detoxification to sulphate. Furthermore, the evolution of excess sulphur-containing compounds in the form of gaseous H_2S has been demonstrated in some species (Taylor and Tingley, 1983). In a similar way, NO_2 dissolves to give equal proportions of nitrate and nitrite ions; although the latter are phytotoxic, they can be transformed by nitrite reductase to ammonia which can contribute to the nitrogen economy of the leaf. The sequence of reactions is less clear for NO, which is relatively insoluble in water (Lenzian and Unsworth, 1982; Koziol and Whatley, 1984).

2. Pollutant Gas Dose Expressions

The discussion in the previous section of the actual quantity of pollutant reaching its "target" within the cell raises the question of how to express this quantity in the most meaningful and useful way. Heck (1984) proposes three different dose expressions: (a) the exposure dose (bulk air concentration × time); (b) the uptake dose (rate of pollutant uptake by the plant × exposure time; this may or may not include pollutant molecules residing on external surfaces) and (c) the effective dose (the fraction of (b) which is actually involved in causing damage). At the present state of understanding, (b) and (c) are not generally practicable, but Fowler and Cape (1982) have suggested an alternative expression "pollutant absorbed dose" (time of exposure-× bulk air concentration × stomatal conductance ($1/r_s$) with reference to the pollutant under study) which could prove more useful.

However, dose expressions are of real value only if there is reciprocity between time and concentration, i.e. a low concentration over a long period giving the same effect as a higher concentration over a shorter period of exposure. This is clearly not the case for acute damage (see below), and although there have been studies in which reciprocity has been observed (e.g. Whitmore, 1985), these are probably the exceptions rather than the rule (Garsed, 1984).

3. Primary Sites of Attack

Once pollutants have entered plant cells, their primary effects will, presumably, be at the molecular or ultrastructural level (although it is not known whether very reactive species such as O_3, whose sites of action appear to be at the plasmalemma, actually penetrate into the cytoplasm to any extent). This assumption is supported by a large number of short-term experiments showing significant changes in, for example, enzyme activities, levels of metabolic intermediates, and membrane permeabilities (Horsman and Wellburn, 1976; Koziol and Whatley, 1984), but the true significance of these findings for the whole plant is not clear because of the multiplicity of effects recorded for each pollutant. However, several primary sites do seem to have been identified.

(i) *Stomata.* As indicated earlier, SO_2, NO_x and O_3 can have a rapid and lasting influence upon stomatal aperture, probably as a consequence of damage to subsidiary and other epidermal cells (Mansfield, 1976; Mansfield and Freer-Smith, 1981). Thus in the case of SO_2, tissue damage can be a consequence of water loss rather than the direct action of the pollutant on metabolism.

(ii) *Chloroplasts.* Exposure of leaves to low doses of SO_2, NO_x and O_3 which do not give visible lesions have been found to cause gross disruption of the thylakoid membrane systems of chloroplasts (e.g. Wellburn *et al.*, 1972). Ozone also causes the breakdown of plasmalemma membranes (Thomson *et al.*, 1966) and it has been suggested that it acts directly upon membranes by reacting with the double bonds of unsaturated lipid molecules, as well as with compounds carrying aromatic rings and sulphydryl groups (Heath, 1975; Koziol and Whatley, 1984).

(iii) *CO_2 fixation.* Exposure of leaves to SO_2 commonly causes a rapid decrease in the rate of CO_2 fixation. Experiments with subcellular fractions have shown that at least part of this inhibition is caused by competition between sulphite and bicarbonate ions for the CO_2-fixing sites on RuBP carboxylase and PEP carboxylase (e.g. Ziegler, 1972), although there will also be alterations in the tertiary structures of enzymes and membranes.

Other studies have indicated that SO_2 acts primarily by blocking sulphydryl groups or through changes in cell pH, whereas NO_x, as nitrite, interferes with chloroplast redox systems (Bleasdale, 1973; Capron and Mansfield, 1976; Mudd and Kozlowski, 1975). It remains to be seen to what extent the damaging effects of SO_2, NO_x, O_3 and

PAN fumigation can be attributed to the initiation of free radical reactions (Lenzian and Unsworth, 1982).

4. Photosynthesis and Dry Matter Production

Since it has been shown that the major pollutants cause changes in stomatal physiology, chloroplast structure, CO_2-fixation reactions and photosynthetic electron transport systems, it is not surprising to find that relatively small doses of air pollution cause rapid (but generally reversible) depressions in the rate of net photosynthesis. Some of the work of Bennett and Hill (1974) in this field is shown in Fig. 7.7, which indicates that the effects of single pollutants on photosynthesis in oat, barley and lucerne follow the series:

$$\text{HF} \quad > \quad \text{O}_3 \quad > \quad \text{SO}_2 \quad > \quad \text{NO}_x$$
$$\left(\begin{array}{c} 0\cdot01 \\ \text{p.p.m.} \end{array}\right) \quad \left(\begin{array}{c} 0\cdot05 \\ \text{p.p.m.} \end{array}\right) \quad \left(\begin{array}{c} 0\cdot2 \\ \text{p.p.m.} \end{array}\right) \quad \left(\begin{array}{c} 0\cdot4{-}0\cdot6 \\ \text{p.p.m.} \end{array}\right)$$

where the concentrations in brackets indicate the minimum pollutant level at which depression in the rate of net photosynthesis could be detected. Subsequent work in the United Kingdom using pea, bean and tomato, has shown that photosynthesis can be sensitive to even lower levels of SO_2 and NO_x (for example, down to 0·03 p.p.m. SO_2 and 0·1 p.p.m. NO_x)(Bull and Mansfield, 1974; Capron and

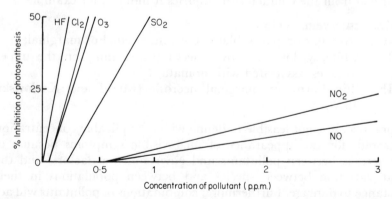

FIG. 7.7. The inhibition of photosynthesis (as measured by CO_2 uptake) in 3–5 week old canopies of barley and oats when exposed to varying levels of gaseous pollutants for two hours. During the treatments, the plants were maintained at 24°C, a wind velocity of 1·2–1·6 m s^{-1}, low humidity (45% RH) and high light (40–50 Klux ≃ 280–350 W m^{-2}) (adapted from Bennett and Hill, 1974).

Mansfield, 1976; Black and Unsworth, 1979). Comparison of these concentrations with the data in Table 7.1 confirms the serious risk of "hidden injury" run by plants growing in urban and industrial areas, and even in rural areas. However, this type of analysis underestimates the risk of damage in most cases as a consequence of the synergistic effects of pollutant combinations (for example most species are relatively insensitive to NO_x, but in combination, it can greatly amplify the effects of SO_2, Table 7.2), and overestimates the hazards to resistant populations (see below).

Where plants are exposed to such levels of pollution over a prolonged period, these reductions in net photosynthesis (in combination with other effects) cause significant reductions in dry matter production, commonly without visible injury, as shown in a number of studies of grass species (Ashenden, 1978, 1979; Ashenden and Mansfield, 1978; Bell and Clough, 1973; Table 7.2). These effects are normally accompanied by reductions in the components of yield (numbers of leaves and tillers, leaf area, etc.) and with accelerated senescence (Table 7.3).

5. Symptoms of Acute Toxicity

Chronic exposure of plants to low concentrations of pollutants can cause progressive leaf chlorosis and premature senescence, symptoms which can be difficult to recognize in the presence of other environmental stresses. In contrast, high concentrations generally cause visible lesions as a consequence of the death, drying out and bleaching of localized areas of leaf tissue. In some cases, the causative agent can be identified from the characteristic injuries it induces; for example:

SO_2 interveinal chlorosis;
NO_x irregular brown or black spots, interveinal or marginal;
O_3 white, yellow or brown flecks (0·1–1 mm) on the upper surfaces, associated with stomata;
HF   "tip burn" or marginal necrosis (Mudd and Kozlowski, 1975).

In practice, it is not feasible to define widely applicable concentration thresholds for the appearance of such visible symptoms because of interactions between pollutants and environmental factors, and the great variation between species and between populations in their resistance to damage. Furthermore, combinations of pollutants will act to give different thresholds and symptoms. However, as a single example, Bennett and Hill (1974) found that, under the conditions of the experiments illustrated in Fig. 7.7, foliar lesions were induced by 0·15 p.p.m. HF, 0·2 p.p.m. O_3, 0·8 p.p.m. SO_2 and by NO_x levels higher than 5 p.p.m.

TABLE 7.3. Components of the yield of S23 *Lolium perenne* grown for 26 weeks in 191 μg m^{-3} SO$_2$ (SO$_2$ plants) or in 9 μg m^{-3} SO$_2$ (control plants), expressed as means per plant (from Bell and Clough, 1973).

	No. of tillers	No. of living leaves	No. of dead leaves	Dry weight of living leaves (g)	Dry weight of "stubble" (g)	Dry weight of dead leaves (g)	Leaf area (cm^2)
SO$_2$ plants	14·84	47·31	12·02	0·388	0·217	0·047	203·6
Control plants	25·18	85·61	6·39	0·791	0·478	0·027	417·2
Difference in productivity between SO$_2$ and control plants[a]	−41%	−45%	+88%	−51%	−55%	+78%	−51%

[a] All differences significant at $p < 0.001$.

6. Ethylene as an Air Pollutant

Although included in Table 7.1, ethylene has been excluded from the foregoing discussion, mainly because its influence on plants strictly as an air pollutant has been largely neglected. However, because of its ubiquity in the atmospheres of all developed countries, its influence upon plant development at very low concentrations (below 0·01 p.p.m.), and the observation that exposure of plant tissues to other air pollutants can stimulate endogenous production of ethylene, much more interest is now being taken in this subject (Koziol and Whatley, 1984; Reid and Watson, 1985).

C. Plant Resistance to Air Pollution

Although it is clearly impossible for plants to be resistant to pollutant gases at all concentrations experienced in the atmosphere, there is ample evidence of species and populations within species which are unusually well-adapted to survive exposure to gaseous pollutants. (In contrast, several very sensitive species are used as biological indicators of air pollution, such as tobacco for ozone; Posthumus, 1982.) Many of the reports of resistance, however, come from unrealistically short investigations using very high concentrations; for example, Taylor et al. (1975) reviewed a number of reports of inter-specific variation in resistance to NO_2 based on fumigation experiments which lasted only a few hours (e.g. 8, 16 and 32 p.p.m. NO_2 for a single hour; note that 1 p.p.m. is representative of the most polluted atmospheres in the United Kingdom, Table 7.1). The problems of interpreting such data are discussed in full by Roose et al. (1982). However, other workers have succeeded in establishing differences in resistance at more realistic dose levels; for example, Taylor and Murdy (1975) demonstrated variation in resistance to the development of SO_2 lesions amongst natural populations of Geranium carolinianum using 0·8 p.p.m. SO_2 for 12 h.

Studies of the mechanisms which lead to reduced damage and improved survival are seriously handicapped by uncertainties about the primary sites of pollutant action (see above). Some progress has been made in the study of stomatal characteristics leading to the avoidance of mesophyll damage (i.e. exclusion of the pollutant molecules). For example, it has been found that the stomata of several field crops are particularly sensitive to O_3, closing rapidly and remaining closed throughout the period of fumigation, thereby denying the gas access to the mesophyll (e.g. onion, Engle and Gabelman, 1966). Such an adaptation, which will interfere with CO_2 uptake and water relations, is appropriate to the episodic nature of ozone pollution, but may not be of value to plants exposed to chronic, long-term pollution. Avoidance of pollutant uptake may also be facilitated by low stomatal density (Dean, 1972). However, in many cases, the relative responses of species or populations are not determined by the amount of pollutant absorbed; for example, an SO_2-resistant population of perennial ryegrass from a highly polluted site in Lancashire actually absorbed more SO_2 than the sensitive S23 cultivar (Bell and Mudd, 1976). In such cases, survival involves the tolerance of elevated tissue levels of pollutant (e.g. the ability of chlorophyll to resist denaturation in the presence of SO_2; Bell and Mudd, 1976) or efficient pathways of detoxification.

One of the most interesting developments over the last decade has been the demonstration that resistant populations of grasses can develop very rapidly, within a few years of the imposition of the pollution stress (although the decline of such populations can be equally rapid; Wilson and Bell, 1985). This implies that there exists considerable variation within grass populations or cultivars in those characteristics contributing to resistance (Roose et al., 1982), but it also implies the existence of powerful selection processes, in spite of the fact that the swards do not show obvious symptoms of acute injury or widespread mortality. These processes remain to be identified, but Mansfield and Freer-Smith (1981) point out that since SO_2 fumigation can interfere with the nitrite reductase activity in cells, the damaging effects of SO_2 on more sensitive genotypes could be amplified considerably if, as is the case in most polluted atmospheres, NO_x is also present.

D. The Influence of Acidic Deposition on Plants

Up to this point, attention has been focused on the effects on plants of air pollutants as gases. However, several of the same pollutants, principally SO_2 and NO_x, can dissolve in water droplets in the atmosphere to give sulphuric and nitric acids, commonly in a ratio of 7:3. Rain falling through unpolluted atmospheres, in equilibrium with CO_2, will have a pH of about 5·6, but values recorded over most of Europe and N. America normally lie in the range 3–4, or even lower, owing to long-range dispersal of SO_2 and NO_x. Thus in addition to the gaseous pollutants taken up through stomata or adsorbed on leaf surfaces, plants can also be intermittently bathed with pollutant solutions of low pH (Last et al., 1986).

Study of this subject can be complicated by the abuse of

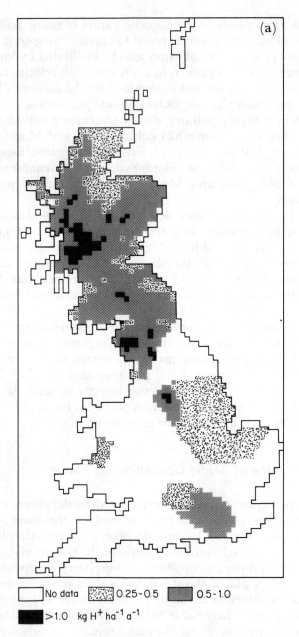

FIG. 7.8. Mean annual acidity (since 1959) received (a) by wet deposition, (b) by dry deposition as SO_2 (redrawn from Fowler *et al.*, 1985).

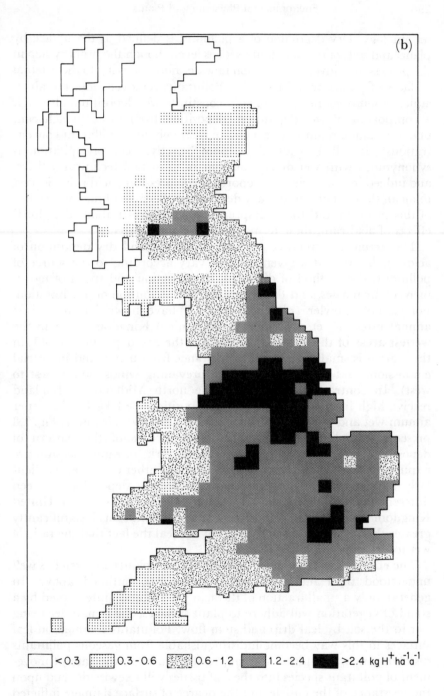

| | <0.3 | | 0.3 - 0.6 | | 0.6 - 1.2 | | 1.2 - 2.4 | | >2.4 kg H$^+$ha^{-1}a^{-1} |

terminology. Dry deposition of a pollutant is defined as the uptake by plants and soils of the pollutant as a gas, even though the primary step in the process may involve dissolution in water films adhering to the external surfaces of plants or soil particles. Pollutants removed from the air in aqueous solution (rain, snow, etc.) constitute wet deposition, although a component of wet deposition, deposition from mists (which can, under certain circumstances, contain relatively high pollutant concentrations), is called occult deposition. Strictly, acid precipitation is synonymous with wet deposition, but in most non-technical articles, and indeed in some scientific reports, the expression "acid rain" is used to mean the sum of wet and dry depositions. Because of the importance of this problem and the widespread public concern about the global effects of acid rain, these terms should be used correctly.

The actual and relative magnitudes of wet and dry deposition of acidic pollutants at a given site will vary according to the source of pollutant, the method of dispersal, the direction and speed of movement of air masses, and the geographical distribution of precipitation. For example, Fowler *et al.* (1985) (Fig. 7.8) have shown that the highest annual rates of wet deposition in the United Kingdom occur in the wettest areas of the west and north, but the dry deposition of SO_2 in these areas is small because of the distance from urban and industrial emmissions and the direction of the prevailing winds (south west to west). In contrast, some parts of the north Midlands of England receive high inputs of both types of deposition (> 1 kg H^+ ha^{-1} per annum wet and $> 2 \cdot 4$ kg H^+ ha^{-1} per annum dry). However, Fig. 7.8 presents a preliminary and incomplete picture of the pattern of deposition, and much more complete networks of sampling sites are required in the United Kingdom and in most other countries, particularly in view of the pronounced differences in deposition between different years. Overall, the impression gained that, over the United Kingdom as a whole, dry deposition of acidic pollutants is significantly greater than wet deposition should not conceal the fact that the ratio of wet to dry deposition varies widely.

The effects of wet deposition of pollutants on plants are even less well understood than those of gaseous pollutants, outlined above. In general, only a small fraction of the acid precipitation intercepted by a stand of vegetation will adhere to plant surfaces, the remainder being lost to the soil by leaf drip and stem flow. Pollutants retained on leaf surfaces in this way become indistinguishable from gaseous pollutants dissolving in water films adhering to leaf surfaces; the rate of movement of pollutant species into the leaf tissues will likewise depend upon the resistance of the cuticle and the degree of surface damage inflicted by the pollutant. Access to the mesophyll will not normally take place via the stomata. However, acid precipitation can have two other important effects on plants. First, the continual flushing of leaf surfaces by acidic solutions during prolonged rainfall can lead to the leaching of mineral ions, particularly Ca^{2+}, from the leaf (Fairfax and Lepp, 1975). The severity of these losses will depend upon leaf age, leaf surface properties and the extent of damage to the cuticle and surface waxes. Secondly, acid precipitation can have indirect effects as a consequence of changes in soil properties. In particular, the acidification of soils can lead to the mobilization of phytotoxic aluminium ions. The effects of aluminium ions on plants are considered in more detail on p. 229. In passing, it should be noted that the mechanisms outlined above tend to concentrate primary (acids) and secondary (aluminium ions, etc.) pollutants in water courses, with serious implications for aquatic plants.

A large number of laboratory and field studies of plant response to simulated acid precipitation has established that this type of pollution can also cause "invisible injuries" (e.g. reduction in dry matter production and in components of yield, without visible lesions, see p. 286) as well as large- and small-scale tissue damage, especially at pH values below 4 (e.g. Paparozzi and Tukey, 1984; Evans et al., 1985). However, little progress has been made towards identifying the mechanisms involved or the primary sites of attack, although results from short-term experiments indicate that pollutant damage can occur even when there are no soil effects. Finally, although much of the research initiated in response to concern about the role of acid rain (sensu lato) in the decline of forest trees in Europe is still at a preliminary stage, it is already clear that (a) aerial pollution is only one of a series of interacting (and highly variable) stresses affecting tree physiology (drought, frost, wind and fungal attack; Pearce, 1986) and (b) that a number of unique "air pollution climates" are involved (Last et al., 1986).

8. Interactions between Organisms

I. INTRODUCTION

All previous chapters have considered physical or chemical aspects of the environment and their effects on plants. In axenic culture such a description of plant–environment interactions might be adequate, but even in the simplest ecological systems other organisms form part of the environment. The effects of these organisms may be of three kinds:

(i) direct influence on the supply or stock of resources;
(ii) indirect effects achieved by alteration of the physical or, more usually, the chemical environment;
(iii) dispersal, as in pollination or seed dispersal.

In this way it is clear that many of the effects of one organism on another can be viewed in terms of the physico-chemical factors discussed in previous chapters. The first category covers situations where organisms compete for resources, where symbiotic associations exist (whether parasitic or mutualistic), or where predation, typically by grazing, occurs; the physiological responses elicited in the plant will be those appropriate to an increase or decrease in the supply or stock of the resource in question. By contrast, alterations to the microclimate or to soil structure, or most strikingly the addition to the environment of toxic or growth-controlling chemicals are of the second type. Again, however, explanations of the effect of one plant on another can be described wholly in terms of physical and chemical characteristics of the environment. Some other plant–plant interactions cannot be explained in this way, such as pollination and reproduction; these fall outside the scope of this book.

Most interactions between plants and other organisms (microbes, fungi, other plants, animals) can, then, be explained in terms of changes in the physical or chemical environment of the affected plant. That is not to imply that such interactions do not involve peculiar ecological and physiological effects and have profound evolutionary implications. Indeed it would be true to say that for most plants the responses shown towards environmental factors can only be understood in the light of the complex interactions that result from the acquistion of the same resources by competing plants, fungi, and microbes, and the modification of other environmental parameters by co-habiting species. It is, for example, of limited value merely to calculate the phosphate-supplying power of a soil and to catalogue in detail the responses of a plant growing on that soil to phosphate supply, if what is required is an understanding of the factors controlling phosphate supply to that plant in the field. The effects of rhizosphere micro-organisms and competing plants, of temperature and moisture on root growth and distribution, of grazing by the soil fauna and of many other factors would also be involved.

This chapter attempts to examine some of the influences of other organisms on plants, regarding them purely as additional environmental factors, and considers in detail three types of interaction.

(i) Competition for resources.
(ii) Predation and parasitism.
(iii) Allelopathic chemicals.

Mutualistic responses have already been considered in some detail in Chapter 3, at least for nitrogen-fixing and mycorrhizal systems.

II. COMPETITION

The definition of competition is one of the longest-running semantic debates in biology. Since we are discussing here the *effects* of competition on plants, we will define it mechanistically rather than in the terms of population dynamics. Competition then represents the situation where the supply of a resource is less than the joint requirements of two organisms, and as a result the performance of one or both is impaired.

Clearly no two individuals of different species can have identical requirements—this would imply a greater degree of affinity than is found within many species. Nevertheless all have some potential competitors when they are exploiting a resource, and survival depends therefore either on partitioning of the resource in some way, thereby avoiding competition, or on the establishment of competitive superiority. Resource partitioning is relatively simple to visualize in animals, where food items, for example, can be categorized on the basis of size or type, but all autotrophic green plants require roughly similar quantities of CO_2, photons, water, and 15 or so inorganic ions. Clearly partitioning, if possible for such resources, must rely on different criteria. If one species possesses competitive superiority over others with respect to all resources, then one might expect that species to eliminate the others in all appropriate habitats. In practice one finds that communities of plants are very diverse and mono-specific stands are rather rare in nature, and many species (e.g. *Pteridium aquilinum*, *Calluna vulgaris*, *Larrea divaricata*) often only manage such dominance as a result of human activity. Clearly then, where species are competing for resources there are processes acting to reduce the intensity of that competition.

A. Competitive Superiority

The simplest competitive situation is one where species A has some advantage over B. This is frequently true in laboratory and greenhouse environments where only one factor is varied. In the oft-quoted experiments of Gause (1934) where two *Paramecium* species competed for food in an otherwise homogeneous environment, one was always competitively superior. Similarly when Idris and Milthorpe (1966) grew barley (*Hordeum vulgare* cv. Union BN) and charlock (*sinapis arvensis*) in pots in an experimental garden, barley showed increased growth rates, net assimilation rates, root growth and nitrogen uptake rates, as its proportion in the mixture decreased. In other words the individual barley plants grew better if their neighbours were charlock than if they were other barleys. This implies that barley was the competitive dominant under these conditions, though under others the situation must be modified, as charlock is an acknowledged weed of barley crops.

The characteristics of plants that convey this superiority are different in the case of competition for photosynthetically active radiation (*PAR*), from that for competition for water or nutrients. Where two plants compete for *PAR*, height and size of photosynthetic organs are critical (Fig. 8.1(a)) and physiological adaptation is only of significance to a subordinate species to compensate for growing in low irradiance. This occurs because radiation is directional; in contrast the supply of nutrients and, to some extent, of water in soil is diffuse and positional effects of the roots are less important. In this case it is the relative quantities of the root systems, and most importantly the ability to lower soil water potential or ionic concentrations at the root surface, which determine competitive superiority. The exception to this is where the surface soil is dry and water is rising from deeper layers—in

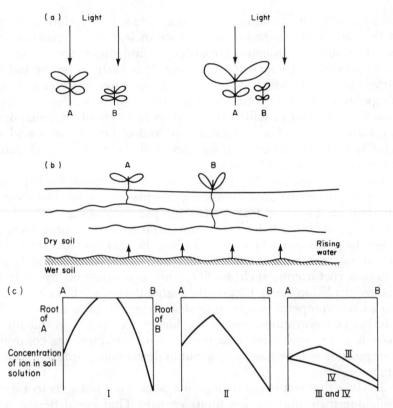

FIG. 8.1.   (a–c) Diagrammatic representation of plant factors controlling the development of competition for light, nutrient ions, and water. In (c) I–IV represent successive time intervals, with plant B attaining the greater supply due to its lower root surface concentration.

that case a deep-rooted plant will intercept the now directional supply of water (Fig. 8.1(b)), or conversely in dry soil, where showers may wet the soil surface and shallow-rooted plants may absorb the water before it can penetrate deeper.

1. Competition for Water and Ions

There is, however, an important distinction between the uptake of water and that of nutrients in the mechanism by which plants lower soil water potential on the one hand and ionic concentration on the other. Water flows into roots and up stems along a gradient of water potential, created by evaporation from the leaves (cf. Chapter 4). The lowering of water potential at the root surface is, therefore a function of the rate of transpiration, and hence of leaf area. Ignoring other variables such as stomatal conductance and differences in the leaf water potential at which stomata close, it is likely to be the leaf:root surface area ratio which determines relative rates of water uptake in competition. Some ions arrive at the root surface in the consequent mass flow of water in sufficient quantities to satisfy plant demand; this is certainly true of Ca^{2+} in calcareous and of Fe^{3+} in very acid soils (cf. Chapter 3). Competition for these will therefore be governed by the same principles as for water.

For other ions, notably $H_2PO_4^-$ mass flow is quite inadequate to satisfy requirements, and diffusion gradients are set up. For these the situation in Figure 8.1(c) arises and the controlling factor is the physiological ability to lower the concentration at the root surface. The lower the diffusion coefficient for the ion, the narrower and steeper will be the depletion zones around the roots. Ions with relatively high diffusion coefficients, such as NO_3^-, are characterized by wide but shallow depletion zones. Clearly the higher the diffusion coefficient, the more likely competition is to ensue since depletion zones are wide; since solution concentrations tend to be higher for such ions, facilitating mass flow or convective transport, shoot characteristics controlling transpiration will be more important in determining uptake in competition.

To show that competition for an ion occurs, it is not sufficient simply to demonstrate that the ion limits growth. That could be the result simply of a very low diffusive flux, which would actually make competition less likely. Two conditions must be satisfied—the ions must not arrive by mass flow at the root surface faster than they are required by the plant, and depletion zones must be wide enough to overlap between adjacent roots. To some extent these conditions counteract one another, since there is a general correlation between ion mobility in soil and soil solution concentration, and so with the contribution of mass flow to uptake (cf. p. 80). They can also be quantified into two dimensionless mathematical expressions which are fully explored by Baldwin and Nye (1974) and Baldwin (1975):

(i) the relationship between mass flow ($V=$ the water flux at the root surface, in cm^3 cm^{-2} s^{-1}) and plant demand ($\alpha=$ root absorbing power, cm s^{-1}: see Nye and Tinker (1969) and p. 86) is simply given by the ratio $V{:}\alpha$. If $V < \alpha$, uptake is greater than supply by mass flow, depletion occurs at the root surface, and competition may ensue. α is itself of course highly dependent on the concentration of the ion at the root surface. The extent of depletion, both in terms of concentration at the root surface and radial spread from the root, depends on D, the diffusion coefficient.

(ii) where depletion occurs, the radius of the depletion zone is approximately given by $2(Dt)^{0.5}$, where D is the diffusion coefficient ($cm^2 s^{-1}$) and t is time (s) (Nye and Tinker, 1977), and the extent and occurrence of overlap is controlled by the function DtL_v, where L_v is root density ($cm\ cm^{-3}$). Baldwin (1975) shows that as DtL_v falls below 2·5 towards zero, the ion becomes increasingly immobile with respect to that particular root system. Uptake of $H_2PO_4^-$ and K^+ in typical soils will in fact be directly related to this function, and for these ions high values of L_v (the only term in the expression determined by the plant) will be advantageous

There emerge from such an analysis, then, four critical parameters—water flux (V), plant demand (α), root density (L_v), and diffusion coefficients (D)—of the soil plant system that control uptake in competition. The situation is greatly complicated, however, by their interactions (Fig. 8.2).

2. Competition for PAR

In contrast to the complexity of competition for nutrients and water, that for *PAR* is governed simply by the directional and transitory nature of the photon as an energy source. If a photon strikes a leaf it may be absorbed, transmitted, or reflected, but both the latter have selective effects, producing radiation of altered spectral quality (see p. 30) and of little photosynthetic value. Competitive superiority for *PAR* therefore resides in the ability to place leaves in illuminated rather than shaded positions, and so will be determined by plant characteristics determining foliage height and the rate at which that height is attained.

Where, as in annual crops and their weeds, both competitors start simultaneously from seed, the critical factors determining rate of growth are seed size, relative growth rate, and elongation rate. Black (1958) sowed together small (4 mg) and large (10 mg) seeds of *Trifolium subterraneum*, which in monoculture produced identical yields, and found that plants developed from large seeds comprised almost the entire sward (92%) after 115 days. This advantage was manifested from the very start, as the large seeds had larger cotyledons, producing a positive feedback, for these captured more light and so

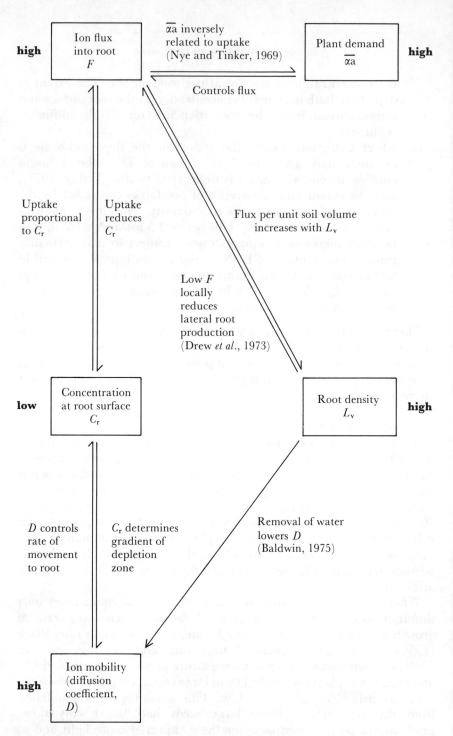

FIG. 8.2. Schematic representation of interactions of major soil and plant variables controlling uptake. Bold type indicates conditions conducive to the onset of competition.

grew faster. Similarly, where the two cultivars Yarloop and Bacchus Marsh, which in monoculture produced equal yields and had modal petiole lengths of 18 and 14 cm, were grown together, Yarloop produced 80% of the final combined yield (Black, 1960).

It is in such circumstances that the significance of etiolation becomes apparent, as a means of growing out of shade, and the importance of the sensitivity of etiolation to far-red radiation, the dominant wavelength of transmitted radiation, is then obvious. McLaren and Smith (1978) have shown that *Rumex obtusifolius* shows greater etiolation if FR increases in proportion to the red flux after germination. Thus seeds germinating under an established canopy, with high FR:R ratio, will respond less than those where the canopy is growing up, gradually increasing the ratio. In the latter situation elongation is likely to bring the plant to the top of the canopy, in the former it would likely be suicidal.

What then of the suppressed plant, confined to sub-canopy layers, or of the suppressed leaf on an otherwise well-illuminated plant? The various adaptations for shade resistance described in Chapter 2 permit survival—namely, decreases in leaf thickness, abscission of leaves below the compensation point (as long as some are above it), and the phenomenon described by Chippindale (1932) as "inanition". In the latter case, plants of *Festuca pratensis* survived for over a year without growing beneath a canopy of *Lolium italicum*; when the canopy was removed and nutrients applied, the *Festuca* resumed growth. Similarly Mahmoud and Grime (1974) have ascribed the ability of *Deschampsia flexuosa* to survive deep shade to its low negative growth rate at very low irradiance.

Leaves in full sun at the top of a canopy run a greater risk of overheating (cf. Chapters 2 and 5). Often they are dissected or more erect, which reduces the heat load they sustain, but also means that they cast less shade on lower leaves. Here, as so often, a compromise is necessary between the conflicting benefits of maximizing capture of *PAR* and minimizing overheating.

3. Interactions in Competition

Following the lead of Donald (1958) a number of workers have attempted to disentangle the effects of competition for *PAR* and nutrients. Using dividers to separate the shoot and root systems, Donald examined the effects of shoot and root competition on the interaction between *Lolium rigidum* and *Phalaris tuberosa*. The figures in Table 8.1 are derived from his paper and show that *P. tuberosa* was clearly the subordinate species, under conditions of both shoot-only

TABLE 8.1.   Growth of two grasses (*Lolium rigidum* and *Phalaris tuberosa*) with all combinations of root and shoot competition and at two nitrogen levels (Donald, 1958).

	Yield (105 days from planting) (g dry matter)							
	Lolium					Phalaris		
	NC	SC	RC	FC		NC	SC	RC	FC
Low N	2·45	2·71	2·12	2·77		2·00	1·63	0·35	0·18
High N	4·71	4·19	4·13	4·72		4·67	3·19	1·17	0·32

	Yield as %age of non-competition yield						
	Lolium				Phalaris		
	SC	RC	FC		SC	RC	FC
Low N	111	87	113		82	18	9 (15)
High N	89	92	100		68	25	7 (17)

NC, SC, RC, FC—no, shoot, root and full competition respectively.
Figures in parentheses represent expected FC values obtained by multiplying SC and RC.

and root-only competition, but particularly in the latter case. Root competition, therefore, appeared to be a more severe treatment. When full competition was permitted (shoot and root—no dividers) yield was further reduced—more so than would be expected by combining the shoot and root competitive effects. The figures in the last column of Table 8.1 represent predicted full competition yields on the assumption that root and shoot effects were simply additive. They are clearly higher than those actually obtained for *P. tuberosa*.

There is here a clear and not unexpected implication of a synergism between the effects of root and shoot competition. Root and shoot perform distinct but almost symbiotic functions. Reduction in nutrient supply to the shoots caused by root competition will lower the efficiency of the shoots and hence the plant's ability to compete for *PAR*; this in turn will reduce the flow of assimilate to the roots, damaging root function, and so on.

Such interactions have led to the suggestion that competitive superiority is largely a function of overshadowing of the leaves of one plant by another, which in turn reduces the competitive ability of the subordinate species for nutrients, producing a "snowball" effect (Newman, 1973). Thus Idris and Milthorpe (1966) sowed barley and charlock (*Sinapis arvensis*) in pure stands and in mixtures with either species in excess. The larger-seeded barley quickly over-shadowed the charlock which was progressively suppressed, even in mixtures in which it was in excess, and this enabled the barley increasingly to dominate the soil and to extract much greater quantities of nitrogen. They were able to show that the nitrogen uptake rate of charlock in the early stages of competition was not reduced by small numbers of barley plants, though the growth rate was much lower (Table 8.2). Their explanation was that the primary factor competed for was *PAR*.

In contrast Aspinall (1960) grew white persicaria (*Polygonum lapathifolium*) with barley and found that, although suppressed plants of *P. lapathifolium* had typical shade adaptations (higher leaf area ratio and specific leaf area for example), in shading experiments significant reductions in relative growth rate were not obtained until irradiance was reduced to 40–45% of full summer daylight. Since such reductions were not obtained in stands of barley until after the commencement of suppression of *P. lapathifolium* by barley, he inferred that nutrient competition resulting from the smaller root system of *P. lapathifolium* was the primary agent. This was confirmed by nutrient addition and by experiments in which the shoots and roots were separated.

The apparent contradiction between these two otherwise similar experiments may possibly be explained by the different growth forms of the competitors. *Sinapis arvensis* is a tall, single-stemmed plant, similar in general habitat to the barley it was grown with. *Polygonum lapathifolium* is a sprawling, weak-stemmed, much-branched plant that cannot compete effectively for *PAR* with the barley canopy. It may well be that selection has proceeded in different directions here; towards competitive ability for *PAR* in the case of *S. arvensis*

TABLE 8.2. Nitrogen uptake and growth rates of competing barley (*Hordeum vulgare*) and charlock (*Sinapis arvensis*) (from Idris and Milthorpe, 1966).

		Nitrogen uptake rate[a] (mg mg^{-1} d^{-1})		Relative growth rate[b] (mg mg^{-1} wk^{-1})	
Mixture		Charlock	Barley	Charlock	Barley
Pure	(95:0)	72·5	9·9	1·03	0·63
Excess	(60:35)	80·5	16·0	0·82	0·68
Deficit	(35:60)	48·9	13·8	0·68	0·81

[a]Uptake rate per unit root dry weight in the first two weeks of the experiment.
[b]Polynomials of the form $\ln W = \ln W_o + bt + 2ct^2$ were fitted to data. The value quoted is of the first-order regression coefficient, b.

and for nutrients with an approach to "shade" physiology for *P. lapathifolium.*

These examples underline the importance of considering the other interacting environmental factors under whose selective pressure the species has evolved. This is beautifully illustrated by the data of Snaydon (1971), who grew acid and calcareous-soil ecotypes of white clover, *Trifolium repens,* on both soils and in all combinations of root and shoot competition. His results (Table 8.3) show that the larger-leaved calcareous-soil plant outyielded the acid-soil ecotype on both soils whenever the shoots were in competition. The nearest approach to the acid-soil plants' native habitat is the full competition/acid-soil combination, yet even here it is outyielded by the calcareous-soil plant: clearly this does not occur in the field. Only where the roots alone compete does the acid-soil plant gain its rightful advantage. The implication is that in the field the only form of effective competition is root competition, and that other factors, probably grazing pressure, prohibit the development of the intensity of shoot competition that occurs in a pot experiment.

The features that confer competitive superiority are relatively simple—leaf canopy architecture and height, rates of transpiration, root morphology and distribution, and physiological parameters of ion uptake—but their operation is complex. Selection may act to minimize these features and so confer competitive incompetence on a plant if their possession carries concomitant disadvantages in a particular field environment. In considering competitive relationships we may not ignore these factors, nor the many other environmental agents and behavioural adaptations that act to reduce the intensity of competition.

TABLE 8.3.   Yield of plants of an acid-soil ecotype of white clover *Trifolium repens* as a percentage of that of a calcareous-soil ecotype, when the two species were grown on both soils in all combinations of root and shoot competition (from Snaydon, 1971).

Competition	Acid soil	Calcareous soil
None	77_b	41_a
Shoots	53_a	38_{ab}
Roots	121_c	36_b
Shoots and roots	72_b	31_c

Within columns figures differently subscripted are significantly different by Duncan's test ($p = 0.05$).

B. The Occurrence, Extent and Ecological Effects of Competition

Until recently, most ecologists would have agreed with the view that competition was an important, if not the major force in structuring ecological communities. Such a view is increasingly controversial, and animal ecologists in particular give more weight to other processes, such as predation and parasitism (Roughgarden and Diamond, 1986). Plant ecologists, too, are tending to pay more attention to these factors, and to others, such as historical accident, climatic fluctuation and disturbance (Pickett and White, 1985). Nevertheless it is intuitively obvious that competition between plants does occur. Perhaps the best experimental evidence for the occurrence of competition in natural communities comes from removal experiments, where one or a few species are removed (physically or by herbicides, for example), and an increase in abundance of others is monitored. Where this has been done, there is normally a rather general improvement in the performance of several other species (Fowler, 1981; Silander and Antonovics, 1982), rather than a specific response by one other species. This suggests that competitive interactions are diffuse: in other words plants compete with their neighbours rather indiscriminately and irrespective of taxonomy.

Nevertheless, ecological theory, and in particular the competitive exclusion principle, would suggest that in any community there would be selection in favour of the differentiation of activity of neighbours, which should lead to a diminution of competitive interaction. Two possible methods of differentiation are in space and in time: it has long been assumed that temporal separation (for example in terms of seasonal development) and spatial separation (in depth of root activity) were common in plant communities and fundamental to their structure. The ecological literature contains many root profile diagrams, such as Fig. 4.9 (p. 159). Often these show differences in depth of root penetration of co-existing species, sometimes coupled with the demonstration that species with the most distinct root patterns are the most frequent neighbours (Cody, 1986).

It is remarkable that, in the three-quarters of a century since Cannon (1911) published his "Root Systems of Desert Plants", so few advances have been made in our understanding of the interactions between root systems. Chemical or radioactive tracers can be used to measure the activity of the root systems of different, co-existing species, and such studies give some evidence that niche differentiation may occur (Fig. 8.3), but it is by no means universal.

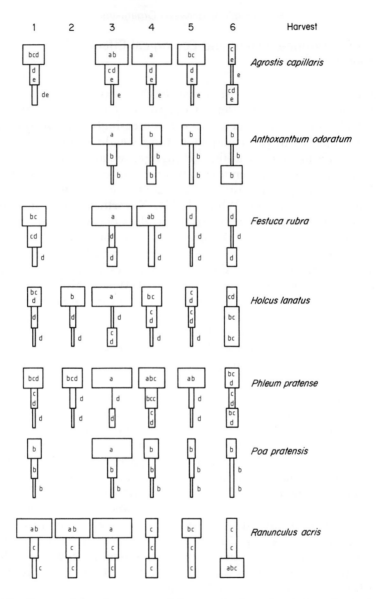

FIG. 8.3. Root activity of seven co-existing grassland species at three depths in an alluvial soil and on six occasions (1–6: 3 and 19 April, 2 and 30 May, and 11 and 26 June 1984) at Wheldrake Ings, Yorkshire, England. Activity is expressed as a proportion of the maximum for each species and was measured using non-radioactive tracers (Li, Rb, Sr) injected into soil. Within species, boxes with different letters differ significantly. Note that activity generally declines with depth, except late in the season when surface soil dries, and that differences between species are not marked. From Fitter, A. H. (1986). *Oecologia* **69**, 594–599.

Niche differentiation based on genetically determined differences in rooting depth, for example, would imply a co-evolutionary process, and this would require a very high degree of predictability in the neighbour relationships of species in communities over many generations. It is just as likely that the apparent differentiation is achieved by the pre-emption of particular soil layers or growth times by dominant species, as can be seen in Fig. 8.3 and in the work of O'Brien et al. (1967). They were able to show that the massive yield depression of *Lolium perenne* by its hybrid with *Festuca pratensis*, × *Festulolium*, was associated with a change in the uptake pattern with depth for phosphate. Phosphorus uptake from a depth of 60 cm, for example, was reduced from 32% of total P uptake to 7%. In contrast, the uptake pattern of the competitive dominant, the hybrid × *Festulolium*, was not altered by either of its parents (Table 8.4).

In many situations, however, it cannot be argued that competition between species is reduced by such avoidance phenomena. Other possible explanations are that other factors are acting to suppress the competitive advantage of one species to some extent, or that co-existence is achieved by differences in competitive ability in relation to

TABLE 8.4. Yield and phosphate uptake from 60 cm depth (as percentage of total) by *Lolium perenne*, *Festuca arundinacea*, and their triploid hybrid, *x Festulolium loliaceum*, when grown for hay (from O'Brien *et al.*, 1967).

| | *Yield (g)* | | |
Competitor	L. perenne	F. arundinacea	x Festulolium
L. perenne	553	377	646
F. arundinacea	199	614	511
x Festulolium	71	164	789

| | *% P uptake from 60 cm* | | |
Competitor	L. perenne	F. arundinacea	x Festulolium
L. perenne	32	32	53
F. arundinacea	56	38	46
x Festulolium	7	16	47

several jointly-limiting resources, as has been advocated by Tilman (1982).

III. PREDATION AND PARASITISM

As Feeny (1976) has pointed out, the continued existence of plants is little short of miraculous. As stationary packages of food for several hundred thousand species of phytophagous insects and several thousand species of parasitic fungi, quite apart from herds of grazing mammals, it is at first sight hard to see how they avoid complete consumption. The devastation caused by locust swarms is of course well known, and occasionally one comes across totally defoliated plants in otherwise healthy vegetation. The winter moth *Operophtera brumata* is able almost to denude oak trees *Quercus robur* under certain conditions, and the chrysomelid beetle *Gastrophysa viridula* can similarly decimate docks *Rumex obtusifolius*. Typically, however, leaves of plants collected from field-grown plants show necrotic patches and nibbled margins, but not a bare skeleton of vascular tissue.

Clearly plants discourage potential pathogens and herbivores from inflicting excessive damage. There appear to be three main mechanisms.

   (i) *Nutritional inadequacy*: most plants have relatively low protein contents. Southwood (1973) quotes figures suggesting that typical values for plant material are 10–30% of dry weight, compared with 30–60% in microbes, 40–60% in insects, and 25–40% in artificial diets for insects. Plant parts have a much greater range, from less than 1% in wood to over 30% in many seeds, and of course many insect herbivores are specialist feeders on the more nutritious parts. It is an intriguing thought that plant breeders, in attempting to improve the food quality of crop plants, may be unwittingly removing this defence. More specific nutritional factors may be important, such as low levels of sterols, of the amino acids tryptophan and methionine in many plant proteins, and even of sodium, which is typically more abundant in animal than in plant tissues.
   (ii) *Physical barriers*: many plant organs have external coverings of scales, hairs, or glandular hairs. Beneath this there may be thick cuticles, thick epidermal cell walls, bands of collenchyma or sclerenchyma, or deposition of substances such as silica (in

grasses) or resins (in conifers). All these present problems for organisms attempting to penetrate plant tissues.

(iii) *Toxins*: perhaps the most dramatic and widely discussed defences are chemical (Levin, 1976). These may be constitutive or induced by attack, and are chemically very diverse, including alkaloids, tannins, cyanogens, and a wide range of phenolic compounds. The reasons for the existence of this diversity are considered below.

These various types of defence have particular value in different situations, depending on the organisms involved and the nature of the attack. It is possible to distinguish two classes of plant-feeding organisms on the basis of the type of material removed. Some organisms obtain essential metabolites by removing whole cells including much structural material, principally cellulose, hemicellulose, and lignin: these are tissue feeders. Others, however, specifically remove cell contents and can be termed metabolite feeders, obtaining a diet containing less low quality material (cellulose, lignin, etc.), though not always well-balanced. It is also less damaging to the plant than tissue feeding; indeed small losses of metabolites may be quite inconsequential, and this is the basis of mutualistic symbiosis.

A. Nature of Attack

1. Metabolite Feeders

Three sources of metabolites are available to plant feeders: xylem contents, phloem contents, and parenchymatous or other cell contents—and two methods which they can use to obtain these metabolites—mechanical insertion of a probe to withdraw material and systemic infection (Table 8.5).

As far as the insect is concerned, in the case of stylet-feeders, the nutritional value declines in the order: cell contents, phloem, xylem. Xylem sap is essentially a dilute solution of mineral salts and, depending on the nitrate reductase activity in the roots, which varies greatly with species, contains a small amount of amino acids. Horsfield (1977) has shown that *Philaenus spumarius*, the well-known spittlebug, is a xylem feeder, and that it preferentially feeds on xylem vessels containing approximately 1 mM amino acids, though xylem may contain less than 0·05 mM *total* N. Indeed when required to utilize xylem containing less than 0·3 mM amino acids, mortality of nymphs was around 90%. Some xylem-feeding insects have extraordinarily long life-cycles: the

TABLE 8.5.   Classification of metabolite feeders by type of food and mode of
attack.

Food source	Systemic infection	Mechanical probe
Xylem	many fungi: e.g. *Verticillium* often cause "wilt" diseases *Ceratostomella ulmi* (Dutch elm disease)	insects; some Cercopidae: e.g. *Philaenus* plants; hemiparasites: e.g. *Rhinanthus*
Phloem	many fungi viruses: e.g. phloem necrosis in elms	insects; most Aphididae: e.g. *Myzus* plants; full parasites: e.g. *Orobanche*
Cytoplasm	many fungi: e.g. *Pythium, Botrytis* ("damping-off") viruses	insects; some Mynidae: e.g. *Leptoterna* some Psyllidae

periodical cicadas, *Magicicada* spp., have a cycle up to 17 years long
and feed on root xylem sap (White and Strehl, 1978). Xylem feeding
clearly requires considerable specialization for an insect, though
for higher plant hemiparasites such as *Euphrasia*, *Rhinanthus*, or *Striga*
it is more straightforward. All these are normally photosynthetic but
have poorly developed root systems. By penetrating xylem vessels of
adjacent plants they ensure a supply of minerals and water, and
although organic compounds are transferred as well (Govier *et al.*,
1968), these are probably not of great significance. The best examples
of such hemiparasites are provided by the annual members of the sub-
family Rhinanthoideae of the Scrophulariaceae.

It has been suggested (Karlsson, 1974) that the hemiparasitic habit
enables these annuals to persist in the closed grassland communities of
which they are characteristic and which are otherwise almost entirely
composed of perennials, since they do not need to use resources in the
production of an extensive root system. It is probably significant that
these hemiparasites are non-mycorrhizal, since mycorrhizal infection
may well enable seedlings to become established by similarly limiting
the need for the early development of roots, in favour of shoot growth.
One further problem of xylem feeding is the very considerable
negative pressures that may build up (Raven, 1983). For higher plant

hemiparasites this is overcome by xylem to xylem continuity and by water secretion from leaf surface glands which increases the water potential gradient (Klaren, 1975; Govier *et al.*, 1968), but it is not known how insects cope, since penetration of the xylem should normally lead to cavitation.

Phloem, by contrast, contains a much more nutritious solution, although its contents fluctuate greatly, both daily and seasonally. Its largest component is sucrose, which is typically $0.2–0.7$ M, and it also contains normally up to 0.5% amino acids and many other substances. For higher plant parasites phloem presumably contains an adequate diet; many such plants, such as *Orobanche* and *Lathraea*, derive all their nutrients from host phloem. For an insect the main problem is the excessive C:N ratio. Aphids, the most successful phloem feeders, overcome this by sugar secretion, the well-known honeydew, and some indeed turn it to advantage by entering a symbiotic relationship with ants, who "farm" the aphids for the sugar. Phloem has the further advantage that it is under positive pressure and so the sucking insect has to control the inflow rather than expend energy on inducing it.

Cell contents are indubitably the best balanced diet for metabolite feeders, but they are not renewable. Whereas an aphid stylet in a phloem vessel will receive a continuous flow of sap, once the contents of a cell have been used up, the organism must move on to a new one, possibly explaining why this mode of feeding has proved most successful in the fungi, whose mode of infection requires them to grow through the host tissue and to pass a series of cells. An alternative is for the parasite to act more as a sink for metabolites without destroying the cell, much in the way that aphids do. Viruses, which to replicate must use cell contents, are renowned for their ability to incorporate themselves into the host genome, replicating in tandem, and most plants can be shown to contain virus particles without exhibiting disease symptoms.

2. Tissue Feeders

Metabolite feeders rarely cause dramatic damage to plants, and then often only incidentally. Tissue feeders, however, particularly leaf feeders, can destroy large proportions of a plant's assimilatory tissue, leading in some cases to total defoliation. Tissue feeders fall mainly into three groups—insects, molluscs and mammals, although birds and reptiles may be important in some habitats. In marine environments molluscs and fish are dominant. The basic qualification for a tissue feeder is the possession of biting or tearing mouthparts, but more subtly they must be able to cope with a cellulose-based diet. To obtain sufficient quantities of nitrogen and other dietary components, they must ingest very large amounts of cellulose, which most animals cannot use. The exceptions are a few molluscs, such as the shipworm *Teredo*, which actually produce their own cellulase, and those animals that have symbiotic bacteria or protozoa in the gut, notably termites (Isoptera) and ruminant mammals. Some insects—leaf-cutting ants and termites—cultivate fungi on leaves brought into the nest, and these break down the cellulose. Larvae of the goat moth, *Cossus cossus*, by contrast, feeding on wood but unable to use cellulose, take three years to reach maturity, and this appears to be due to poor nutritional quality, since on artificial diets they pupate within one year. Most tissue feeders are therefore faced with the problem of having to ingest an enormous quantity of material to obtain their requirements. Quantitatively this is measured as the ratio of the energy content of the ingested food to that of the biomass produced, the production–consumption ratio or efficiency of conversion of ingested food (*ECI*). *ECI* is typically around 10% for detritivores feeding on very N-poor materials, 10–20% for most phytophagous insects, and as high as 40% (or rarely 60%) for those feeding on N-rich sources (seeds, pollen) and for carnivores.

B. Plant Defences

As briefly discussed above, plants can defend themselves by nutritional inadequacy, by physical barriers, and by chemical toxins. In addition the activities of predators may be important, though the plant usually has little or no control over these. An exception is the symbiosis between ants and plants with extrafloral nectaries (i.e. nectaries outside the flowers and not connected with pollination). This relationship is exemplified by some Central American *Acacia* species which have hollow thorns used as nests and extra-floral nectaries and protein-rich appendages used as food by colonies of ants which deter potential herbivores. If the ants are removed, the plant is quickly defoliated (Janzen, 1975). Such symbioses are widespread if often looser and less well-developed; even some non-flowering plants such as bracken fern seem to rely on them to a small extent.

1. Nutritional Inadequacy
It appears to be generally true that plant tissue is only barely adequate as a diet for most insects (Southwood, 1973), although more usable for the herbivorous mammals and for fungi, which benefit from

symbiotic associations on the one hand and very versatile metabolic systems on the other. The failure of most animals to evolve a cellulase, even more striking in view of the fact that a handful of species have achieved it, is almost certainly explained by the very low protein contents of plant material, making cellulose unnecessary as an energy source since it is protein and not energy that is limiting. Many animal species avoid this problem by feeding on sap, but as mentioned above, this too may have severe nutritional imbalances. The only species to avoid this problem to any great extent are the cytoplasm feeders, since cytoplasm has a much higher N content than whole plant tissue. There are, however, other problems associated with feeding on cytoplasm. Generally increases in N content are likely to improve growth and survival of the feeder, as has been shown for the aphid *Myzus persicae* by van Emden *et al.* (1969) and for many other insects. It is a striking consequence of this that nutrient deficiency for a plant may actually be an important defence mechanism, and that there may have been selection for low N contents. However, low N content may not be a wholly adequate defence for a plant, since it could lead to a grazing animal eating more tissue, so as to obtain sufficient, and this would be more damaging to the plant. There is therefore a dilemma: high N content increases palatability and so the likelihood of attack, while low N content may make individual attacks more severe. However, low N content may also reduce both the individual and the population growth rates of the herbivore and so reduce its equilibrium population size, so that the total effect on the plant population is less (Lawton and McNeill, 1979). Nevertheless, other defences may be required.

2. Physical Barriers

Despite their nutritional inadequacy to most herbivores, plants are the primary producers and are eaten. All plants present, to a greater or lesser extent, physical problems to at least some potential predators or pathogens. Tissue feeders can be deterred by a variety of external morphological traits, whose scale depends on the size of the animal. The ferocious spines of arid-zone trees, such as *Acacia karoo*, have clearly evolved in relation to grazing by large mammals, but less dramatic than these, there exists a wide range of smaller structures. Spines are widespread, particularly on woody plants, where they are sufficiently strong to give protection. Some are detachable, as in *Rosa*, which presumably reduces the danger of stem breakage by passing animals catching them; others are outgrowths of the wood. Leaf-spines are also common in thistles (*Cirsium*, *Carduus*, etc.) and these grade into hairs. Indeed, most angiosperms are hairy in some part.

Hairs are found on higher plants in a bewildering variety of forms (Fig. 8.4). Although their main effect was once considered to be in reducing transpiration by thickening the leaf boundary layer, numerous studies have failed to show any such general relationship (but see Chapters 4 and 5, p. 223). Undoubtedly in some cases they serve this function, and also increase leaf reflectivity (Chapter 5), but hairiness appears to be at least as strongly associated with defence against insect attack. This was shown clearly for leaf spines by Merz (1959), who offered leaves of holly, *Ilex aquifolium*, to larvae of the moth, *Lasiocampa quercus*, which normally feed along the edges of leaves. The larvae refused to eat the holly. The same leaves with the spines removed were avidly devoured. Leaf spines can therefore be an effective deterrent,

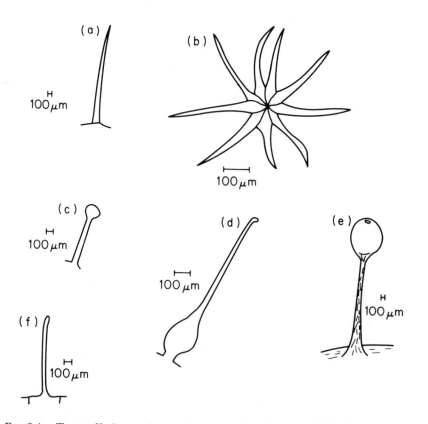

FIG. 8.4. Types of hairs on plant surfaces. (a) simple hair from capsule of *Silene dioica*; (b) stellate hair from *Sida* (after Esau); (c) glandular hair from *Epilobium hirsutum*; (d) stinging hair from *Urtica dioica*; (e) glandular hair from *Drosera capensis*; (f) root hair from *Lolium perenne*.

and spiny or hooked hairs can also prevent insects feeding. Gilbert (1971) has shown that the hooked hairs of *Passiflora adenopoda* prohibit movement of and may even wound the caterpillars of Heliconid butterflies, damaging predators of other *Passiflora* species. Even soft hairs may be effective against smaller insects. Mutant lines of soybean varieties differing only in hairiness have been shown to have leafhopper (*Empoasca fabae*) infestations closely related to hair density (Singh *et al.*, 1971). In these experiments height growth of the soybeans was closely related to the infestation of *E. fabae* suggesting that powerful selection pressures may be involved. The whole field of plant pubescence has been reviewed by Johnson (1975).

In addition to hairs, many other physical structures have been implicated in defence, including thick cuticles (important against fungal pathogens), silicious walls, oxalate raphides (crystals) in cells, corky layers, resin and latex ducts and so on. Clearly the borderline between chemical and physical defences is hard to draw. Many hairs are glandular and can secrete either sticky substances, which immobilize insects, or toxins, and the full development of this line is seen in the stinging hairs of nettles (*Urtica*) and other families such as Loasaceae.

3. Chemical Defences

All plants contain compounds usually known as "secondary plant products", once considered to be waste products, but now generally accepted as largely defensive. There are at least 30 000 such compounds known from the relatively small sample of plants so far analysed (Harborne, 1977). The waste product theory seems in retrospect somewhat far-fetched, since organisms with a primarily anabolic biochemistry should be able to avoid producing wastes, but the clear toxicity of many of these compounds to fungi, insects, or other enemies is a clear pointer to their function.

One problem for a plant of adopting a chemical defence against predation or parasitism is that any chemical toxic enough to be effective is likely to be autotoxic—so where can it be stored? There are two compartments—the cell wall and the vacuole—which are remote from metabolism, and these are indeed found to contain many toxins. Alternatively two compounds (typically a glycoside and glycosidase enzyme), that are themselves innocuous but that if mixed produce a toxin, can be stored in separate compartments. Finally, special organs may occur such as glandular hairs, which are wholly devoted to the storage and excretion of toxins. The extreme development of this can perhaps be seen on the leaf surface of sundews *Drosera*, which is covered with stalked glands, and whose secretion of digestive enzymes turns a defensive into an offensive mechanism.

For a long time it was held, on rather slender evidence, that chemical defences were of two types: specific toxins which poisoned particular metabolic pathways, and rendered a plant unusable except to a few herbivores capable of detoxifying them; and quantitative defences, which were generally deleterious chemicals that reduced food value in a way that herbivores could not easily counteract. The example of the latter case usually given was the tannins, which form complexes with proteins and can make up several per cent of the dry weight of a leaf. Feeny (1976) argued that the metabolic cost of producing large quantities of such general feeding deterrents was large and that they were therefore characteristic of "apparent" species, that is those that are conspicuous either by being abundant in space or persistent in time. Long-lived climax species, such as oak, would therefore be expected to use this defence (and indeed oak leaves may contain up to 5% by weight of tannins in September), whereas more fugitive or sparsely-dispersed species would contain the metabolically less expensive specific toxins.

The hypothesis in this simple form is no longer tenable. Tannins are recognized as being chemically diverse and it seems probable that the two main classes perform different functions (Zucker, 1983). Condensed tannins, which build up in deciduous leaves in late summer and are abundant in wood, may act to inhibit microbial cellulases and other degradative enzymes. Their role may be partly posthumous, regulating the rate of nutrient release from litter. Hydrolysable tannins, of which different plants contain many different chemical species, are more likely candidates for a defensive role against insects, but each probably interacts with a specific target molecule, often an enzyme. Tannins seem therefore to be no different to other chemical defences in this respect.

The range of these defences is large. They may be as simple as a low cell pH, achieved by the storage of organic acids, which is thought to explain the immunity of *Oxalis* spp. to attacks by dodders *Cuscuta* (Kuijt, 1969), or they may be represented by any of a variety of glycosides, non-protein amino acids, terpenes, alkaloids, hormone mimics and others. The chemistry and mode of action of many are discussed by Harborne (1977). Such defences are very powerful and often represent a relatively trivial metabolic cost, but in many cases some herbivore or group of herbivores has evolved a means of detoxifying them. Particular herbivores thus become specialized to feed on one plant species or group, which is why most insect larvae have well-defined host plant ranges. A neat illustration of this was provided by Erickson and Feeny(1974), who showed that the swallow-tail butterfly *Papilio polyxenes*, which normally feeds on the leaves of various species of the Umbelliferae, was unable to complete its development successfully if fed on celery leaves (*Apium graveolens*, Umbelliferae) cultured in sinigrin, a mustard oil glucoside characteristic of the Cruciferae. This may well explain the inability of many insects to eat crucifers. On the other hand, sinigrin is a specific feeding attractant for other species, such as Pierid butterflies, which are adapted to feeding on crucifers and capable of detoxifying it.

All defences involve the plant in some diversion or expenditure of resources, although it is not yet possible to calculate the costs and benefits involved. The costs are reduced if the defence system is inducible—synthesized only when the plant is attacked. Several workers have shown that insect damage to some leaves on a plant may reduce the palatability of other, undamaged leaves (e.g. Wratten *et al.*, 1984; Edwards and Wratten, 1982; Haukioja and Niemela, 1977). More ambitious notions of inter-plant communication (the so-called "talking trees" phenomenon) are highly controversial (Fowler and Lawton, 1985).

Where attack is systemic, however, the plant is able to predict the infection, since fungi, in particular, need actually to grow through tissue. Resistance to parasitic fungi often involves the phenomenon of hypersensitivity, where cells around the point of infection die. In such cases it is frequently found that chemicals capable of inhibiting fungal growth (phytoalexins) accumulate in the dying cells. Their role as agents limiting fungal growth in infected tissue is as yet not fully established, an alternative, though less probable explanation being that they accumulate as a consequence of fungal attack and necrosis (Deverall, 1977). However, the fact that their production declines after successful systemic infection, and that fungi capable of this often have enzymes that can detoxify them, strongly suggests that phytoalexins are indeed a form of inducible chemical defence in response to a predictable attack.

When the necrotrophic fungus *Botrytis fabae* attacks broad bean leaves it causes initially small brown lesions of necrotic tissue. This is the hypersensitive reaction. A phytoalexin, wyerone acid, accumulates in these patches, but *B. fabae* reduces this to a less active form, grows through into living tissue, and starts the systemic infection (Mansfield and Deverall, 1974a, b). *B. cinerea*, by contrast, a much less specific necrotroph, is unable to reduce wyerone or to grow through the necrotic tissue. It is possible, therefore, to recognize both general and specific forms of defence, the balance between the two being controlled by life-history and abundance, and constitutive and inducible defences. Because of the unpredictable nature of most insect and other animal predation on plants, inducible defence compounds are only found to any great extent in the plant: fungus relationship.

There is, however, a whole group of organisms feeding on plants for which chemical forms of defence would appear to be inadequate—the stylet feeders. Aphids and stylet-feeding nematodes only insert a relatively inert tube into the plant tissue and typically feed on phloem fluid which does not normally contain toxins. Nevertheless, some aphids at least show a high degree of host specificity, suggesting a chemical or possibly a physical relationship. It is known that the glandular hairs of some plants secrete toxins, and these may be important in controlling aphid infestations. It seems possible that the primary defence of some plants against aphid attack is simply that in obtaining their food aphids make themselves very vulnerable to predators.

C. Plant Responses

Since productivity is a function of photosynthetic rate and leaf area, it can be severely limited by grazing of photosynthetic tissue or by pathogen damage. Plants can compensate for such loss in two ways: physiologically by an increase in photosynthetic rate in surviving leaves, both where other leaves are grazed (Nowak and Caldwell, 1984) and where they are infected by biotrophic fungi (Roberts and Walters, 1986); and also morphologically by new leaf growth. These new leaves become important sinks for photosynthate, and their demand is partially responsible for increased photosynthetic activity, which is mediated by a reduction in leaf carbohydrate content. It is often the case, too, that roots become much weaker sinks for carbohydrate after loss of leaves, and may actually become carbon sources (Bokhari, 1977). In a comparison of two sedges—cotton-grass *Eriophorum angustifolium* and *Carex aquatilis*, which has a larger and longer-lived root system—Chapin and Slack (1979) showed that *C. aquatilis* could continue to finance new leaf growth over several cycles of defoliation from carbohydrate reserves stored in the roots. In contrast, defoliation of *E. angustifolium* rapidly inhibited root growth, as it had no reserves and received no carbohydrate from the shoot because of competition from the developing new leaves.

If the loss of tissue is severe and the plant has neither adequate reserves nor sufficient surviving leaf area to resynthesize carbohydrates for leaf growth, there may be breakdown of proteins and other non-storage substrates. This profoundly affects a range of other activities, including nutrient uptake (Davidson and Milthorpe, 1966).

There is also the morphological problem that regrowth requires meristems. When a dicot leaf is damaged, some limited meristematic activity may occur around the site of damage, but the leaf cannot regenerate the lost tissue. Grasses, in contrast, have continuously active basal meristems and if the leaf is cut, new tissue is produced from the base. This is of course why grasses make good lawns and it also gives them a unique ability to withstand grazing. Grasses and grazing animals have certainly co-evolved (Stebbins, 1981); indeed some would go so far as to say that they had a symbiotic, even mutualistic, relationship (Owen and Wiegert, 1981). In the sense that grasses are adapted to withstand grazing, whereas many dicots suffer much more severe damage, this may be true; but there is little or no evidence that grasses get any direct benefit from being grazed, such as would make the relationship mutualistic (Belsky, 1986).

IV. ALLELOPATHY

Allelopathy is the production of substances by one plant injurious to another plant (Müller, 1970) or to microbes as well (Rice, 1984). It is a most controversial topic, as can be judged from reading Harper's (1975) review of the first edition (1974) of Rice's book. The problem is that plants contain an enormous range of more or less toxic substances, and many experimenters have attempted to demonstrate allelopathic effects by applying extracts of one plant to the seeds or seedlings of another. Quite apart from the fact that the extracts are not appropriate experimental material, since they do not occur in nature, frequently they are not sterile so that bacterial transformations may have taken place, and usually the plants have no ecological relationship. Such experiments are hard to interpret. The critical question is whether some plants have a toxic influence on others growing in the field, and this must be separate from any competitive effects for light, water, and nutrients. Thus one of the classic instances of allelopathy has always been held to be the effect of the crucifer *Camelina sativa* on flax *Linum usitatissimum* (Grümmer, 1958). Kranz and Jacob (1977) have, however, demonstrated that the interaction can be interpreted almost entirely in terms of competition for nutrients.

To demonstrate an allelopathic effect in the field it would be necessary at least to show:

 (i) that the source plant contains a toxin;
 (ii) that the toxin can move from source to target;
 (iii) that it is present in the environment of the target plant at a
        sufficient concentration to produce toxic effects; and
 (iv) that there are no simpler explanations.

In addition there is the severe problem that since the toxin presumably moves by physical processes, this requires a concentration gradient
and it will be present at lower (probably much lower) concentrations
around the target than the source plant. The source plant must
therefore either be very much more resistant than the target or release
the toxin in an inactive form. Even this still leaves the question of
ecological significance unanswered.

One very detailed investigation of an apparent case of allelopathy is
the work of Müller on the Californian chaparral (Müller, 1966). This
vegetation comprises areas of annual grasses and extensive clumps of
aromatic shrubs, particularly *Salvia leucophylla* and *Artemisia californica*.
Around each patch of shrubs there occurs characteristically a bare
zone one to two metres wide, and beyond that a zone of stunted growth
three to eight metres wide. Such zones extend equally uphill or
downhill. Muller and his co-workers systematically eliminated a range
of possible explanations—competition for water or nutrients, grazing
by animals sheltering in the shrubs (though Bartholomew (1970)
produced evidence that this might nevertheless be the explanation),
and the washing out of water-soluble toxins—and were left with the
conclusion that volatile toxins were responsible. They demonstrated
that a range of terpenes were given off by the shrubs, including α-
pinene, β-pinene, camphene, cineole and camphor, and that these were
capable of severely inhibiting seedling growth in native grasses, *Festuca
megalura*, *Bromus* spp. and *Stipa pulchra*, more so than in a standard test
plant, cucumber. Finally, they showed that the terpenes were adsorbed
by soil, remained toxic after at least two months storage in soil, and
could dissolve in cuticular waxes.

This investigation satisfies three of the requirements (above) for a
complete demonstration of an allelopathic effect, but the uncertainty
over the role of grazing animals leaves requirement (iv) open. Müller
has since described similar situations for a range of Californian arid-
zone plants, but bare zones can also be found around plants not
containing these toxins. Even if such phenomena are allelopathic, they
do not resolve the question of the importance of allelopathy in more
mesic environments where no such dramatic patterning exists. Many
workers have found inhibitory substances in plants, particularly in

species such as *Mercurialis perennis* and *Allium ursinum*, which tend to form mono-specific stands, but few have carried out all the necessary tests to demonstrate allelopathy.

Newman and Rovira (1975) found that when eight common British grassland plants were subjected to leachate of each of the other species, there were pronounced effects on yield. Leachates from all species caused yield reductions in at least some others, and the mean effect of each species as donor of leachate was a reduction of yield, averaged over all receiver species. Of the eight species tested, six had reduced yields from all donors, although of 48 combinations of these, only 10 were statistically significant. Strikingly *Trifolium repens* had its yield increased by leachate from all but one (*Holcus lanatus*) of the other species, though not from other clover plants. Again only one effect was significant: when *Rumex acetosa* increased its yield by 25% (Table 8.6). These results are important as the species were not chosen because they showed any apparent allelopathic interactions. Indeed, all commonly co-habit and all but one were collected from a single meadow. It is possible therefore that allelopathic influences are a normal phenomenon, but that their effect is generally small. Clearly a yield depression of a competitor of only a few per cent could be highly beneficial.

TABLE 8.6.   Effect of leachate from various species (donors) on growth of those species in separate pots (receivers). Figures are percentage differences from the control donor with no plant leachates (from Newman and Rovira, 1975).

Receivers		Donors				
		Ao	*Hl*	*Lp*	*Hr*	*Tr*
Anthoxanthum odoratum	(Ao)	+4	−9	−1	−5	−26[a]
Holcus lanatus	(Hl)	−23[a]	−12	−22[a]	−18	−6
Lolium perenne	(Lp)	−22[a]	−24	−25[a]	−18	−10
Hypochaeris radicata	(Hr)	−22[a]	−30[a]	−16	−38[a]	−23[a]
Trifolium repens	(Tr)	+10	−8	+14	+9	0

[a] Effect significant at p = 0·05.
Plantago lanceolata showed no effects as donor or receiver.
Cynosurus cristatus showed no effects as receiver, but a negative effect (−20%[a]) as donor to *Rumex acetosa*.
Rumex acetosa also significantly increased yield (+26%[a]) of *Trifolium repens*.

A. Mechanisms

Plants certainly produce toxic chemicals, though the significance of this is still unclear. Still more obscure are the mechanisms of action. It seems likely that many effects are indirect: perhaps the stimulation or depression of micro-organisms might be responsible, for this certainly occurs (Newman *et al.*, 1977). The role of phenolic acids is of particular interest, since these are common breakdown products, for example of lignins, and have been identified as active agents in leaves of *Adenostoma fasciculatum* and *Arctostaphylos glandulosa*, two more Californian chaparral shrubs (McPherson and Müller, 1969). Phenolic acids are abundant in soils and have been identified in rhizosphere samples (Pareek and Gaur, 1973); they have been shown by Glass (1973, 1974) massively to inhibit uptake of phosphorus and potassium by barley (Fig. 6.2). The implications of this work are unexplored, but they might well help to explain the results of Newman and Miller (1977), who, following up the earlier work of Newman and Rovira (1975), found considerable effects of leachates on P uptake.

It seems possible, therefore, that phenolic acids may act by interfering with membrane function. W. H. Müller (1965) has suggested that volatile terpenes are inhibitors of cell division, and this may be true of many toxins, since inhibition of germination is one of the commonest effects. One would expect, however, to find that target plants had evolved resistance to allelopathic toxins, much as they have to heavy metals (Chapter 5) or as insects have to plant defence compounds. Unfortunately our knowledge is far too piecemeal for it yet to be possible to investigate this.

B. Allelopathy in Perspective

The effects considered so far have involved toxic interactions between species mediated largely by terpenoids or phenolic compounds. Such effects are, however, both chemically and functionally only part of a wider spectrum. Plants exude from all parts a truly astonishing range of chemicals—sugars and scent compounds from flowers, terpenoids and soluble leachates from leaves, and an enormous range from roots. Rovira (1969) compiled reports of 10 sugars, 19 amino acids, 10 organic acids, 2 flavonones, 3 enzymes, and a nucleotide in root exudates. A similar list now would be much longer. These various chemicals may have any number of functions.

(i) *Simple excretion*: it is widely assumed that plants do not require to excrete waste products but the salt produced by halophytes such as mangroves from salt glands (Chapter 6) is certainly excreted, as are protons and hydroxyl ions from roots, and the anthocyanin which turns some nutrient-deficient plants purple is apparently a waste product. Others may be too.

(ii) *Discouragement of predators and pathogens*: this function is well established and has already been discussed.

(iii) *Encouragement of symbionts*: establishment of a symbiotic relationship requires some recognition process. Roots have defences against pathogens and these defences have to be circumnavigated for nodule or mycorrhizal symbiosis to occur. The legume–*Rhizobium* interaction is known to involve secretion of tryptophan by the legume, its conversion by the bacterium to IAA, which causes curling of the root, and enzyme degradation of root cell walls. Other such situations may involve exudates.

(iv) *Direct effects on competitors*: these are discussed above. Their existence can be taken as proven, but their significance is unclear. Potentially they could be highly potent in situations where the competitive balance is delicate.

(v) *Autotoxicity*: may investigators have found that some plants produce toxins with specific effects on their own seedlings. One of the classic investigations is that of Bonner and Galston (1944), who noticed that, in plantations of guayule (*Parthenium argentatum*–Compositae), edge plants were always larger than central ones. They eventually ascribed this to root-exuded *trans*-cinnamic acid, which they found to be toxic to *Parthenium* at 1 p.p.m., but to tomato only at 100 p.p.m. Such effects can perhaps be explained on the grounds that, for a competitive dominant, the most severe form of competition is intra-specific. It was noticeable that in the experiments of Newman and Rovira (1975) species that were autotoxic were those that tended to grow naturally as isolated individuals, such as *Plantago lanceolata*, while species found naturally in pure stands, for example *Holcus lanatus*, were not autotoxic, and were more inhibited by other species.

Interactions between plants and other organisms, then, involve both the removal of resources from the environment (as in competition) and the addition of substances to it (as in defence and allelopathy). The range of compounds produced by plants is vast and we have as yet only scratched the surface in this field.

9. An Ecological Perspective

I. ALLOCATION

Environmental physiology is in effect the physiology of adaptation, and so it involves a consideration of optimal patterns of response, as described in Chapter 1. The operation of any physiological process requires resources, whether these be carbon for respiration or for molecular skeletons, nitrogen for enzyme construction, phosphorus for nucleotides or water for the transport of materials. This is why the first part of this book was devoted to describing the ways in which plants respond to limitations in the supply of these resources.

Selection will act on a plant in relation to these resources in two ways, in terms of both the amount the plant acquires and the way in which it uses each resource. For example, a plant growing beneath a leaf canopy is limited in its growth by the low photon flux density (cf. p. 82), and therefore by the small amount of carbon it can fix. Increased fitness is likely (but not certain) to result both from mechanisms which increase quantum efficiency and from patterns of carbon allocation which favour leaf growth over roots, thereby increasing the interception of radiation. Some extreme shade plants from tropical rainforests have curious leaf properties (such as red undersurfaces or blue iridescence) which apparently increase absorption (Lee and Graham, 1986), and it seems reasonable to propose that these have been selected precisely as a result of that. A more general response to low irradiation, however, is an increase in allocation of biomass to leaves as opposed to roots, just as plants in nutrient-deficient or very dry environments tend to display the opposite allocation pattern.

For any given set of environmental conditions, one particular pattern of allocation of resources to various structures and activities will result in maximum carbon gain; this pattern is defined as optimal. Of course, such a plant might be so vulnerable to herbivores, for example, that this pattern would fail to maximize fitness and would be selected against. Nevertheless, this approach is valuable because it

enables us to quantify the physiological and morphological characteristics which contribute to fitness. Iwasa and Roughgarden (1984)
modelled resource allocation in a simple way and showed that when
the allocation ratio between roots and shoots deviated from the
optimum, all future allocation should be to the part that could restore
the optimum (leaves in shade, roots in drought) until that was regained
(Fig. 9.1).

Although resource allocation is normally perceived in terms of
carbon, there is no reason why other plant resources such as nitrogen
should not be the critical currency in particular conditions. For
example, there is a very close relationship between the maximum
photosynthetic rate that a leaf can achieve (A_{max}) and its nitrogen
concentration (Fig. 9.2). This appears to be because the single greatest
limitation to the maximum rate of photosynthesis in C_3 plants is the
amount of ribulose bisphosphate carboxylase, the CO_2-fixing enzyme.
This enzyme is the most abundant protein in the world and typically

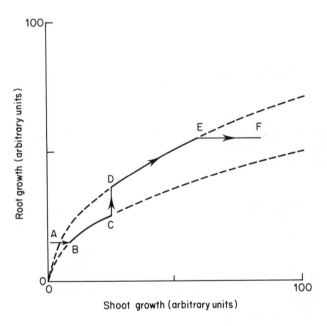

FIG. 9.1. Simulation of root and shoot growth under changing environmental
conditions. The two dashed lines represent optimal growth paths, the upper for
conditions where water limits growth more than PAR, the lower for PAR-limited
growth. In each case the appropriate organ receives greater allocation of biomass. The
thick solid line is a growth trajectory for a plant limited by PAR from A→C and from
E→F, and by water from C→E. From Iwasa and Roughgarden (1984).

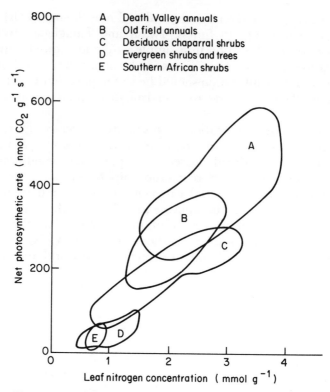

FIG. 9.2. The relationship between photosynthetic rate and leaf nitrogen concentration, for a wide range of species, shows that investment of N into leaves is a major constraint. Each line encircles an area containing many data points for each category. From Field and Mooney (1986).

comprises around half the total leaf protein. The relationship of Fig. 9.2 implies a severe allocation problem for plants: maximizing carbon gain is promoted by high allocation of both carbon and nitrogen to leaves. The former allows the development of large leaf areas which increase the interception of radiation, while the latter provides the biochemical machinery for CO_2 assimilation. The acquisition of the nitrogen, however, is a root function, and maximizing that may involve preferential allocation of resources to roots.

Such conflicts lead to non-linear responses. In other words, there is no simple linear relationship between pattern and response; rather there is an optimum, a pattern of resource allocation either side of which growth (or some correlate of growth) is reduced. Another example of such an optimization model is that proposed by Givnish

(1986) to relate resource allocation and stomatal behaviour. Models of optimal stomatal function have been suggested by Cowan (1982) and Cowan and Farquhar (1977) (see Chapter 1), which depend on the slope of the relationship between transpiration and CO_2 assimilation. Givnish also takes into account the cost of the roots required to provide the water for transpiration and the effects on leaf water potential that changes in the ratio of transpiring to absorbing surface will have, and in so doing derives a remarkably simple relationship:

$$\frac{r_m}{r_s} = \frac{f}{1-f} \qquad (9.1)$$

where r_m and r_s are the mesophyll (residual) and stomatal resistances in the leaves, and f is the fraction of biomass allocated to the leaves (so that $1-f$ is the fraction allocated to roots). This relationship must be maintained if carbon gain is to be maximized, and it relates in a strikingly simple way the physiological and the morphological phenomena. The two resistances represent on the one hand the efficiency of the biochemical machinery and on the other the control of CO_2 acquisition and water loss, while the morphological ratio is a familiar one that crops up in many guises and in many situations. Where the "biochemical" resistance predominates, leafy plants should be favoured, but where stomatal resistance is high, suggesting that water may be limiting, more resources must go to roots.

II. STRATEGIES

Defining an optimum response pattern as that which maximizes carbon gain is a highly simplistic approach, albeit useful in pinpointing the consequences of particular patterns of behaviour. In practice the biological optimum is that which maximizes fitness, and that means that the reproductive success of the individual must be taken into account as well. Reproduction is another activity to be included in any budget of plant resources, along with shoot and root vegetative growth, respiration, defences against animals and pathogens, storage and so on. Resources allocated to reproduction are not available for the acquisition of further resources, as can be seen simply by considering the classical growth equation (Chapter 1) relating relative growth rate ($\mathbf{R}$) to leaf area ratio ($\mathbf{F}$) and net assimilation rate ($\mathbf{E}$):

$$\mathbf{R} = \mathbf{E} \times \mathbf{F} \quad \text{or} \qquad (9.2)$$

$$\frac{1}{W} \cdot \frac{\mathrm{d}W}{\mathrm{d}t} = \frac{1}{A_{\mathrm{L}}} \cdot \frac{\mathrm{d}W}{\mathrm{d}t} \cdot \frac{A_{\mathrm{L}}}{W} \tag{9.3}$$

This clearly demonstrates that growth rate depends on the ratio of leaf area (A_{L}) to plant weight (W), so that diversion of resources to reproductive structures reduces growth rate. In this sense therefore reproduction and growth are alternatives. In many plants it is also true that their branching structure or architecture restricts the number of growing points they can bear, and here the conversion of a meristem to a flower means that it cannot produce leaves (Fig. 9.3). Of course, many reproductive structures are green, particularly bracts and sepals around flowers, and they may make some contribution to growth (or at least to the carbon requirements of the flower they surround: Bazzaz *et al.*, 1979), but this does not contradict the essential point.

Some plants are annual or ephemeral: they complete their life-cycle

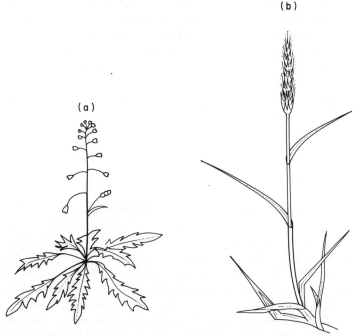

Fig. 9.3. Two plant architectures which involve constraints on growth following initiation of reproduction. (a) *Capsella bursa-pastoris* (Cruciferae) is a typical rosette plant in which leaf production ceases after the production of the almost leafless flowering stem. (b) *Phleum arenarium* (Gramineae) is an annual grass and grows by producing new tillers at the stem base. Each tiller is partially independent and has determinate growth: it produces a few leaves and ends in a flowering spike.

within a single growing season, or in the extreme (*Arabidopsis thaliana*, a temperate weed, and some desert plants) within a few weeks. They may then devote a very high proportion of their resources, often as high as 50–60%, to seed production. As a result they are capable of rapid changes in population size, and are particularly suited to growth in environments where the adult population is likely to be drastically reduced by hazards outside their range of resistance. In other words, selection acts on their ability to achieve high rates of population increase, described by the parameter r in the classical logistic equation for population growth:

$$\frac{\mathrm{d}N}{\mathrm{d}t} = \frac{rN(K-N)}{K} \tag{9.4}$$

Here N is population size and t time; K is the equilibrium population size. Such species are often referred to as r-selected. At the other extreme are plants which persist for a long time—they are perennial. Often they do not reproduce at all for the first few years of their life; many trees, for example, have such a juvenile period of from 5–15 years, and even in monocarpic species (those which only reproduce once in their life-time), the pre-reproductive phase can be remarkably long (e.g. *Agave deserti*, cf. Chapter 4). Even when they start to produce seeds, trees rarely devote more than a few per cent of their annual productivity to sexual reproduction. Perennial herbs tend to have higher reproductive allocation rates, largely because they produce less supporting tissue, but maximum values of around 30% are typical. Of course, many perennials reproduce both vegetatively and sexually, and in such cases the total reproductive allocation again seems to have a similar ceiling, though it may be difficult to distinguish vegetative reproduction (e.g. by a stolon which produces new plants at its tip) from growth, particularly in grasses, which normally grow in just this way by producing tillers, which are effectively new grass plants, though physically linked to the parent.

In all such plants, selection appears to have acted more on their ability to occupy space, resist competition, and maintain activity over a long period. They will tend therefore to exhibit only slight changes in population size, and often this is maintained at or near some equilibrium value denoted by K in equation 9.4 (often ineptly termed the "carrying capacity"). They are therefore referred to as K-selected. This dichotomy between growth-oriented and reproduction-oriented, between K-selected and r-selected plants offers a useful evolutionary viewpoint on the constraints which determine plant growth and form. We expect (and find) that plants of disturbed environments are

r-selected, those of undisturbed and productive sites, such as temperate forests, are K-selected.

But such a one-dimensional ordination of plant form is inevitably a caricature of its diversity. In particular it fails totally to distinguish plants of fertile, productive sites where resources provide little limit to growth, from those of extreme environments, such as deserts, mountains and toxic soils. Both groups apparently fall into the K-selected category. They are however separated in the alternative, functional classification suggested by Grime (1977, 1979), which has three, rather than two poles. It derives from a two-dimensional ordination of the favourability of the environment and the level of disturbance, in which one extreme (unfavourable, disturbed sites) is uninhabitable, leaving a triangle (Fig. 9.4).

Grime refers to the positions that plants occupy in this triangle as strategies, a term that has provoked much controversy, because to some

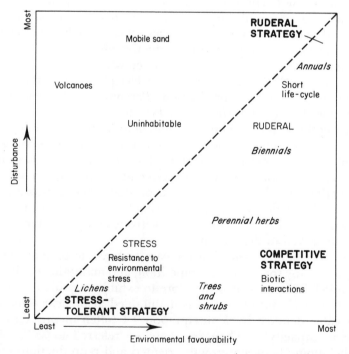

FIG. 9.4.   Grime's (1977) triangular ordination derives from an ordination using the two axes of disturbance and environmental favourability. Bold type represents the three primary strategies, italics the general position occupied by some major life-forms, and roman type the main selective forces.

it suggests connotations of planning by plants of their goals, or in other words of teleology. Nevertheless the word is useful, as it indicates that taxonomically unrelated, but ecologically similar species of plant may possess common characteristics that enable them to survive in particular habitats. There are three underlying strategies in this scheme, at the three poles of the triangle:

  (i) in favourable, disturbed environments the *ruderal* strategy is favoured, corresponding to *r*-selection;
 (ii) in favourable, undisturbed sites *competitive* plants occur;
(iii) and in unfavourable, but undisturbed habitats are found *stress-tolerators*.

Various intermediate categories are possible within the scheme, but the relationship of these three strategies to the theme of this book is very close. Ruderals are plants of rapid growth potential, devoting a large proportion of their resources to reproduction, and growing in environments which, because they are recently disturbed, offer little competition for resources from other individuals. In consequence, ruderals can obtain resources without difficulty, and have little recourse to symbioses for nutrient acquisition, such as nitrogen-fixing bacteria or mycorrhizas. Equally, they are short-lived, unlikely to suffer severe pathogen damage, and so devote little to defence.

The competitive strategy, in contrast, though still often associated with rapid growth rate, is found in more persistent plants, which use resources less for seed production and more for support tissue to provide height growth which increases competitive effectiveness for radiation; for root systems to obtain water and ions in competition; for symbioses and defence, and so on (cf. Chapter 8). They often have high turn-over rates, for example, for leaves (cf. Chapter 3), since in a competitive situation the ability to redistribute resources from old leaves to new and better-placed ones may be crucial.

Finally, and perhaps most interestingly, there are stress-tolerators. These are plants of unfavourable environments, where resistance to environmental extremes may be the dominant selective force. In deserts, adaptations which increase water use efficiency are obvious, but may require diversion of resources or directly reduce growth potential. Very high root weight ratios imply low growth rates, as do morphological adaptations such as reduced leaf blade dimensions (Chapters 4 and 5) or metabolic phenomena such as crassulacean acid metabolism (CAM; Chapters 2 and 4).

The concept of strategies, in the sense that plants which have similar ecological behaviour share common sets of characteristics, is not new.

The Danish botanist Raunkiaer (1907) produced a scheme of plant life-forms, based upon the position of the overwintering organs (seeds, buds below, at or above ground level, in water, etc.), that proved successful and was widely used for comparing floras from different habitats. Another Dane, Warming (1909), developed a strictly habitat-based classification, including categories such as lithophytes (plants on rock) and psammophytes (plants on sand), as well as more familiar terms such as halophytes, for plants of saline soils.

The success of Raunkiaer's scheme derived from its ability to relate other aspects of a plant's biology to its habitat preference; in this sense it is a set of strategies. In deserts, for example, therophytes (annuals) are particularly numerous, while in grasslands it is hemicryptophytes (buds at ground level) and in the tropics phanerophytes (buds > 25 cm above the ground—trees and shrubs) which are most abundant. It is easy to produce arguments based on adaptation to explain these distributions, but they say little about physiology. Schemes like those of Grime (1979) offer a greater possibility of explaining ecological phenomena on a physiological basis.

III. COMMUNITIES

Concepts such as strategies and life-forms imply a recognition that the selective effects of environment on a wide taxonomic range of species will be sufficiently similar and powerful that they will come to possess many similar features, both physiological and morphological. When this similarity is perceived on a wide enough geographic scale we call it *convergent evolution*. In Mediterranean climates throughout the world (for example in California, Chile, Western Australia and southern Africa as well as around the Mediterranean itself) a community of evergreen, sclerophyllous, often aromatic, drought-resistant shrubs occurs, in which the component species are very similar in appearance and ecology, but often unrelated taxonomically.

The same phenomenon can be seen in a less striking fashion, however, in any plant community. A typical north temperate decidu-ous forest will have in its canopy a number of different tree species, perhaps in the genera *Quercus, Tilia, Betula, Ulmus*, and *Acer*, all in different taxonomic families and yet all clearly possessing many common features. Whether they have converged from very different ancestors or have all descended from a common ancestor, it is certainly the case that part of the reason why they are able to occur together in the same habitat is that they are all suited ("adapted") to it.

This rather unremarkable statement leads, however, to a paradox. It is an axiom of ecology (Gause's principle) that no two species can co-exist if they have identical requirements, or in other words share the same niche. There is, therefore, a conflict: for two species to have the potential to co-exist they must be sufficiently similar to survive in the same environment, yet if they are too similar, their requirements may overlap to the extent that competition between them may eliminate one or other (Chapter 8). The fundamental question, therefore, is whether a greater increase in fitness is achieved by a phenotypic change which leads to greater adaptation to the physical environment or by one which reduces competitive overlap. For example, in the native vegetation of the north American prairies, the grass *Agropyron spicatum* is a common and important species, highly competitive towards the other native grasses it grows with. It shows virtually no root growth in winter, when soil temperatures are very low, but then undergoes rapid elongation in spring, which enables it to exploit soil moisture before the period of summer drought (Harris, 1967). If this behaviour were analysed in terms of optimization, it might well be possible to demonstrate that this was an optimum growth pattern. Over much of its former native area, however, it has been replaced by the introduced Eurasian grass *Bromus tectorum*, whose roots elongate at lower temperatures and are already established in the soil profile before *Agropyron* becomes active (Fig. 9.5). It is therefore able to outcompete *Agropyron* for water, by absorbing the available water in any given soil layer before the *Agropyron* roots can reach it.

This example emphasizes the importance of considering the total environment of a plant, which is why we have included organisms as environmental factors in this survey of environmental physiology. What may appear to be the optimum for a process in controlled conditions in the laboratory (the *physiological optimum*), may be very far from it in the field. In the same way one can describe the fundamental niche (Hutchinson, 1957) of a species in terms of its ability to withstand physico-chemical features of the environment by, for example, growing plants in pots in a glasshouse in a range of soils and measuring their performance. Invariably, one finds that plants will grow (and often grow well) under such conditions in soils which they appear totally unable to colonize in nature. This has been demonstrated many times (e.g. by Tansley, 1917 and Ellenberg, 1953; Fig. 9.6), and Hutchinson (1957) called the often very small part of the fundamental niche in which a plant can grow in the wild, its *realized niche*.

Interactions between species then are an integral part of the

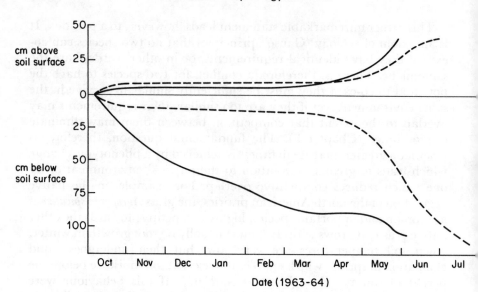

FIG. 9.5. Above- and below-ground growth of two co-occurring grasses —the native *Agropyron spicatum* (– – –) and the introduced *Bromus tectorum* (——)—in the North American prairies. *Bromus* starts root growth earlier in the season and so always has roots at greater depth than *Agropyron*. This seems to explain its competitive superiority for water. From Harris (1967).

environmental physiology of a plant. They lead to large modifications in optimal behaviour, and may result in a species normally growing under conditions very different from those to which it is apparently adapted. One should beware, however, of taking too deterministic a view of a plant community. Competitive effects on a plant in a community are typically the result of the actions of many other species, and not specifically of one frequently associated neighbour. Clearly, if two species normally grew in association and interacted in such a way that, for example, one always shaded the leaves of the other, one would expect that there would be powerful selection pressure on the subordinate species for the development of a shade physiology, taller stems, earlier growth or some other "adaptation". In practice, however, such precise association is rare and the sort of co-evolution that this implies is probably unusual in plant communities.

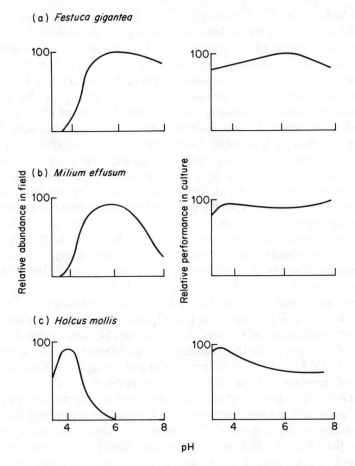

Fig. 9.6.   Fundamental and realized niches of three common woodland grasses in relation to soil pH. On the left is the distribution of the species in the field; on the right the relative performance at different pH values in culture. The best performance in the field is usually at a more extreme pH than in culture. From Ellenberg (1953).

IV. DYNAMICS

One of the most powerful factors mitigating against the sort of predictability of association that might lead to co-evolution, is the dynamic nature of plant communities. Communities are made up of individuals, which have finite life-spans. If, on average, each death is replaced by an individual of the same species, the community is stable, but it is probably more common for different species to invade. This is particularly true if a community is developing from some recent disturbance, in the process known as ecological succession.

The environment of early successional communities differs in many ways from that of communities which have not been disturbed for a long period. Where the disturbance has been simply the removal of the vegetation, which then typically regenerates rapidly in the process known as secondary succession, the main environmental effect is on the microclimate, since the soil is often more or less undisturbed. If the mature community is of trees, these create a sub-canopy environment of low radiation flux density, low windspeed, little temperature fluctuation, and normally little risk of water deficit. In these conditions, seedlings of many trees can establish much more readily than is sometimes possible in exposed, recently disturbed environments. The effects can sometimes be unexpected: Finegan (1984) found that oak *Quercus robur* seeds failed to colonize a bare chalk quarry floor because they were too visible and so eaten by birds. Once hawthorn *Crataegus monogyna* became established, however, the acorns could be hidden in the hawthorn leaf litter, and oaks appeared.

If the environmental disturbance is more severe and a new, undeveloped substrate, such as moraine gravel, lava or sand, is colonized, then there are enormous differences in soil factors between early and late stages of succession. In the classic study at Glacier Bay, Crocker and Major (1955) found a 10-fold increase in soil nitrogen over the first 100 years of succession (Fig. 9.7), largely because of the activity of nitrogen-fixing alders. This was accompanied by a reduction in pH of over 3 units and a large increase in soil organic matter and therefore soil water-holding capacity. Obviously the initial colonists—plants such as fireweed *Epilobium angustifolium* and later mountain avens *Dryas drummondii*—experienced a totally different environment from the spruces and firs that eventually came to occupy the site.

At one level these changes can be seen in terms of Grime's strategies (see p. 336, above). In secondary successions, the initial colonists are typically fast-growing plants, capable of taking advantage of the relatively fertile soil and lack of competition; in other words, they are *ruderals*. In primary successions, however, where soil water and nutrient levels are initially low and other environmental factors, such as temperature, are often extreme, it is slow-growing plants resistant to such stresses that predominate. The classic primary colonists are lichens, which are excellent examples of *stress-tolerant* plants. Whichever starting point a succession may derive from, however, the mature communities that result are ones in which the level of utilization of resources, and hence competition for those resources, is high. The

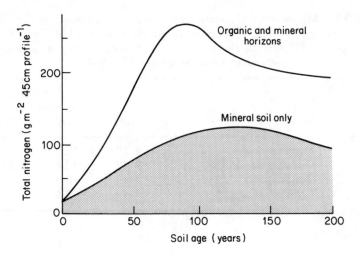

Fɪɢ. 9.7. Soil nitrogen concentration increased markedly in the period from 50–100 years after exposure on glacial moraine at Glacier Bay, Alaska, corresponding with the period of dominance of N-fixing alders. The later decline occurs when spruce replaces alder. From Crocker and Major (1955).

plants that dominate these therefore typify the *competitive* strategy.

At the physiological level of analysis similar patterns emerge. Early successional species are by definition shorter-lived than later species, and this implies, as we have seen, faster growth rates. These in turn require higher photosynthetic rates, a phenomenon which is generally well supported by experiments (Bazzaz, 1979). In a detailed study of succession in central European hedgerows, Küppers (1982: described in Schulze *et al.*, 1986) measured maximum CO_2 assimilation rate and annual productivity for several woody species. The early successional blackthorn *Prunus spinosa* and the mid-successional hawthorn *Crataegus macrocarpa* and field maple *Acer campestre* had much higher rates for both parameters than the late-successional beech *Fagus sylvatica* (Table 9.1). Many late-successional plants have a typical "shade-plant" physiology (Fig. 9.8; cf. p. 53).

The necessity for the early species to be highly productive has already been stressed. Later species may have to devote more resources to support, defence and other non-growth functions. In addition, there is an architectural problem to be considered. Horn's (1971) theory of monolayer and multilayer tree canopies has already been described (Chapter 2, p. 41). As he points out, the greatest productivity is attained by exposing the largest leaf area to radiation, and this is

TABLE 9.1. Maximum CO_2 assimilation rate (MCAR/μmol m^{-2} s^{-1}), leaf productivity (g C g^{-1} leaf y^{-1}) and stand productivity (g C m^{-2} y^{-1}) for four woody species differing in the stage at which they appear in a hedgerow succession. From Schulze *et al.* (1986).

| Stage | Species | MCAR | Productivity | |
			Leaf	Stand
Early	*Prunus spinosa*	9–12	12·6	372
Mid	*Crataegus macrocarpa*	9–12	12·1	516
	Acer campestre	9–12	9·7	324
Late	*Fagus sylvatica*	3–4	2·4	108

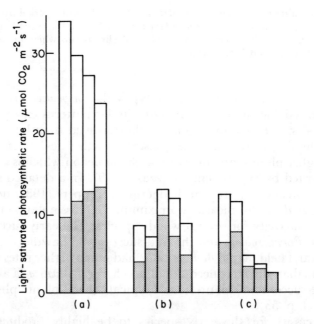

FIG. 9.8. Light-saturated rate of photosynthesis in (a) early, (b) mid and (c) late-successional species, from a successional gradient in Illinois. Open bars are for plants grown in sun, the hatched parts for plants grown in shade (1% full daylight). The early species are herbs (*Ambrosia trifida, Datura stramonium, Chenopodium album, Polygonum pennsylvanicum*); the mid species are fast-growing trees and shrubs (*Rhus typhina, Prunus pennsylvanica, Fraxinus americana, Platanus occidentalis, Quercus imbricaria*); and the late species are shade-resistant herbs (*Hydrophyllum virginianum*) or trees (*Quercus rubra, Aesculus glabra, Tilia americana, Acer saccharum*). From Bazzaz and Carlson (1982).

achieved by the multilayer canopy, which lets a high proportion of the incident radiation penetrate to the lower layers. In doing so, however, it also permits shade-resistant saplings of other species to grow beneath the canopy, and so is susceptible to invasion and replacement. The monolayer trees, with a low total leaf area produced by a poorly penetrable canopy, are therefore much less productive but relatively immune to invasion, until a tree dies.

Similar differences can be found in relation to other environmental factors. Nitrogen, as shown in Fig. 9.7, usually builds up in soil as succession and soil development proceeds, as a result of microbial dinitrogen fixation. Tilman (1985) showed that soil nitrogen increased from 20 to 850 mg kg^{-1} during succession on prairie areas in Minnesota. When nine species were ranked in order of the time of their development in this successional sequence, it can be seen that their ability to extract soil nitrogen declines accordingly (Fig. 9.9). One

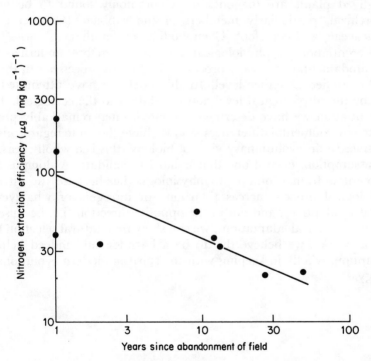

FIG. 9.9. Nitrogen extraction efficiency of nine plant species as a function of the time each takes to colonize abandoned agricultural fields. Extraction efficiency, measured as the quantity of nitrogen (µg) extracted by the plant, as a function of the N concentration in soil (mg kg^{-1}), is lower in later colonists. N in soil increases from 20 to 850 mg kg^{-1} during the sequence. From Tilman (1985).

reason for the success of the early successional species in this system, therefore, may be their greater ability to obtain this limiting resource.

An alternative response to resource limitation is the adoption of symbioses. Nitrogen-fixing associations (see Chapter 3, p. 110) are commonly found in developing communities, at middle stages of successions. Indeed, the invasion of a nitrogen-fixing species may be a critical step in the process, as with alder *Alnus crispa* at Glacier Bay (see Fig. 9.7). As N concentrations in soil increase, however, the benefits to the fixing species decline and the costs of the symbiosis presumably remain, rendering the species less competitive. N-fixers are therefore rare in mature communities. In contrast, the availability of phosphorus almost invariably declines with soil and community age, as P is lost by erosion, chemical inactivation and transfer to intractable organic complexes. The benefits of being mycorrhizal, therefore, would be expected to increase as the community matures, and it is the case that short-lived plants are the ones most commonly found to be non-mycorrhizal, particularly members of the cabbage (Cruciferae or Brassicaceae) and goosefoot (Chenopodiaceae) families.

Differentiation of physiological response therefore underlies the most fundamental ecological processes, and those processes can therefore be studied at either level. In this book we have attempted to describe the physiological level and to relate it to the ecological. The differentiation we have described is related, often remarkably precisely, to environmental differences, and we have chosen to regard this as adaptive. In an evolutionary view of biology, this is a wholly reasonable assumption, even if one that is hard to validate. As long as the argument is from evolution to physiology (i.e. because the altered physiological process increases carbon—or nitrogen, or whatever—gain, it is adaptive), and not in the opposite direction (i.e. because of these physiological adaptations, we can show that natural selection has been at work), we believe this to be a harmless and indeed helpful assumption, wholly in keeping with the spirit of modern evolutionary biology.

References

ABBAS AL-ANI, M. K. and HAY, R. K. M. (1983). The influence of growing temperature on the growth and morphology of cereal seedling root systems. *J. exp. Bot.* **34,** 1720–1730.

ABELES, F. B. (1973). "Ethylene in Plant Biology". Academic Press, New York and London.

ABEYAKOON, K. F. and PIGOTT, C. D. (1975). The inability of *Brachypodium sylvaticum* and other species to utilise apatite or organically bound phosphate in calcareous soil. *New Phytol.* **74,** 147–154.

ABRAHAMSON, W. G. and CASWELL, H. (1982). On the comparative allocation of biomass, energy and nutrients in plants. *Ecology* **63,** 982–991.

ADDICOTT, F. T. and LYON, J. L. (1973). Physiological ecology of abscission. *In* "Shedding of Plant Parts" (Ed. T. T. Kozlowski), pp. 85–124. Academic Press, New York and London.

ADDISON, P. A. and BLISS, L. C. (1984). Adaptations of *Luzula confusa* to the polar semi-desert environment. *Arctic* **37,** 121–132.

AHMAD, I. and WAINWRIGHT, S. J. (1976). Ecotype differences in leaf surface properties of *Agrostis stolonifera* from salt marsh, spray zone, and inland habitats. *New Phytol.* **76,** 361–366.

ALBERDA, T. (1954). Growth and root development of lowland rice and its relation to oxygen supply. *Plant and Soil* **5,** 1–28.

ALBERT, R. (1975). Salt regulation in halophytes. *Oecologia* **21,** 57–71.

ALEXANDER, C. and HADLEY, G. (1984). The effect of mycorrhizal infection of *Goodyera repens* and its control by fungicide. *New Phytol.* **97,** 391–400.

ALEXANDER, I. J. and HOGBERG, P. (1986). Ectomycorrhizas of tropical angiospermous trees. *New Phytol.* **102,** 541–549.

AMEN, R. D. (1966). The extent and role of seed dormancy in alpine plants. *Quart. Rev. Biol.* **41,** 271–281.

ANDERSON, M. C. (1964). Studies of the woodland light climate. I. The photographic computation of light conditions. *J. Ecol.* **52,** 27–42.

ANDERSON, M. C. (1970). Radiation climate, crop architecture and photosynthesis. *In* "Prediction and Measurement of Photosynthetic Productivity" (Ed. I. Setlik), pp. 71–78. Centre for Agricultural Publishing and Documentation, Wageningen.

ANDERSON, M. C. and MILLER, E. E. (1974). Forest cover as a solar camera: penumbra effect in plant canopies. *J. appl. Ecol.* **1,** 691–698.

ANDERSON, W. P. (1976). Transport through roots. *In* "Transport in Plants IIB". *Encyclopaedia of Plant Physiology* **2**, 129–156. Springer Verlag, Berlin.

ANDREW, C. S. and VANDEN BERG, P. J. (1973). The influence of aluminium on phosphate sorption by whole plants and excised roots of some pasture legumes. *Austr. J. agric. Res.* **24**, 341–351.

ANTONOVICS, J., BRADSHAW, A. D. and TURNER, R. G. (1971). Heavy metal tolerance in plants. *Adv. Ecol. Res.* **7**, 8–25.

AP REES, T. and WILSON, P. M. (1984). Effects of reduced supply of oxygen on the metabolism of roots of *Glyceria maxima* and *Pisum sativum*. *Z. Pflanzenphysiol.* **114**, 493–503.

ARMSTRONG, W. (1964). Oxygen diffusion from the roots of some British bog species. *Nature* **204**, 801–802.

ARMSTRONG, W. (1967). The oxidising activity of roots in waterlogged soils. *Physiol. Plant.* **20**, 920–926.

ARMSTRONG, W. (1975). Waterlogged soils. *In* "Environment and Plant Ecology". (Ed. J. R. Etherington), pp. 181–218. Wiley, London.

ARMSTRONG, W. (1979). Aeration in higher plants. *Adv. Bot. Res.* **7**, 226–332.

ARMSTRONG, W. and BOATMAN, D. J. (1967). Some field observations relating the growth of bog plants to conditions of soil aeration. *J. Ecol.* **55**, 101–110.

ARMSTRONG, W., BOOTH, T. C., PRIESTLY, P. and READ, D. J. (1976). The relationship between soil aeration, stability and growth of Sitka spruce (*Picea sitchensis* (Bong.) Carr.) on upland peaty gleys. *J. appl. Ecol.* **13**, 585–591.

ARMSTRONG, W. and GAYNARD, T. J. (1976). The critical oxygen pressures for respiration in intact plants. *Physiol. Plant.* **37**, 200–206.

ARMSTRONG, W. and WEBB, T. (1985). A critical oxygen pressure for root extension in rice. *J. exp. Bot.* **36**, 1573-1582.

ASHENDEN, T. W. (1978). Growth reductions in cocksfoot (*Dactylis glomerata* L.) as a result of SO_2 pollution. *Environ. Pollut.* **15**, 161–166.

ASHENDEN, T. W. (1979). The effects of long-term exposures to SO_2 and NO_2 pollution on the growth of *Dactylis glomerata* L. and *Poa pratensis* L. *Environ. Pollut.* **18**, 249–258.

ASHENDEN, T. W. and MANSFIELD, T. A. (1977). Influence of wind speed on the sensitivity of ryegrass to SO_2. *J. exp. Bot.* **28**, 729–735.

ASHENDEN, T. W. and MANSFIELD, T. A. (1978). Extreme pollution sensitivity of grasses when SO_2 and NO_2 are present in the atmosphere together. *Nature* **273**, 142–143.

ASHER, C. J. and LONERAGAN, J. F. (1967). Response of plants to phosphate concentration in solution culture. 1. Growth and phosphate content. *Soil Sci.* **103**, 225–233.

ASPINALL, D. (1960). An analysis of competition between barley and white persicaria. II Factors determining the course of competition. *Ann. appl. Biol.* **48**, 637–654.

AUNG, L. H. (1974). Root-shoot relationships. *In* "The Plant Root and Its Environment" (Ed. E. W. Carson). pp. 29–61. University Press, Virginia.

AUSTENFELD, F-A. (1974a). Der Einfluss des NaCl und anderer Alkalisalze auf die Nitratreduktaseaktivität von *Salicornia europaea* L. *Z. Pflanzenphys.* **71**, 288–296.

AUSTENFELD, F-A. (1974b). Untersuchungen zum Ionenaushalt von *Salicornia europaea* L. unter besonderer Berücksichtigung des Oxalats in Abhängigkeit von der Substratsalinität. *Biochem. Physiol. Pflanzen* **165**, 303–316.

BAGSHAW, R., VAIDYANATHAN, L. V. and NYE, P. H. (1972). The supply of nutrient ions to plant roots in soil V. *Plant and Soil* **37**, 617–626.

BALDWIN, J. P. (1975). A quantitative analysis of the factors affecting plant nutrient uptake from some soils. *J. Soil Sci.* **26**, 195–206.

BALDWIN, J. P. and NYE, P. H. (1974). Uptake of solutes by multiple root systems from soil IV. A model to calculate the uptake by a developing root system or root hair system of solutes with concentration variable diffusion coefficients. *Plant and Soil* **40**, 703–706.

BANGE, G. G. J. (1953). On the quantitative explanation of stomatal transpiration. *Acta Bot. Neerlandica,* **2**, 255–296.

BANNISTER, P. (1971). The water relations of heath plants from open and shaded habitats. *J. Ecol.* **59**, 51–64.

BANNISTER, P. (1976). "Introduction to Physiological Plant Ecology". Blackwell, Oxford.

BARBER, D. A., BOWEN, G. D. and ROVIRA, A. D. (1976). Effects of microorganisms on absorption and distribution of phosphate in barley. *Aust. J. Plant Physiol.* **3**, 801–808.

BARBER, S. A. (1974). Influence of the plant root on ion movement in soil. *In* "The Plant Root and its Environment" (Ed. E. W. Carson), pp. 525–563. University Press, Virginia.

BARCLAY, A. M. and CRAWFORD, R. M. M. (1982). Plant growth and survival under strict anaerobiosis. *J. exp. Bot.* **33**, 541–549.

BARLEY, K. (1970). The configuration of the root system in relation to nutrient uptake. *Adv. Agron.* **22**, 159–201.

BARLEY, K. and ROVIRA, A. D. (1970). The influence of root hairs on the uptake of phosphate. *Comm. Soil. Sci. Pl. Anal.* **1**, 287–292.

BAROSS, J. A. and MORITA, R. Y. (1978). Microbial life at low temperatures: Ecological aspects. *In* "Microbial Life in Extreme Environments" (Ed D. J. Kushner), pp. 9–71. Academic Press, London and New York.

BARROW, N. J. (1973). On the displacement of adsorbed anions from soil. I. Displacement of molybdate by phosphate and by hydroxide. *Soil Sci.* **116**, 423–431.

BARROW, N. J. (1975). The response to phosphate of two annual pasture species. II. The specific rate of uptake of phosphate, its distribution, and use for growth. *Austr. J. Agric. Res.* **26**, 145–156.

BARRS, H. D. (1968). Determination of water deficits in plant tissues. *In* "Water Deficits and Plant Growth" (Ed. T. T. Kozlowski), vol. 1, pp. 235–368. Academic Press, New York and London.

BARTHOLOMEW, B. (1970). Bare zone between California shrub and grassland communities: the role of animals. *Science* **170**, 1210–1212.

BAUMEISTER, W. and KLOOS, G. (1974). Über die Salzsekretion bei *Halimione portulacoides* (L.) Aellen. *Flora* **163**, 310–326.

BAYLIS, G. T. S. (1975). The magnolioid mycorrhiza and mycotrophy in root systems derived from it. *In* "Endomycorrhizas" (Eds. F. E. Sanders, B. Mosse and P. B. Tinker), pp. 373–390. Academic Press, London and New York.

BAZZAZ, F. A. (1979). The physiological ecology of plant succession. *Ann. Rev. Ecol. Syst.* **11**, 287–310.

BAZZAZ, F. A. and CARLSON, R. W. (1982). Photosynthetic acclimation to variability in the light climate of early and late successional plants. *Oecologia* **54**, 313–316.

BAZZAZ, F. A., CARLSON, R. W. and HARPER, J. L. (1979). Contribution to reproductive effort by photosynthesis of flowers and fruits. *Nature* **279**, 554–555.

BECKETT, P. H. T. (1964). Potassium–calcium exchange equilibria in soils: specific adsorption sites for potassium. *Soil Sci.* **97**, 376–383.

BEGG, J. E. (1980). Morphological adaptations of leaves to water stress. In "Adaptation of Plants to Water and High Temperature Stress" (Eds. N. C. Turner and P. J. Kramer), pp. 33–42. Wiley-Interscience, New York.

BELL, J. N. B. (1984). Air pollution problems in western Europe. ln "Gaseous Air Pollutants and Plant Metabolism" (Eds. M. J. Koziol and F. R. Whatley), pp. 3–24. Butterworth, London.

BELL, J. N. B. and CLOUGH, W. S. (1973). Depression of yield in ryegrass exposed to SO_2. *Nature Lond.* **241**, 47–94.

BELL, J. N. B. and MUDD, C. H. (1976). Sulphur dioxide resistance in plants: a case study of *Lolium perenne*. *In* "Effects of Air Pollutants on Plants" (Ed. T. A. Mansfield), pp. 87–103. Cambridge University Press, Cambridge.

BELSKY, A. J. (1986). Does herbivory benefit plants? A review of the evidence. *Am. Nat.* **127**, 870-892.

BEN-AMOTZ, A. and AVRON, M. (1973). The role of glycerol in the osmotic regulation of the halophilic alga, *Dunialella parva*. *Plant Physiol.* **51**, 875–887.

BENNERT, W. H. and MOONEY, H. A. (1979). The water relations of some desert plants in Death Valley, California. *Flora* **168**, 405–427.

BENNETT, J. H. and HILL, A. C. (1974). Acute inhibition of apparent photosynthesis by phytotoxic air pollutants. *In* "Air Pollution Effects on Plant Growth" (Ed. M. Dugger). *American Chemical Society Symposium Series* **3**, 115–127.

BENSINK, J. (1971). On morphogenesis of lettuce leaves in relation to light and temperature. *Meded. Landbouw. Wageningen* **71** (15), 1–93.

BERNAL, J. D. (1965). The structure of water and its biological implications. *Symp. Soc. Exp. Biol.* **19**, 17–32.

BERRY, J. and BJÖRKMAN, O. (1980). Photosynthetic response and adaptation to temperature in higher plants. *Ann. Rev. Plant Physiol.* **31**, 491–543.

BERRY, J. A. and RAISON, J. K. (1981). Responses of macrophytes to temperature. *Encyclopaedia of Plant Physiology* **12A**, 278–338.

BEWLEY, J. D. and KROCHKO, J. E. (1982). Desiccation-tolerance. *Encyclopaedia of Plant Physiology* **12B**, 325–378.

BHAT, K. K. S. and NYE, P. H. (1973). Diffusion of phosphate to plant roots in soil. I. Quantitative autoradiography of the depletion zone. *Plant and Soil* **38,** 161–175.

BILLINGS, W. D. (1974). Arctic and alpine vegetation: plant adaptations to cold summer climates. *In* "Arctic and Alpine Environments" (Eds. J. D. Ives and R. G. Barry), pp. 403–443. Methuen, London.

BILLINGS, W. D. and BLISS, L. C. (1959). An alpine snowbank environment and its effect on vegetation, plant development and productivity. *Ecology* **40,** 388–397.

BILLINGS, W. D., GODFREY, P. J., CHABOT, B. F. and BOURQUE, D. P. (1971). Metabolic acclimation to temperature in arctic and alpine ecotypes of *Oxyria digyna. Arc. Alp. Res.* **3,** 277–289.

BILLINGS, W. D. and MOONEY, H. A. (1968). The ecology of arctic and alpine plants. *Biol. Rev.* **43,** 481–529.

BISCOE, P. V., UNSWORTH, M. H. and PINCKNEY, H. R. (1973). The effects of low concentrations of sulphur dioxide on stomatal behaviour in *Vicia faba. New Phytol.* **72,** 1299–1306.

BJÖRKMAN, O. (1968). Further studies on differentiation of photosynthetic properties in sun and shade ecotypes of *Solidago virgaurea. Physiol. Plant.* **21,** 84–99.

BJÖRKMAN, O. (1980). The response of photosynthesis to temperature. *In* "Plants and their Atmospheric Environment" (Eds. J. Grace, E. D. Ford and P. G. Jarvis), pp. 273–301. Blackwell Scientific Publications, Oxford.

BJÖRKMAN, O. (1981). Responses to different quantum flux densities. *Encyclopaedia of Plant Physiology* **12A,** 57–108.

BJÖRKMAN, O., BOARDMAN, N. K., ANDERSON, J. M., GOODCHILD, D. J. and PYLIOTIS, N. A. (1972). Effect of light intensity during growth of *Atriplex patula* on the capacity of photosynthetic reactions, chloroplast components and structure. *Carnegie Inst. Wash. Yearbook* **71,** 115–135.

BJÖRKMAN, O. and HOLMGREN, P. (1963). Adaptability of the photosynthetic apparatus to light intensity in ecotypes from exposed and shaded habitats. *Physiol. Plant.* **16,** 889–914.

BJÖRKMAN, O., NOBS, M., TROUGHTON, J., BERRY, J., NICHOLSON, F. and WARD, W. (1973/74). Transplant studies in the Death Valley and the Bodega Head experimental gardens. *Carnegie Inst. Wash. Yearbook* **73,** 748–757.

BJÖRKMAN, O., PEARCY, R. and NOBS, M. (1971). Hybrids between *Atriplex* species with and without β-carboxylation photosynthesis. *Carnegie Inst. Wash. Yearbook* **69,** 640–648.

BLACK, C. C. (1971). Ecological implications of dividing plants into groups with distinct photosynthetic production capacities. *Adv. Ecol. Res.* **7,** 87–114.

BLACK, J. N. (1958). Competition between plants of different initial seed size in swards of subterranean clover (*Trifolium subterraneum*) with particular reference to leaf area and micro-climate. *Austr. J. agric. Res.* **9,** 299–318.

BLACK, J. N. (1960). The significance of petiole length, leaf area, and light interception in competition between strains of subterranean clover (*Trifolium subterraneum* L.) grown in swards. *Austr. J. agric. Res.* **11**, 277–291.

BLACK, M. (1969). Light controlled germination of seeds. *In* "Dormancy and Survival" *Symp. Soc. Exp. Biol.* **23**, 193–218.

BLACK, M. and WAREING, P. F. (1955). Growth studies in woody species. VII. Photoperiodic control of germination in *Betula pubescens* Ehrh. *Physiol. Plant.* **8**, 300–316.

BLACK, V. J. and UNSWORTH, M. H. (1979). A system for measuring effects of sulphur dioxide on gas exchange of plants. *J. exp. Bot.* **30**, 81–88.

BLACKMAN, G. E. and RUTTER, A. J. (1946). Physiological and ecological studies in the analysis of plant environment. I. The light factor and the distribution of the bluebell (*Scilla non-scripta*) in woodland communities. *Ann. Bot.*(N.S.) **10**, 361–390.

BLACKMAN, G. E. and WILSON, G. L. (1951). Ibid. II. The constancy for different species of a logarithmic relationship between net assimilation rate and light intensity, and its ecological significance. *Ann. Bot.*(N.S.) **15**, 63–94.

BLACKMAN, V. H. (1919). The compound interest law and plant growth. *Ann. Bot.* **33**, 353–360.

BLAIR, G. J., MAMARIL, C. P. AND MILLER, M. H. (1972). Effect of nitrogen status on short-term phosphate uptake. *Comm. Soil. Sci. Pl. Anal.* **3**, 23–27.

BLEASDALE, J. K. A. (1973). Effects of coal smoke pollution gases on the growth of ryegrass (*Lolium perenne* L.). *Environ. Pollut.* **5**, 275–285.

BLISS, L. C. (1962). Adaptations of arctic and alpine plants to environmental conditions. *Arctic* **15**, 117-144.

BLISS, L. C. (1975). Devon Island, Canada. *In* "Structure and Function of Tundra Ecosystems" (Eds. T. Rosswall and O. W. Heal). *Ecol. Bull.* **20**, 17–60. Swedish Natural Science Research Council, Stockholm.

BOGGIE, R. (1972). Effect of water-table height on root development of *Pinus contorta* on deep peat in Scotland. *Oikos* **23**, 304–312.

BOKHARI, U. G. (1977). Regrowth of western wheatgrass utilising ^{14}C-labelled assimilates stored in below-ground parts. *Plant and Soil* **48**, 115–127.

BOLE, J. B. (1973). Influence of root hairs in supplying soil phosphorus to wheat. *Can. J. Soil Sci.* **53**, 169–175.

BOND, G. (1951). The fixation of nitrogen associated with the root nodules of *Myrica gale* L. with special reference to its pH relation and ecological significance. *Ann. Bot.*(N. S.) **15**, 447–459.

BOND, G. (1976). The results of the IBP survey of root nodule formation in non-leguminous angiosperms. *In* "Symbiotic Nitrogen Fixation in Plants" (Ed. P. S. Nutman), **7**, 443–474. Cambridge University Press, Cambridge.

BONNER, J. and GALSTON, A. W. (1944). Toxic substances from the culture medium of Guayule which may inhibit growth. *Bot. Gaz.* **106**, 185–198.

BOORMAN, L. (1967). Biological flora of the British Isles: *Limonium vulgare* Mill. and *L. humile* Mill. *J. Ecol.* **55**, 221–232.

BOORMAN, L. and FULLER, R. M. (1982). Effects of added nutrients on dune swards grazed by rabbits. *J. Ecol.* **70**, 345–356.

BOROWITZKA, L. J. and BROWN, A. D. (1974). The salt relations of marine and halophytic species of the unicellular green alga *Dunaliella*. The role of glycerol as a compatible solute. *Arch. Microbiol.* **96**, 37–52.

BOYER, J. S. (1970). Leaf enlargement and metabolic rates in corn, soybean and sunflower at various leaf water potentials. *Plant Physiol.* **46**, 233–235.

BOYSEN-JENSEN, P. and MÜLLER, D. (1929). Die maximale Ausbeute und der tägliche Verlauf der Kohlensäureassimilation. *Jahrb. Wiss. Bot.* **70**, 493–502.

BRADSHAW, A. D. (1965). Evolutionary significance of phenotypic plasticity in plants. *Adv. Genetics* **13**, 115–155.

BRADSHAW, A. D. (1972). Some of the evolutionary consequences of being a plant. *Evol. Biol.* **5**, 25–47.

BRADSHAW, A. D. (1973). Environment and phenotypic plasticity. *Brookhaven Symp. Biol.* **25**, 75–94.

BRADSHAW, A. D., CHADWICK, M. J., JOWETT, D. LODGE, R. W. and SNAYDON, R. W. (1960a). Experimental investigations into the mineral nutrition of several grass species III. Phosphate level. *J. Ecol.* **48**, 631–637.

BRADSHAW, A. D., LODGE, R. W., JOWETT, D. and CHADWICK, M. J. (1960b). Experimental investigations into the mineral nutrition of several grass species II. Calcium and pH. *J. Ecol.* **48**, 143–150.

BRADSHAW, A. D., CHADWICK, M. J., JOWETT, D. and SNAYDON, R. W. (1964). Experimental investigations into the mineral nutrition of several grass species IV. Nitrogen level. *J. Ecol.* **52**, 665–676.

BRADSHAW, A. D., McNEILLY, T. S. and GREGORY, R. P. G. (1965). Industrialisation, evolution, and the development of heavy metal tolerance in plants. *In* "Ecology and the Industrial Society". *Brit. Ecol. Soc. Symp.* **5**, 327–343.

BRADY, N. C. (1974). "The Nature and Properties of Soils" 8th Edn. Macmillan, New York.

BROOKES, P. C., POWLSON, D. S. and JENKINSON, D. S. (1985). The microbial biomass in soil. In "Ecological Interactions in Soil" (Eds. A. H. Fitter, D. Atkinson, D. J. Read and M. B. Usher), pp. 123–125. Blackwell Scientific Publications, Oxford.

BROWN, A. D. (1976). Microbial water stress. *Bact. Rev.* **40**, 803–846.

BROWN, A. D. H., MARSHALL, D. R. and MUNDAY, J. (1976). Adaptedness of variants at an alcohol dehydrogenase locus in *Bromus mollis* L. (Soft bromegrass). *Aust. J. biol. Sci.* **29**, 389–396.

BROWN, J. C. (1972). Competition between phosphate and the plant for Fe from Fe^{++} ferrozine. *Agron. J.* **64**, 240–243.

BROWN, M. E. (1975). Rhizosphere micro-organisms—opportunist, bandits, or benefactors. *In* "Soil Microbiology" (Ed. N. Walker), pp. 21–38. Butterworth, London.

BROWNELL, P. F. and CROSLAND, C. J. (1972). Requirement for sodium as a micro-nutrient for species having the C_4 dicarboxylic photosynthetic pathway. *Plant Physiol.* **49**, 794–797.

BULL, J. N. and MANSFIELD, T. A. (1974). Photosynthesis in leaves exposed to SO_2 and NO_2. *Nature* **250**, 443–444.

BURKE, M. J., GUSTA, L. V., QUAMME, H. A., WEISER, C. J. and LI, P. H. (1976). Freezing and injury in plants. *Ann. Rev. Pl. Physiol.* **27**, 507–528.

BURSTRÖM, H. G. (1968). Calcium and plant growth. *Biol. Rev.* **43**, 298–316.

CALDWELL, M. M. (1968). Solar ultraviolet radiation as an ecological factor for alpine plants. *Ecol. Monogr.* **38**, 243–268.

CALDWELL, M. M. (1981). Plant response to solar ultraviolet radiation. *Encyclopaedia of Plant Physiology* **12A**, 170–197.

CALLAGHAN, T. V. and LEWIS, M. C. (1971). The growth of *Phleum alpinum* L. in contrasting habitats at a sub-antarctic station. *New Phytol.* **70**, 1143–1154.

CANNON, W. A. (1911). The root habits of desert plants. *Carnegie Inst. Wash. Publ.* 131.

CAPRON, T. M. and MANSFIELD, T. A. (1976). Inhibition of net photosynthesis in tomato in air polluted with NO and NO_2. *J. exp. Bot.* **27**, 1181–1186.

CARADUS, J. R. (1981). Effect of root hair length on white clover *Trifolium repens* cv Tamar grown over a range of phosphorus levels. *N.Z. J. Agric. Res.* **24**, 353–358.

CARGILL, S. M. and JEFFERIES, R. L. (1984). Nutrient limitation of primary production in a sub-arctic salt marsh. *J. appl. Ecol.* **21**, 657–668.

CARLSON, P. S. (1980). "The Biology of Crop Productivity". Academic Press, New York.

CARTWRIGHT, B. (1972). The effect of phosphate deficiency on the kinetics of phosphate absorption by sterile excised barley roots, and some factors affecting the ion uptake efficiency of roots. *Comm. Soil. Sci. Pl. Anal.* **3**, 313–322.

CASWELL, H., REED, F., STEPHENSON, F. N. and WERNER, P. A. (1973). Photosynthetic pathways and selective herbivory: a hypothesis. *Amer. Nat.* **107**, 465–480.

CHABOT, B. F. (1979). Metabolic and enzymatic adaptations to low temperatures. *In* "Comparative Mechanisms of Cold Adaptation" (Eds. L. S. Underwood, L. L. Tieszen, A. B. Callahan and G. E. Folk), pp. 283–301. Academic Press, New York.

CHAPIN, F. (1974). Morphological and physiological mechanisms of temperature compensation in phosphate absorption along a latitudinal gradient. *Ecology* **55**, 1180–1198.

CHAPIN, F. S. (1980). The mineral nutrition of wild plants. *Ann. Rev. Ecol. Syst.* **11**, 233–260.

CHAPIN, F. S. (1983). Direct and indirect effects of temperature on arctic plants. *Polar Biol.* **2**, 47–52.

CHAPIN, F. S., FOLLETT, J. M. and O'CONNOR, K. F. (1982). Growth, phosphate absorption and phosphorus chemical fractions in two *Chionochloa* species. *J. Ecol.* **70**, 305–321.

CHAPIN, F. S. and SLACK, M. (1979). Effect of defoliation upon root growth, phosphate absorption and respiration in nutrient-limited tundra grami-noids. *Oecologia* **42,** 67–79.

CHAZDON, R. L. and PEARCY, R. W. (1986a, b). Photosynthetic responses to light variation in rainforest species. 1. Induction under constant and fluctuating light conditions. II. Carbon gain and photosynthetic efficiency during lightflecks. *Oecologia* **69,** 517–523, 524–531.

CHENERY, E. A. and SPORNE, K. P. (1976). A note on the evolutionary status of aluminium accumulation among dicotyledons. *New Phytol.* **76,** 551–554.

CHIPPINDALE, H. G. (1932). The operation of interspecific competition in causing delayed growth of grasses. *Ann. appl. Biol.* **19,** 221–242.

CHIRKOVA, T. V. (1978). Some regulatory mechanisms of plant adaptation to temporal anaerobiosis. *In* "Plant Life in Anaerobic Environments" (Eds. D. D. Hook and R. M. M. Crawford), pp. 137–158. Ann Arbor Science Publishers Inc., Michigan.

CHRISTIE, E. K. and MOORBY, J. (1975). Physiological responses of arid grasses. I. The influence of phosphorus supply on growth and phosphorus absorption. *Austr. J. Agric. Res.* **26,** 423–436.

CHRISTIE, P., NEWMAN, E. I. and CAMPBELL, R. (1974). Grassland species can influence the abundance of microbes on each other's roots. *Nature* **250,** 570.

CLARKSON, D. T. (1966). Aluminium tolerance within the genus *Agrostis*. *J. Ecol.* **54,** 167–178.

CLARKSON, D. T. (1967). Phosphorus supply and growth rates in species of *Agrostis* L. *J. Ecol.* **55,** 707–731.

CLARKSON, D. T. (1985). Factors affecting mineral nutrient acquisition by plants. *Ann. Rev. Plant Physiol.* **36,** 77–115.

CLARKSON, D. T., HALL, K. C. and ROBERTS, J. K. M. (1980). Phospholipid composition and fatty acid desaturation in the roots of rye during acclimatization to low temperature: positional analysis of fatty acids. *Planta* **149,** 464–471.

CLARKSON, D. T., ROBARDS, A. W. and SANDERSON, J. (1971). The tertiary endodermis in barley roots: fine structure in relation to radial transport of ions and water. *Planta* **96,** 296–305.

CLARKSON, D. T. and SANDERSON, J. (1969). The uptake of a polyvalent cation and its distribution in the root apices of *Allium cepa*: tracer autoradio-graphic studies. *Planta* **89,** 136–154.

CLARKSON, D. T. and SANDERSON, J. (1971). Inhibition of the uptake and long-distance transport of calcium by aluminium and other polyvalent cations. *J. exp. Bot.* **23,** 837–851.

CLARKSON, D. T., SANDERSON, J. and RUSSELL, R. S. (1968). Ion uptake and root age. *Nature* **220,** 805–806.

COCKBURN, W. (1985). Variation in photosynthetic acid metabolism in vascular plants: CAM and related phenomena. *New Phytol.* **101,** 3–24.

CODY, M. L. (1986). Structural niches in plant communities. *In* "Community Ecology" (Eds. J. Diamond and T. J. Case), pp. 381–405. Harper & Row, New York.

COE, M. D. (1964). The chinampas of Mexico. *Sci. Amer.* **211**, 90–98.

CONNOR, D. J., LEGGE, N. J. and TURNER, N. C. (1977). Water relations of Mountain ash (*Eucalyptus regnans* F. Muell.) forests. *Aust. J. Plant Physiol.* **4**, 735–762.

COOK, C. D. K. (1972). Phenotypic plasticity with particular reference to three amphibious plant species. *In* "Taxonomy and Ecology" (Ed. V. H. Heywood), pp. 97–111. Systematics Association Special Volume No. 5. Academic Press, London and New York.

COOMBE, D. E. (1966). The seasonal light climate and plant growth in a Cambridgeshire wood. *In* "Light as an Ecological Factor". *Symp. Brit. Ecol. Soc.* **6**, 148–166.

COOPER, A. J. (1973). "Root Temperature and Plant Growth". Res. Rev. 4, Commonwealth Bureau of Horticulture and Plantation Crops.

COOPER, J. P. and BREEZE, A. L. (1971). Plant Breeding: forage grasses and legumes. *In* "Potential Crop Production" (Eds. P. F. Wareing and J. P. Cooper), pp. 295–318. Heinemann, London.

COULT, D. A. and VALLANCE, K. B. (1958). Observations on the gaseous exchange which takes place between *Menyanthes trifoliata* and its environment. *J. exp. Bot.* **9**, 384–402.

COUPLAND, R. T. and JOHNSON, R. E. (1965). Rooting characteristics of native grassland species in Saskatchewan. *J. Ecol.* **53**, 475–507.

COWAN, I. R. (1977). Stomatal behaviour and environment. *Adv. Bot. Res.* **4**, 117–228.

COWAN, I. R. (1982). Regulation of water use in relation to carbon gain in higher plants. *Encyclopaedia of Plant Physiology* **12B**, 589–613.

COWAN, I. R. and FARQUHAR, G. B. (1977). Stomatal function in relation to leaf metabolism and environment. *In* "Integration of Activity in the Higher Plant" (Ed. D. H. Jennings), pp. 471–505. Cambridge University Press, Cambridge.

COWAN, I. R. and MILTHORPE, F. L. (1968). Plant factors influencing the water status of plant tissues. *In* "Water Deficits and Plant Growth" (Ed. T. T. Kozlowski), vol. 1, pp. 137–193. Academic Press, New York and London.

COWLING, D. W., JONES, L. H. P. and LOCKYER, D. R. (1973). Increased yield through correction of sulphur deficiency in ryegrass exposed to sulphur dioxide. *Nature* **243**, 479–480.

COXSON, D. S. and KERSHAW, K. A. (1983). The ecology of *Rhizocarpon superficiale*. 1. The rock surface boundary layer microclimate. *Can. J. Bot.* **61**, 3009–3018.

CRAFTS, A. S. (1968a). Water structure and water in the plant body. *In* "Water Deficits and Plant Growth" (Ed. T. T Kozlowski), vol. 1, pp. 23–47, Academic Press, New York and London.

CRAFTS, A. S. (1968b). Water deficits and physiological processes. In "Water Deficits and Plant Growth" (Ed. T. T. Kozlowski), vol. 2, pp. 85–133, Academic Press, New York and London.

CRAWFORD, R. M. M. (1972). Some metabolic aspects of ecology. *Trans. Bot. Soc. Edin.* **41**, 309–322.

CRAWFORD, R. M. M. (1977). Tolerance of anoxia and ethanol metabolism in germinating seeds. *New Phytol.* **79**, 511–517.

CRAWFORD, R. M. M. (1978). Metabolic adaptations to anoxia. *In* "Plant Life in Anaerobic Environments" (Eds. D. D. Hook and R. M. M. Crawford), pp. 119–136. Ann Arbor Science Publishers Inc., Michigan.

CRAWFORD, R. M. M. (1982). Physiological responses to flooding. *Encyclopaedia of Plant Physiology* **12B**, 453–477.

CRAWFORD, R. M. M. and MCMANMON, R. M. (1968). Inductive responses of alcohol and malic dehydrogenases in relation to flooding tolerance in roots. *J. exp. Bot.* **19**, 435–441.

CRAWFORD, R. M. M. and PALIN, M. A. (1981). Root respiration and temperature limits to the north-south distribution of four perennial maritime plants. *Flora* **171**, 338–354.

CROCKER, R. L. and MAJOR, J. (1955). Soil development in relation to vegetation and surface age, Glacier Bay, Alaska. *J. Ecol.* **43**, 427–448.

CROSS, J. R. (1975). Biological flora of the British Isles: *Rhododendron ponticum. J. Ecol.* **63**, 345–359.

CROSSLEY, G. K. and BRADSHAW, A. D. (1968). Differences in response to mineral nutrients of populations of ryegrass, *Lolium perenne* L. and orchard grass *Dactylis glomerata* L. *Crop Sci.* **8**, 383–387.

CRESSWELL, E. and GRIME, J. P. (1981). Induction of a light requirement during seed development and its ecological consequences. *Nature* **291**, 583–585.

CRUSH, J. R. (1974). Plant growth responses to vesicular arbuscular mycorrhiza VII. Growth and nodulation of some herbage legumes. *New Phytol.* **73**, 743–749.

CUMMING, B. G. (1963). Dependence of germination on photoperiod, light quality, and temperature in *Chenopodium. Can. J. Bot.* **41**, 1211–1233.

DAHL, E. (1951). On the relation between summer temperature and the distribution of alpine vascular plants in the lowlands of Fennoscandia. *Oikos* **3**, 22–52.

DANIEL, P. P., WOODWARD, F. I., BRYANT, J. A. and ETHERINGTON, J. A. (1985). Nocturnal accumulation of acid in leaves of wall pennywort (*Umbilicus rupestris*) following exposure to water stress. *Ann. Bot.* **55**, 217–223.

DARWIN, F. (1876). On the hygroscopic mechanism by which certain seeds are enabled to bury themselves in the ground. *Trans. Linn. Soc. London. 2nd Series. Bot.* **1**, 149–167.

DAUBENMIRE, R. F. (1947). "Plants and Environment". Wiley, New York.

DAUM, C. R. (1967). A method for determining water transport in trees. *Ecology* **48**, 425–431.

DAVIDSON, J. L. and MILTHORPE, F. L. (1966). The effect of temperature on the regrowth of cocksfoot (*Dactylis glomerata* L.). *Ann. Bot.* **29**, 407–417.

DAVIDSON, R. L. (1969). Effect of root-leaf temperature differentials on root-shoot ratios in some pasture grasses and clover. *Ann. Bot.* (N.S.) **33**, 561–569.

DAVIES, D. D., GREGO, S. and KENWORTHY, P. (1974). The control of the production of lactate and ethanol by higher plants. *Planta* **118**, 297–310.

DAVIES, D. D., NASCIMENTO, K. H. and PATIL, K. D. (1974). The distribution and properties of NADP malic enzyme in flowering plants. *Phytochem.* **13**, 2417–2425.

DAVIES, M. S. and SNAYDON, R. W. (1973a, b), (1974). Physiological differences among populations of *Anthoxanthum odoratum* L. collected from the Park Grass Experiment, Rothamsted. I. Response to calcium. *J. appl. Ecol.* **10**, 33–45; II. Response to aluminium. *J. appl. Ecol.* **10**, 47–55; III. Response to phosphorus. *J. appl. Ecol.* **11**, 699–670.

DAVIES, T. (1980). Grasses more sensitive to SO$_2$ pollution in conditions of low irradiance and short days. *Nature* **284**, 488–485.

DAVISON, A. W. and BAILEY, I. F. (1982). SO$_2$ pollution reduces the freezing resistance of ryegrass. *Nature* **297**, 400–402.

DAVIS, K. S. and DAY, J. A. (1961). "Water". Heinemann, London.

DAXER, H. (1934). Über die Assimilations-ökologie der Waldbodenflora. *Jahrb. Wiss. Bot.* **80**, 363–420.

DEAN, C. E. (1972). Stomate density and size as related to ozone-induced weather fleck in tobacco. *Crop Sci.* **12**, 547–548.

DEL MORAL, R. (1972). On the variability of chlorogenic acid concentration. *Oecologia* **9**, 289–300.

DEVERALL, B. J. (1977). "Defence mechanisms of plants". Cambridge University Press, Cambridge.

DIRR, M. A., BARKER, A. V. and MAYNARD, D. M. (1973). Extraction of nitrate reductase from leaves of Ericaceae. *Phytochem.* **12**, 1261–1264.

DITTMER, H. J. (1940). A quantitative study of the subterranean members of soybean. *Soil Conserv.* **6**, 33–34.

DÖBEREINER, J. and DAY, J. M. (1976). Associative symbioses and free-living systems. "Proceedings of the 1st International Symposium on Nitrogen Fixation" (Eds. W. E. Newton and C. J. Nyman), pp. 518–538. Washington State University Press, Pullman.

DODDEMA, H., TELKAMP, G. P. and OTTEN, H. (1979). Uptake of nitrate by mutants of *Arabidopsis thaliana*, disturbed in uptake or reduction of nitrate. *Physiol. Plant.* **45**, 297–346.

DONALD, C. M. (1958). The interaction of competition for light and nutrients. *Aust. J. agric. Res.* **9**, 421–435.

DOWDELL, R. J., SMITH, K. A., CREES, R. and RESTALL, S. W. F. (1972). Field studies of ethylene in the soil atmosphere—equipment and preliminary results. *Soil Biol. Biochem.* **4**, 325–331.

DREW, M. C. (1975). Comparison of the effects of a localized supply of phosphate, nitrate, ammonium and potassium on the growth of the seminal root system, and the shoot, in barley. *New Phytol.* **75**, 479–490.

DREW, M. C. (1979). Properties of roots which influence rates of absorption. *In* "The Soil-Root Interface" (Eds. J. L. Harley and R. S. Russell), pp. 21–38. Academic Press, London and New York.

DREW, M. C., JACKSON, M. B., GIFFARD, S. C. and CAMPBELL, R. (1981). Inhibition by silver ions of gas space (aerenchyma) formation in adventitious roots of *Zea mays* L. subjected to exogenous ethylene or to oxygen deficiency. *Planta* **153**, 217–224.

DREW, M. C. and LYNCH, J. M. (1980). Soil anaerobiosis, micro-organisms and root function. *Ann. Rev. Phytopath.* **18**, 37–66.

DREW, M. C. and NYE, P. H. (1969). The supply of nutrient ions by diffusion to plant roots in soil. II. *Plant and Soil* **31**, 407–424.

DREW, M. C. and SAKER, L. R (1975). Nutrient supply and the growth of the seminal root system in barley. II. *J. exp. Bot.* **26**, 79–90.

DREW, M. C., SAKER, L. R. and ASHLEY, T. W. (1973). Nutrient supply and the growth of the seminal root system in barley. I. *J. exp. Bot.* **24**, 1189–1202.

DREW, M. C., SAKER, L. R., BARBER, S. A. and JENKINS, W. (1984). Changes in the kinetics of phosphate and potassium absorption in nutrient-deficient barley roots measured by a solution-depletion technique. *Planta* **160**, 490–499.

DUDNEY, P. J. (1973). An approach to the growth analysis of perennial plants. *Proc. Roy. Soc. London.* **B184**, 217–220.

DUMBROFF, E. B. and COOPER, A. W. (1974). Effects of salt stress applied in balanced nutrient solutions at several stages during growth of tomato. *Bot. Gaz.* **135**, 219–224.

DUNCAN, W. G. and OHLROGGE, A. J. (1958). Principles of nutrient uptake from fertiliser bands. II. *Agron. J.* **50**, 605–608.

DUSS, F. and BRÄNDLE, R. (1982). Die Überflutungstoleranz der Teichsimse (*Schoenoplectus lacustris* (L.) Palla): 5. Die bildung von verschiedenen Gärungsprodukten und Transportsubstanzen im Rhizomgewebe bei Sauerstoffmangel. *Flora* **172**, 217–222.

DUVIGNEAUD, P. and DENAYER DE SMET, S. (1963). Cuivre et végétation au Katanga. *Bull. Soc. Roy. Bot., Belgique* **96**, 93–231.

EDWARDS, P. J. and WRATTEN, S. D. (1982). Wound-induced changes in palatability of birch (*Betula pubescens* Ehrh. ssp *pubescens*). *Am. Nat.* **120**, 816–818.

EHLERINGER, J. R. (1978) Implications of quantum yield differences on the distributions of C_3 and C_4 grasses. *Oecologia* **31**, 255–267.

EHLERINGER, J. R. (1980). Leaf morphology and reflectance in relation to water and temperature stress. *In* "Adaptations of Plants to Water and High Temperature Stress" (Eds. N. C. Turner and P. J. Kramer), pp. 295–308. Wiley–Interscience, New York.

EHLERINGER, J. R. and FORSETH, I. (1980). Solar tracking by plants. *Science* **210**, 1094–1098.

EHLERINGER, J. R. and WERK, K. S. (1986). Modifications of solar radiation absorption patterns and implications for carbon gain at the the leaf level.

In "On the Economy of Plant Form and Function" (Ed. T. J. Givnish). Cambridge University Press, Cambridge.

ELKINGTON, T. T. and JONES, B. M. G. (1974). Biomass and primary productivity of birch (*Betula pubescens* S. Lat.) in south-west Greenland. *J. Ecol.* **62**, 821–830.

ELLENBERG, H. (1953). Physiologisches und ökologisches Verhalten derselben Pflanzenarten. *Ber. Deut. Botan. Ges.* **65**, 351–361.

EMERSON, R. and ARNOLD, W. (1932). A separation of the reactions in photosynthesis by means of intermittent light. *J. Gen. Physiol.* **19**, 391–420.

ENGLE, R. L. and GABELMAN, W. H. (1966). Inheritance and mechanism for resistance to ozone damage in onion, *Allium cepa* L. *Proc. Amer. Soc. Hort. Sci.* **89**, 423–430.

EPSTEIN, E. (1961) The essential role of calcium in selective cation transport by plant cells. *Plant Physiol.* **37**, 682–685.

EPSTEIN, E. (1969). Mineral metabolism of halophytes. *In* "Ecological Aspects of the Mineral Nutrition of Plants". *Brit. Ecol. Soc. Symp.* **9**, 345–355.

EPSTEIN, E. (1973). Mechanisms of ion transport through plant cell membranes. *Int. Rev. Cytol.* **34**, 123–168.

EPSTEIN, E. and HAGEN, C. E. (1952). A kinetic study of the absorption of alkali cations by barley roots. *Plant Physiol.* **27**, 457–474.

ERICKSON, J. M. and FEENY, P. (1974). A chemical barrier to the black swallow-tail butterfly *Papilio polyxenes*. *Ecology* **55**, 103–111.

ESAU, K. (1965). "Plant Anatomy", 2nd Edn. Wiley, New York.

ETHERINGTON, J. R. (1967). Soil water and the growth of grasses. II. Effects of soil water potential on growth and photosynthesis of *Alopecurus pratensis*. *J. Ecol.* **55**, 373–380.

ETHERINGTON, J. R. (1983). Control of germination and seedling morphology by ethene: differential responses, related to the habitat of *Epilobium hirsutum* L. and *Chamaenerion angustifolium* (L.) J. Holub. *Ann. Bot.* **52**, 653–658.

EVANS, G. C. (1972). "The Quantitative Analysis of Plant Growth". Blackwells, Oxford.

EVANS, G. C. and HUGHES, A. P. (1961). Plant growth and the aerial environment. I. Effect of artificial shading on *Impatiens parviflora*. *New Phytol.* **60**, 150–180.

EVANS, L. S., LEWIN, K. F., SANTUCCI, K. A. and PATTI, M. J. (1985). Effects of frequency and duration of simulated acidic rainfalls on soybean yields. *New Phytol.* **100**, 199–208.

FAIRFAX, J. A. W. and LEPP, N. W. (1975). Effect of simulated acid rain on cation loss from leaves. *Nature* **255**, 324–325.

FEENY, P. (1976). Plant apparency and chemical defence. *In* "Recent Advances in Phytochemistry. 10. Biochemical Interactions between Plants and Insects" (Eds. J. W. Wallace and R. L. Mansell). pp. 1–40. Plenum Press, New York.

FERGUS, I. F., MARTIN, A. E., LITTLE, I. P. and HAYDOCK, K. P. (1972). Studies on soil potassium. II. The Q/I relation and other parameters compared with plant uptake of potassium. *Aust. J. Soil Res.* **10**, 95–111.

FERRARI, G. and RENOSTO, F. (1972). Comparative studies on the active transport by excised roots of inbred and hybrid maize. *J. agric. Sci.* **79,** 105–108.

FIELD, C. and MOONEY, H. A. (1986). The photosynthesis-nitrogen relationship in wild plants. *In* "On the Economy of Plant Form and Function" (Ed. T. J. Givnish), pp. 25–55. Cambridge University Press, Cambridge.

FINEGAN, B. (1984). Forest succession. *Nature* **312,** 109–114.

FINLAY, R. D. (1985). Interactions between soil micro-arthropods and endomycorrhizal associations of higher plants. *In* "Ecological Interactions in Soil" (Eds. A. H. Fitter, D. Atkinson, D. J. Read and M. B. Usher), pp. 319–331. Blackwell Scientific Publications, Oxford.

FIRN, R. D. (1986). Phototropism. *In* "Photomorphogenesis in Plants" (Eds. R. E. Kendrick and G. H. M. Kronenberg), pp. 369–390. Junk. The Hague, Netherlands.

FIRTH, J. N. M. (1978). The origin and exploitation of non-ferrous metals. *In* "Environmental Management of Mineral Wastes" (Eds. G. S. Goodman and M. J. Chadwick), pp. 259–272. Sijthoff and Noordhoff, Alphen.

FITTER, A. H. (1976). Effects of nutrient supply and competition from other species on root growth of *Lolium perenne* in soil. *Plant and Soil* **45,** 177–189.

FITTER, A. H. (1977). Influence of mycorrhizal infection on competition for phosphorus and potassium by two grasses. *New Phytol.* **79,** 119–125.

FITTER, A. H. (1985). Functional significance of root morphology and root system architecture. *In* "Ecological Interactions in Soil" (Eds. A. H. Fitter, D. Atkinson, D. J. Read and M. B. Usher), pp. 87–106. Blackwell Scientific Publications, Oxford.

FITTER, A. H. (1986). Effect of benomyl on leaf phosphorus concentration in alpine grasslands: a test of mycorrhizal benefit. *New Phytol.* **103,** 767–776.

FITTER, A. H. (1987). An architectural approach to the comparative ecology of plant root systems. *New Phytol.* **106,** 61–77.

FITTER, A. H. and ASHMORE, C. J. (1974). Response of *Veronica* species to a simulated woodland light climate. *New Phytol.* **73,** 997–1001.

FITTER, A. H. and BRADSHAW, A. D. (1974). Root penetration of *Lolium perenne* on colliery shale in response to reclamation treatments. *J. appl. Ecol.* **11,** 609–616.

FITTER, A. H. and SMITH, C. J. (1979). "A Wood in Askham". Ebor Press, York.

FLOWERS, T. J. and LÄUCHLI, A. (1983). Sodium versus potassium: substitution and compartmentation. *Encyclopaedia of Plant Physiology* **15B,** 651–681.

FOGEL, R. (1985). Roots as primary producers in below-ground ecosystems. *In* "Ecological Interactions in Soil" (Eds. A. H. Fitter, D. Atkinson, D. J. Read and M. B. Usher), Blackwell Scientific Publications, Oxford.

FOOTE, B. D. and HOWELL, R. W. (1964). Phosphorus tolerance and sensitivity of soybeans as related to uptake and translocation. *Plant Physiol.* **39,** 610–613.

FOWLER, D. and CAPE, J. N. (1982). Air pollutants in agriculture and horticulture. *In* "Effects of Gaseous Air Pollution in Agriculture and Horticulture" (Eds. M. H. Unsworth and D. P. Ormrod), pp. 1–26. Butterworth, London.

FOWLER, D., CAPE, J. N. and LEITH, I. D. (1985). Acid inputs from the atmosphere in the United Kingdom. *Soil Use and Management* **1,** 3–5.

FOWLER, N. (1981). Competition and coexistence in a North Carolina grassland. 2. The effects of the experimental removal of species. *J. Ecol.* **69,** 843–854.

FOWLER, S. V. and LAWTON, J. H. (1985). Rapidly induced defences and talking trees: the devil's advocate position. *Am. Nat.* **126,** 181–195.

FOX R. L., HASAN, S. M. and JONES, R. C. (1971). Phosphate and sulphate sorption by latosols. *Proc. Int. Symp. Soil Fert. Evaln. New Delhi* **1,** 857–864.

FOY, C. D., CHANEY, K. L. and WHITE, M. C. (1978). The physiology of metal toxicity in plants. *Ann. Rev. Plant Physiol.* **29,** 511–566.

FOYER, C. H. and HALL, D. O. (1980). Oxygen metabolism in the active chloroplast. *Trends in Biochemical Sciences 1980,* 188–191.

FRIED, M. and BROESHART, H. (1967). "The Soil-Plant System". Academic Press, New York and London.

GAFF, D. F. (1980). Protoplasmic tolerance of extreme water stress. *In* "Adaptation of Plants to Water and High Temperature Stress" (Eds. N. C. Turner and P. J. Kramer), pp. 207–230. Wiley-Interscience, New York.

GALLAGHER, J. N. and BISCOE, P V. (1979). Field studies of cereal leaf growth. III. *J. exp. Bot.* **117,** 645–655.

GARDNER, W. R. (1960). Dynamic aspects of water availability to plants. *Soil Sci.* **89,** 63–73.

GARNER, W. W. and ALLARD, H. A. (1920). Effect of length of day on plant growth. *J. agr. Res.* **18,** 553–606.

GARSED, S. G. (1984). Uptake and distribution of pollutants in the plant and residence time of active species. *In* "Gaseous Air Pollutants and Plant Metabolism"'(Eds. M. J. Koziol and F. R. Whatley), pp. 83–103. Butterworth, London.

GATES, C. T., HAYDOCK, K. P. and ROBINS, M. F. (1970). Response to salt in *Glycine*: 4. Salt concentration and the content of phosphorus, potassium, sodium and chloride in cultivars of *G. wightii* (*G. javanica*). *Aust. J. exp. Agric. Anim. Husb.* **10,** 99–110.

GATES, D. M. (1968). Transpiration and leaf temperature. *Ann. Rev. Pl. Physiol.* **19,** 211–238.

GATES, D. M. (1976). Energy exchange and transpiration. *In* "Water and Plant Life" (Eds. O. L. Lange, L. Kappen and E.-D. Schulze). *Ecological Studies* **19,** 137–147. Springer Verlag, Berlin.

GATES, D. M. and PAPIAN, L. E. (1971). "Atlas of Energy Budgets of Plant Leaves". Academic Press, London and New York.

GAUHL, E. (1969). Leaf factors affecting the rate of light-saturated photosynthesis in ecotypes of *Solanum dulcamara. Carnegie Inst. Wash. Yrbk* **68,** 633–636.

GAUHL, E. (1976). Photosynthetic response to varying light intensity in ecotypes of *Solanum dulcamara* L. from shaded and exposed habitats. *Oecologia* **22**, 275–286.

GAUHL, E. (1979). Sun and shade ecotypes in *Solanum dulcamara* L.: Photosynthetic light dependence characteristics in relation to mild water stress. *Oecologia* **39**, 61–70.

GAUSE, G. F. (1934). "The Struggle for Existence" (Reprinted 1964). Hafner Publishing Company, New York.

GAUSLAA, Y. (1984). Heat resistance and energy budget in different Scandinavian plants. *Holarctic Ecol.* **7**, 1–78.

GEIGER, R. (1965). "Die Atmosphäre der Erde". Darmstadt, Perthes.

GEORGE, M. F., HONG, S. G. and BURKE, M. J. (1977). Cold hardiness and deep supercooling of hardwoods: its occurrence in provenance collections of red oak, yellow birch, black walnut and black cherry. *Ecology* **58**, 674–680.

GERDEMANN, J. W. and TRAPPE, J. M. (1975). Taxonomy of the Endogonaceae. *In* "Endomycorrhizas" (Ed. F. E. Sanders *et al.*), pp. 35–51. Academic Press, London and New York.

GERRETSEN, F. C. (1948). The influence of micro-organisms on the phosphate intake by the plant. *Plant and Soil* **1**, 51–81.

GIGON, A. and RORISON, I. H. (1972). The response of some ecologically distinct plant species to nitrate and ammonium nitrogen. *J. Ecol.* **60**, 93–102.

GILBERT, L. E. (1971). Butterfly: plant co-evolution: has *Passiflora adenopoda* won the selectional race with Heliconid butterflies? *Science* **172**, 585–586.

GILLER, K. E. and DAY, J. M. (1985). Nitrogen fixation in the rhizosphere: significance in natural and agricultural systems. *In* "Ecological Interactions in Soil" (Eds. A. H. Fitter, D. Atkinson, D. J. Read and M. B. Usher), pp. 127–147. Blackwell Scientific Publications, Oxford.

GIVNISH, T. J. (1986). Optimal stomatal conductance, allocation of energy between leaves and roots, and the marginal cost of transpiration. *In* "On the Economy of Plant Form and Function" (Ed. T. J. Givnish), pp. 171–214. Cambridge University Press, Cambridge.

GLASS, A. D. M. (1973). Influence of phenolic acids on ion uptake. I. Inhibition of phosphate uptake. *Plant Physiol.* **51**, 1037–1041.

GLASS, A. D. M. (1974). Influence of phenolic acids on ion uptake. III. Inhibition of potassium absorption. *J. exp. Bot.* **25**, 1104–1113.

GLASS, A. D. M. (1978). Regulation of potassium influx into intact roots of barley by internal potassium levels. *Can. J. Bot.* **56**, 1759–1764.

GLEAVES, T. J. (1973). Gene flow mediated by wind-borne pollen. *Heredity* **31**, 355-366.

GODWIN, H., CLOWES, D. R. and HUNTLEY, B. (1974). Studies in the ecology of Wicken Fen. V. Development of fen carr. *J. Ecol.* **62**, 197–214.

GOLDBERG, D. E. (1982). The distribution of evergreen and deciduous trees in relation to soil type: an example from the Sierra Madre, Mexico, and a general model. *Ecology* **63**, 942–951.

GOLDSTEIN, G. and MEINZER, F. (1983). Influence of insulating dead leaves and low temperatures on water balance in an Andean giant rosette plant. *Plant, Cell and Env.* **6,** 649–656.

GOOD, R. (1964). "The Geography of Flowering Plants". Longman, London.

GOODLASS, G. and SMITH, K. A. (1978). Effect of pH, organic matter content and nitrate on the evolution of ethylene from soils. *Soil Biol. Biochem.* **10,** 193–199.

GOODMAN, P. J., FOTHERGILL, M. and HUGHES, D. M. (1974). Variation in nitrate reductase, nitrite and nitrite reductase in some grasses and cereals. *Ann. Bot.* (N.S.) **38,** 31–37.

GORSKI, T. (1975). Germination of seeds in the shadow of plants. *Physiol. Plant.* **34,** 342–346.

GOULD, S. J. and VRBA, E. S. (1982). Exaptation—a missing term in the science of form. *Paleobiology* **8,** 4–15.

GOVIER, R. N., BROWN, J. G. and PATE, J. S. (1968). Hemiparasitic nutrition in Angiosperms. II. Root haustoria and leaf glands of *Odontites verna* (Bell) Dum. and their relevance to the abstraction of solutes from the host. *New Phytol.* **67,** 963–972.

GRABHERR, G. and CERNUSCA, A. (1977). Influence of radiation, wind and temperature on the CO_2 gas exchange of the alpine dwarf shrub community *Loiseleurietum centrariosum. Photosynthetica* **11,** 22–28.

GRACE, J. (1983). "Plant–Atmosphere Relations". Chapman & Hall, London.

GRAY, J. T. (1983). Nutrient use by evergreen and deciduous shrubs in southern California. 1. Community nutrient cycling and nutrient use efficiency. *J. Ecol.* **71,** 21–41.

GREAVES, M. P. and DARBYSHIRE, J. F. (1972). The ultrastructure of the mucilaginous layer on plant roots. *Soil Biol. Biochem.* **4,** 443–449.

GREEN, D. G. and WARDER, F. G. (1973). Accumulation of damaging concentrations of phosphorus by leaves of Selkirk Wheat. *Plant and Soil* **38,** 567–572.

GREEN, M. S. and ETHERINGTON, J. R. (1977). Oxidation of ferrous iron by rice (*Oryza sativa* L.) roots: a mechanism for waterlogging tolerance? *J. exp. Bot.* **28,** 678–690.

GREENWAY, H. (1962). Plant response to saline substrates. I. Growth and ion uptake of several varieties of *Hordeum* during and after sodium chloride treatment. *Aust. J. Biol. Sci.* **15,** 16–38.

GREENWAY, H. (1973). Salinity, plant growth, and metabolism. *J. Aust. Inst. Agr. Sci.* March 1973, 24–34.

GREENWOOD, D. J. (1967). Studies on the transport of oxygen through the stems and roots of vegetable seedlings. *New Phytol.* **66,** 337–347.

GRIEVE, P. W. and POVEY, M. J. W. (1981). Evidence for the osmotic dehydration theory of freeze damage. *J. Sci. Fd. Agric.* **32,** 96–98.

GRIME, J. P. (1966). Shade avoidance and shade tolerance in flowering plants. *In* "Light as an Ecological Factor". *Symp. Brit. Ecol. Soc.* **6,** 187–207.

GRIME, J. P. (1977). Evidence for the existence of three primary strategies in plants and animals and its relevance to ecological and evolutionary theory. *Am. Nat.* **111,** 1169–1194.

GRIME, J. P. (1979). "Plant Strategies and Vegetation Processes", Wiley, London.

GRIME, J. P., CRICK, J. C. and RINCON, J. A. (1986). The ecological significance of plasticity. In "Plasticity in Plants" (Eds. D. H. Jennings and A. Trewavas), pp. 7–29. Cambridge University Press, Cambridge.

GRIME, J. P. and HODGSON, J. G. (1969). An investigation of the ecological significance of lime-chlorosis by means of large-scale comparative experiments. In "Ecological Aspects of the Mineral Nutrition of Plants". Brit. Ecol. Soc. Symp. 9, 67–100.

GRIME, J. P. and HUNT, R. (1975). Relative growth rate: its range and adaptive significance in a local flora. J. Ecol. 63, 393–422.

GROBBELAAR, N., CLARKE, B. and HOUGH, M. C. (1971). The nodulation and nitrogen fixation of isolated roots of Phaseolus vulgaris L. III. The effect of carbon dioxide and ethylene. Plant and Soil (Special Volume), 215–223.

GRUBB, P. J. and SUTER, M. B. (1971). The mechanism of acidification of soil by Calluna and Ulex and the significance for conservation. In "The Scientific Management of Animal and Plant Communities for Conservation". Brit. Ecol. Soc. Symp. 11, 115–135.

GRÜMMER, G. (1958). Die Beeinflussung des Leinertrages durch Camelina-Arten. Flora 146, 158–177.

GUDERIAN, R. (1977). "Air Pollution". Ecological Studies 22. Springer Verlag, Berlin.

GULMON, S. L. and MOONEY, H. A. (1977). Spatial and temporal relationships between two desert shrubs Atriplex hymenelytra and Tidestromia oblongifolia. J. Ecol. 65, 831–838.

GUTTERMAN, Y. (1980/81). Annual rhythm and position effects in the germinability of Mesembryanthemum nodiflorum. Israel J. Bot. 29, 93–97.

GUPTA, P. L. and RORISON, I. H. (1975). Seasonal differences in the availability of nutrients down a podzolic profile. J. Ecol. 63, 521–534.

HÅBJØRG, A. (1978a). Photoperiod ecotypes in Scandinavian trees and shrubs. Meld. Norges Landbrukshøgskole 57(33) 2–20.

HÅBJØRG, A. (1978b). Climatic control of floral differentiation and development in selected latitudinal and altitudinal ecotypes of Poa pratensis. Meld. Norges Landbrukshøgskole 57(17), 1–21.

HAINSWORTH, J. M. and AYLMORE, L. A. G. (1986). Water extraction by single plant roots. Soil Sci. Soc. Am. J. 50, 841–848.

HALL, I. R (1984). Taxonomy of VA mycorrhizal fungi. In "VA Mycorrhiza" (Eds. C. L. Powell and D. J. Bagyaraj), pp. 57–94. CRC Press, Boca Raton.

HARBERD, D. J. (1961). Observations on population structure and longevity of Festuca rubra L. New Phytol. 60, 184–206.

HARBORNE, J. (1977). "Introduction to Ecological Biochemistry". Academic Press, London and New York.

HARLEY, J. and SMITH, S. E. (1983). Mycorrhizal Symbiosis. Academic Press, London.

HARLEY, J. L. and LEWIS, D. H. (1969). The physiology of ectotrophic mycorrhizas. Adv. Microbiol. Physiol. 3, 53–58.

HARPER, J. L. (1969). The role of predation in vegetational diversity. *In* "Diversity and Stability in Ecological Systems". *Brookhaven Symp. Biol.* **22**, 48–62.

HARPER, J. L. (1975). Review of "Allelopathy" by E. L. Rice. *Quart. Rev. Biol.* **50**, 493–495.

HARPER, J. L. (1982). After description. *In* "The Plant Community as a Working Mechanism" (Ed. E. I. Newman), pp. 11–25. Blackwell Scientific Publications, Oxford.

HARPER, J. L. (1986). Modules, branches and the capture of resources. *In* "Population Biology and the Evolution of Clonal Organisms" (Eds. J. B. C. Jackson, L. E. Buss and R. E. Cook), pp. 1–34. Yale University Press, New Haven.

HARPER, J. L. and BENTON, R. A. (1966). The behaviour of seeds in soil. II. The germination of seeds on the surface of a water supplying substrate. *J. Ecol.* **54**, 151–166.

HARPER, J. L. and SAGAR, G. R. (1953). Some aspects of the ecology of buttercups in permanent grassland. *Proc. Br. Weed Control Conf.* **1**, 256–265.

HARPER, J. L., WILLIAMS, J. T. and SAGAR, G. R. (1965). The behaviour of seeds in soil. 1. The heterogeneity of soil surfaces and its role in determining the establishment of plants from seed. *J. Ecol.* **53**, 273–286.

HARRIS, G. A. (1967). Some competitive relationships between *Agropyron spicatum* and *Bromus tectorum*. *Ecological Monographs* **37**, 89–111.

HARRIS, G. A. and WILSON, A. M. (1970). Competition for moisture among seedlings of annual and perennial grasses as influenced by root elongation at low temperature. *Ecology* **51**, 530–534.

HARRISON-MURRAY, R. S. and CLARKSON, D. T. (1973). Relationships between structural development and absorption of ions by the root system of *Cucurbita pepo*. *Planta* **114**, 1–16.

HART, M. G. R. (1963). Observations on the source of acid in empoldered mangrove soils. II. Oxidation of soil polysulphides. *Plant and Soil* **19**, 106–114.

HARVEY, D. M. R., HALL, J. L., FLOWERS, T. J. and KENT, B. (1981). Quantitative ion localisation within *Suaeda maritima* leaf mesophyll cells. *Planta* **151**, 550–560.

HASSAN-PORATH, E and POLJAKOFF-MAYBER, A. (1969). The effect of salinity on the malic dehydrogenase of pea roots. *Plant Physiol.* **44**, 103–104.

HAUKIOJA, E. and NIEMELA, P. (1977). Retarded growth of a geometrid larva after mechanical damage to leaves of its host tree. *Ann. Zool. Fennica* **14**, 48–52.

HAVILL, D. C., INGOLD, A. and PEARSON, J. (1985). Sulphide tolerance in coastal halophytes. *Vegetatio* **62**, 279–285.

HAVILL, D. C., LEE, J. A. and STEWART, G. R. (1974). Nitrate utilisation by species from acidic and calcareous soils. *New Phytol.* **73**, 1221–1232.

HAY, R. K. M. (1977). Effects of tillage and direct drilling on soil temperature in winter. *J. Soil Sci.* **28**, 403–409.

HAY, R. K. M. (1978). Seasonal changes in the position of the shoot apex of winter wheat and spring barley in relation to the soil surface. *J. agric. Sci.* **91**, 245–248.

HAY, R. K. M. (1981a). "Chemistry for Agriculture and Ecology". Blackwell Scientific Publications, Oxford.

HAY, R. K. M. (1981b). Timely planting of maize: a case history from the Lilongwe Plain. *Trop. Agric.* **58**, 147–155.

HAY, R. K. M. and ALLEN, E. J. (1978). Tuber initiation and bulking in the potato (*Solanum tuberosum*) under tropical conditions: the importance of soil and air temperature. *Trop. Agric.* **55**, 289–295.

HAY, R. K. M. and TUNNICLIFFE WILSON, G. (1982). Leaf appearance and extension in field-grown winter wheat plants: the importance of soil temperature during vegetative growth. *J. agric. Sci. Camb.* **99**, 403–410.

HAYMAN, D. S. and MOSSE, B. (1972). Plant growth responses to vesicular-arbuscular mycorrhiza. III. Increased uptake of labile P from soil. *New Phytol.* **71**, 41–47.

HEATH, R. L. (1975). Ozone. *In* "Responses of Plants to Air Pollution" (Eds. J. B. Mudd and T. T. Kozlowski), pp. 23–55. Academic Press, New York and London.

HEATH, R. L. (1980). Initial events in injury to plants by air pollutants. *Ann. Rev. Plant Physiol.* **31**, 395–431.

HECK, W. W. (1984). Defining gaseous pollution problems in north America. *In* "Gaseous Air Pollutants and Plant Metabolism" (Eds. M. J. Koziol and F. R. Whatley), pp. 35–48. Butterworth, London.

HEDBERG, I. and HEDBERG, O. (1979). Tropical-alpine life forms of vascular plants. *Oikos* **33**, 297–307.

HEDLEY, M. J., NYE, P. H. and WHITE, R. E. (1982). Plant-induced changes in the rhizosphere of rape (*Brassica napus* var. Emerald) seedlings. 2. Origin of pH change. *New Phytol.* **91**, 31–44.

HEGARTY, T. W. and ROSS, H. A. (1980/81). Investigations of control mechanisms of germination under water stress. *Israel J. Bot.* **29**, 83–92.

HEIDE, O. M., HAY, R. K. M. and BAUGERÖD, H. (1985). Specific daylength effects on leaf growth and dry-matter production in high-latitude grasses. *Ann. Bot.* **55**, 579–586.

HEIMER, Y. M. (1973). The effects of sodium chloride, potassium chloride and glycerol on the activity of nitrate reductase of a salt-tolerant and two non-tolerant plants. *Planta* **113**, 279–281.

HELLKVIST, J., RICHARDS, G. P. and JARVIS, P. G. (1974). Vertical gradients of water potential and tissue water relations in Sitka Spruce trees measured with the pressure chamber. *J. appl. Ecol.* **11**, 637–667.

HENDERSON, L. J. (1913). "The Fitness of the Environment". Macmillan, London.

HENDRY, G. A. F. and BROCKLEBANK, K. J. (1985). Iron-induced oxygen radical metabolism in waterlogged plants. *New Phytol.* **101**, 199–206.

HEPLER, P. K. and WAYNE, R. O. (1985). Calcium and plant development. *Ann. Rev. Plant Physiol.* **36**, 397–439.

HEWITT, E. J. (1967). "Sand and Water Culture Methods Used in the Study of Plant Nutrition". C.A.B., London.

HIESEY, W. M., NOBS, M. A. and BJÖRKMAN, O. (1971). Experimental studies on the nature of species. V. Biosystematics, genetics, and physiological ecology of the *Erythranthe* section of *Mimulus*. *Carnegie Inst. Wash. Publ.* **628**.

HIGGS, D. E. B. and JAMES, D. B. (1969). Comparative studies on the biology of upland grasses. I. Rate of dry matter production and its control in four grass species. *J. Ecol.* **57**, 553–564.

HILL, B. S. and HILL, A. E. (1973). Enzymatic approaches to chloride transport in the *Limonium* salt gland. *In* "Ion Transport in Plants" (Ed. W. P. Anderson), pp. 379–384. Academic Press, London and New York.

HILLEL, D. (1972). Soil moisture and seed germination. *In* "Water Deficits and Plant Growth" (Ed. T. T. Kozlowski), vol. 3, pp. 65–89. Academic Press, New York and London.

HODGSON, D. R. and BUCKLEY, G. P. (1975). A practical approach towards the establishment of trees and shrubs on pulverised fuel ash. *In* "The Ecology of Resource Degradation and Renewal" (Eds. M. J. Chadwick and G. T. Goodman). *Brit. Ecol. Soc. Symp.* **15**, 305–330.

HOLFORD, I. C. R. (1976). Effects of phosphate buffer capacity of soil on the phosphate requirements of plants. *Plant and Soil* **45**, 433–444.

HOLMES, M. G. and SMITH, H. (1975). The function of phytochrome in plants growing in the natural environment. *Nature* **254**, 512–514.

HOLMES, M. G. and SMITH, H. (1977). The function of phytochrome in the natural environment. II. The influence of vegetation canopies on the spectral energy distribution of natural daylight. *Photochem. Photobiol.* **25**, 539–546.

HOLMGREN, P. (1968). Leaf factors affecting light-saturated photosynthesis in ecotypes of *Solidago virgaurea* from exposed and shaded habitats. *Physiol. Plant.* **21**, 676–698.

HOLMGREN, P., JARVIS, P. G. and JARVIS, M. S. (1965). Resistances to carbon dioxide and water vapour transfer in leaves of different plant species. *Physiol. Plant.* **18**, 557–573.

HOOK, D. D (1984). Adaptations to flooding with fresh water. *In* "Flooding and Plant Growth" (Eds. T. T. Kozlowski), pp. 265–294. Academic Press, New York.

HOOK, D. D. and SCHOLTENS, J. R. (1978). Adaptations and flood tolerance of tree species. *In* "Plant Life in Anaerobic Environments" (Eds. D. D. Hook and R. M. M. Crawford), pp. 299–331. Ann Arbor Science Publishers Inc., Michigan.

HOOK, D. D., BROWN, C. L. and KORMANIK, P. P. (1971). Inductive flood tolerance in swamp tupelo (*Nyssa sylvatica* var *biflora* (Walt.) Sarg.). *J. exp. Bot.* **22**, 78–89.

HOPE-SIMPSON, J. F. (1938). A chalk flora of the Lower Greensand, and its use in determining the calcicole habit. *J. Ecol.* **26**, 218–235.

HORAK, O. and KINZEL, H. (1971). Typen des Mineralstoffwechsels bei den höheren Pflanzen. *Oster. bot. Z.* **119**, 475–495.

HORN, H. (1971). "The Adaptive Geometry of Trees". University, Princeton.

HORSFIELD, D. (1977). Relationships between feeding of *Philaenus spumarius* (L.) and the amino acid concentration in the xylem sap. *Ecol. Entomol.* **2**, 259–266.

HORSMAN, D. C. and WELLBURN, A. R. (1976). Guide to the metabolic and biochemical effects of air pollutants on higher plants. *In* "Effects of Air Pollutants on Plants" (Ed. T. A. Mansfield), pp. 185–199. Cambridge University Press, Cambridge.

HSIAO, T. C. (1973). Plant responses to water stress. *Ann. Rev. Pl. Physiol.* **24**, 519–570.

HSIAO, T. C., ACEVEDO, E., FERERES, E. and HENDERSON, D. W. (1976). Water stress, growth and osmotic adjustment. *Phil. Trans. R. Soc. London* **B273**, 479–500.

HSIAO, T. C., SILK, W. K. and JING, J. (1985). Leaf growth and water deficits: biophysical effects. *In* "Control of Leaf Growth" (Eds. N. R. Baker, W. J. Davies and C. K. Ong), pp. 239–266. Cambridge University Press, Cambridge.

HUBER, B. (1928). Weitere quantitative Untersuchungen über das Wasserleitungssystem der Pflanzen. *Jahrb. Wiss. Bot.* **67**, 877–959.

HUGHES, A. P. (1959). Effects of the environment on leaf development in *Impatiens parviflora* D.C. *J. Linn. Soc. (Bot)* **56**, 161–165.

HULTÉN, E. (1962). "The Circumpolar Plants, II, Dicotyledons". *K. svenska. Vetensk. Akad. Handl. ser.* **5**, 13(1).

HUNT, R. (1970). "Relative growth rate: its range and adaptive significance in a local flora". Ph.D. Thesis, University of Sheffield.

HUTCHINSON, G. E. (1957). Concluding remarks. *In* "Population Studies: Animal Ecology and Demography". *Cold Spring Harbour Symposium on Quantitative Biology* **22**, 415–427.

HUTCHINSON, T. C. (1967). Comparative studies of the ability of species to withstand prolonged periods of darkness. *J. Ecol.* **55**, 291–299.

HYAMS, E. (1952). "Soil and Civilization". Thames and Hudson, London.

IDRIS, H. and MILTHORPE, F. L. (1966). Light and nutrient supplies in the competition between barley and charlock. *Oecologia Plant.* **1**, 143–164.

ISHIKAWA, M. and SAKAI, A. (1981). Freezing avoidance mechanisms by supercooling in some *Rhododendron* flower buds with reference to water relations. *Plant Cell Physiol.* **22**, 953–967.

IWASA, Y. and ROUGHGARDEN, J. (1984). Shoot/root balance of plants: optimal growth of a system with many vegetative organs. *Theoretical Population Biology* **25**, 78–105.

JACKSON, L. W. R. (1967). Effect of shade on leaf structure of deciduous tree species. *Ecology* **48**, 498–499.

JACKSON, M. B. (1985). Ethylene and the responses of plants to excess water in their environment—a review. *In* "Ethylene and Plant Development" (Eds. J. A. Roberts and G. A. Tucker), pp. 241-265. Butterworth, London.

JACKSON, M. B. and DREW, M. C. (1984). Effects of flooding on growth and metabolism of herbaceous plants. *In* "Flooding and Plant Growth" (Ed. T. T. Kozlowski), pp. 47–128. Academic Press, New York.

JACKSON, M. B., HERMAN, B. and GOODENOUGH, A. (1982). An examination of the importance of ethanol in causing injury to flooded plants. *Plant Cell Env.* **5**, 163–172.

JACKSON, M. B. and KOWALESKA, A. K. B. (1983). Positive and negative messages from roots induce foliar desiccation and stomatal closure in flooded pea plants. *J. exp. Bot.* **34**, 493–506.

JACOBSEN, T. AND ADAMS, R. M. (1958). Salt and silt in ancient Mesopotamian agriculture. *Science* **128**, 1251–1258.

JAEGER, E. C. (1940). Desert Wild Flowers. Stanford University Press, Stanford.

JAFFRE, T., BROOKS, R. R., LEE, J. and REEVES, D. (1976). *Serbetia acuminata*: a nickel-accumulating plant from New Caledonia. *Science* **193**, 579–580.

JANIESCH, P. (1973). Beitrag zur Physiologie der Nitrophyten: Nitratspeicherung und Nitratassimilation bei *Anthriscus sylvestris* Hoffm. *Flora* **162**, 479–491.

JANZEN, D. H. (1975). *Pseudomyrmex nigropilosa*: a parasite of a mutualism. *Science* **188**, 936–937.

JARVIS, P. G. (1964). The adaptability to light intensity of seedlings of *Quercus petraea* (Matt.) Liebl. *J. Ecol.* **52**, 545–571.

JARVIS, P. G. and MANSFIELD, T. A. (EDS.) (1981). "Stomatal Physiology". Cambridge University Press, Cambridge.

JEFFREE, C. E., JOHNSON, R. P. C. and JARVIS, P. G. (1971). Epicuticular wax in the stomatal antechamber of Sitka Spruce and its effects on the diffusion of water vapour and carbon dioxide. *Planta* **98**, 1–10.

JEFFREY, D. W. (1964). The formation of polyphosphate in *Banksia ornata*, an Australian heath plant. *Aust. J. biol. Sci.* **17**, 845–854.

JEFFREY, D. W. and PIGOTT, C. D. (1973). The response of grasslands on sugar limestone in Teesdale to application of phosphorus and nitrogen. *J. Ecol.* **61**, 85–92.

JENSEN, R. D., TAYLOR, S. A. and WIEBE, H. H. (1961). Negative transport and resistance to water flow through plants. *Plant. Physiol.* **36**, 633–638.

JESCHKE, W. D. (1973). K^+-stimulated Na^+ efflux and selective transport in barley roots. *In* "Ion Transport in Plants" (Ed. W. P Anderson), pp. 285–296. Academic Press, London and New York.

JOHN, C. D (1977). The structure of rice roots grown in aerobic and anaerobic environments. *Plant and Soil* **47**, 269–274.

JOHN, C. D. and GREENWAY, H. (1976). Alcoholic fermentation and activity of some enzymes in rice roots under anaerobiosis. *Aust. J. Pl. Physiol.* **3**, 325–336.

JOHN, M. K. (1972). Cadmium adsorption maxima of soils as measured by the Langmuir isotherm. *Can. J. Soil. Sci.* **52**, 343–350.

JOHNSON, D. A. and TIESZEN, L. L. (1976). Aboveground biomass allocation, leaf growth and photosynthesis patterns in tundra plant forms in arctic Alaska. *Oecologia* **24**, 159–173.

JOHNSON, H. B. (1975). Plant pubescence: an ecological perspective. *Bot. Rev.* **41**, 233–258.

JONES, H. E. (1971). Comparative studies of plant growth and distribution in relation to waterlogging. II. An experimental study of the relationship between transpiration and the uptake of iron in *Erica cinerea* L. and *E. tetralix* L. *J. Ecol.* **59**, 167–178.

JONES, H. E. and ETHERINGTON, J. R. (1970). Comparative studies of plant growth and distribution in relation to waterlogging. I. The survival of *Erica cinerea* L. and *E. tetralix* L. and its apparent relationship to iron and manganese uptake in waterlogged soil. *J. Ecol.* **58**, 487–496.

JONES, R. (1972). Comparative studies of plant growth and distribution in relation to waterlogging. V. The uptake of iron and manganese by dune and slack plants. *J. Ecol.* **60**, 131–140.

JORDAN, C. F. and KLINE, J. R. (1977). Transpiration of trees in tropical rainforests. *J. appl. Ecol.* **14**, 853–860.

JUNIPER, B. E. and ROBERTS, R. M. (1966). Polysaccharide synthesis and the fine structure of root cap cells. *J. Roy. Micro. Soc.* **85**, 63–72.

KAPPEN, L. (1981). Biological significance of resistance to high temperatures. *Encyclopaedia of Plant Physiology* **12A**, 439–474.

KARLSSON, T. (1974). Recurrent ecotypic variation in Rhinantheae and Gentianaceae in relation to hemi-parasitism and mycotrophy. *Bot. Not.* **127**, 527–539.

KASPERBAUER, M. J. (1971). Spectral distribution of light in a tobacco canopy and effects of end-of-day light quality on growth and development. *Plant Physiol.* **47**, 775–778.

KASPERBAUER, M. J. and PEASLEE, D. E. (1973). Morphology and photosynthetic efficiency of tobacco leaves that received end-of-day red or far-red light during development. *Plant Physiol.* **52**, 440–442.

KATZNELSON, H. (1946) The 'rhizosphere effect' of mangels on certain groups of soil micro-organisms. *Soil Sci.* **62**, 343–354.

KAUFMANN, M. R. (1972). Water deficits and reproductive growth. *In* "Water Deficits and Plant Growth" (Ed. T. T. Kozlowski), vol. 3, pp. 91–124. Academic Press, New York and London.

KAY, Q. O. N. (1971). Biological Flora of the British Isles: *Anthemis cotula* L. and *A. arvensis* L. *J. Ecol.* **59**, 623–648.

KEELEY, J. E. (1978) Malic acid accumulation in roots in response to flooding: evidence contrary to its role as an alternative to ethanol. *J. exp. Bot.* **29**, 1345–1349.

KEELEY, J. E. (1983). Crassulacean acid metabolism in the seasonally submerged aquatic *Isoetes howellii. Oecologia* **58**, 57–62.

KEELEY, J. E. and FRANZ, E. H. (1979). Alcoholic fermentation in swamp and upland populations of *Nyssa sylvatica*: temporal changes in adaptive strategy. *Amer. Nat.* **113**, 587–592.

KEVAN, P. G. (1975). Sun-tracking solar furnaces in high-arctic flowers: significance for pollination and insects. *Science* **189**, 723–726.

KIMBALL, S. L., BENNETT, B. D. and SALISBURY, F. B. (1973). The growth and development of montane species at near-freezing temperatures. *Ecology* **54**, 168–173.

King, T. J. (1975). Inhibition of seed germination under leaf canopies in *Arenaria serpyllifolia, Veronica arvensis,* and *Cerastium holosteoides. New Phytol.* **75,** 87–90.

Kinzel, H. (1969). Ansätze zu einer vergleichenden Physiologie der Mineralstoffwechsel und ihre ökologischen Konsequenzen. *Ber. deutsch. Bot. Ges.* **82,** 143–158.

Kinzel, H. (1982). Pflanzenökologie und Mineralstoffwechsel. Ulmer, Stuttgart.

Kjellberg, B., Karlsson, S. and Kerstensson, I. (1982). Effects of heliotropic movements of flowers of *Dryas octopetala* L. on gynoecium temperature and seed development. *Oecologia* **54,** 11–13.

Klaren, C. H. (1975). "Physiological Aspects of the Hemiparasite *Rhinanthus serotinus*". Groningen.

Klein, R. M. (1978). Plants and near-ultraviolet radiation. *Bot. Rev.* **44,** 1–127.

Kleinkopf, G. E. and Wallace, A. (1974). Physiological basis for salt tolerance in *Tamarix ramosissima. Plant Sci. Lett.* **3,** 157–163.

Klikoff, L. G. (1969). Temperature dependence of mitochondrial oxidative rates in relation to plant distribution. *In* "Physiological Systems in Semi-Arid Environments" (Eds. C. C. Hoft and M. L. Riedesell), pp. 263–269. University of New Mexico Press, Albquerque.

Koch, K. E. and Johnson, C. R. (1984). Photosynthetic partitioning in split-root citrus seedlings with mycorrhizal and non-mycorrhizal root systems. *Plant Physiol.* **75,** 26–30.

Koch, K. and Mengel, K. (1974) The influence of the level of potassium supply to young tobacco plants on short-term uptake and utilisation of nitrate nitrogen. *J. Sci. Fd Agric.* **25,** 465–471.

Kochenderfer, J. N. (1973) Root distribution under some forest types native to West Virginia. *Ecology* **54,** 445–449.

Koller, D. (1972) Environmental control of seed germination. *In* "Seed Biology" (Ed. T. T. Kozlowski), vol. 2, pp. 1–101. Academic Press, New York and London.

Koller, D. and Hadas, A. (1982). Water relations in the germination of seeds. *Encyclopaedia of Plant Physiology* **12B,** 401–431.

Koller, D. and Negbi, M. (1966). Germination of seeds of desert plants. Final report to USDA, project no AlO-FS-6, Department of Botany, Hebrew University, Jerusalem, Israel.

Kondo, J. (Ed.) (1984). Studies on effects of air pollutant mixtures on plants. *Res. Rep. Nat. Inst. Env. Studies, Japan* **66.**

Koziol, M. J. and Whatley, F. R. (Eds.) (1984). "Gaseous Air Pollutants and Plant Metabolism". Butterworth, London.

Kozlowski, T. T. (1964). "Water Metabolism in Plants". Harper and Row, New York.

Kozlowski, T. T. (1972). Shrinking and swelling of plant tissues. *In* "Water Deficits and Plant Growth" (Ed. T. T. Kozlowski), vol. 3, pp. 1–64. Academic Press, New York and London.

KOZLOWSKI, T. T. (1976). Water relations and tree improvement. *In* "Tree Physiology and Yield Improvement" (Eds. M. G. R. Cannell and F. T. Last), pp. 307–327. Academic Press, London and New York.

KOZLOWSKI, T. T. (1984). Responses of woody plants to flooding. *In* "Flooding and Plant Growth" (Ed. T. T. Kozlowski), pp. 129–163. Academic Press, New York.

KRAMER, P. J. (1969). "Plant and Soil Water Relationships". McGraw-Hill, New York

KRAMER, P. J. and KOZLOWSKI, T. T. (1979). "Physiology of Woody Plants". Academic Press, New York.

KRANZ, E. and JACOB, F. (1977). Zur Mineralstoff-Konkurrenz zwischen *Linum* und *Camelina* II. Aufnahme von ^{32}P-Phosphat und ^{86}Rubidium. *Flora* **166,** 505–516.

KRAUS, H. (1969). Osmoregulation mit α-galactosyl-glyceriden bei *Ochromonas malhamensis. Ber. deutsch. bot. Ges.* **82,** 115–125.

KROG, J. (1955). Notes on temperature measurements indicative of special organisation in arctic and subarctic plants for utilisation of radiated heat from the sun. *Physiol. Plant.* **8,** 836–839.

KROH, G. C. and BEAVER, D. L. (1978). Insect response to mixture and monoculture patches of Michigan old-field annual herbs. *Oecologia* **31,** 269–275.

KU, H. S., SUGE, H., RAPPAPORT, L. and PRATT, H. K. (1970). Stimulation of rice coleoptile growth by ethylene. *Planta* **90,** 333–339.

KUIJT, J. (1969). "The Biology of Parasitic Flowering Plants". University of California Press, Berkeley.

KUMMEROW, J. (1980). Adaptation of roots in water-stressed native vegetation. *In* "Adaptation of Plants to Water and High Temperature Stress" (Eds. N. C. Turner and P. J. Kramer), pp. 57–73. Wiley-Interscience, New York.

KYLIN, A. and KÄHR, M. (1973) The effect of magnesium and calcium on adenosine triphosphatases from wheat and oat roots at different pH. *Physiol. Plant.* **28,** 452–457.

LAMM, C. G., TJELL, J. C., MØLLER, O. and CHRISTIANSEN, T. F. (1969). Plant nutrient availability in soils. II. Quantity-intensity relationships of phosphorus and manganese as influenced by soil pH. *Acta agric. Scand.* **19,** 135–140.

LAMBERS, H. (1979). Energy Metabolism in Higher Plants in Different Environments. Rijksuniversiteit te Groningen, Netherlands.

LANCE, J. C. and PEARSON, R. W. (1969). Effects of low concentrations of aluminium on growth and water and nutrient uptake by cotton roots. *Soil Sci. Soc. Amer. Proc.* **33,** 95–98.

LANE, S. D., MARTIN, E. S. and GARROD, J. F. (1978). Lead toxicity effects on indole-3-ylacetic acid induced cell elongation. *Planta* **144,** 79–84.

LANGE, O. L., NOBEL, P. S., OSMOND, C. B. and ZEIGLER, H. (EDS.) (1982). Physiological Plant Ecology 2. Water Relations and Carbon Assimilation. *Encyclopaedia of Plant Physiology* **12B.**

Lange, O. L., Kappen, L. and Schulze, E.-D. (1976). "Water and Plant Life". *Ecological Studies* **19**. Springer-Verlag, Berlin.

Lange, O. L. and Zuber, M. (1977). *Frerea indica*, a stem succulent CAM plant with deciduous C_3 leaves. *Oecologia* **31**, 67–72.

Langridge, J. and McWilliam, J. R. (1967) Heat responses of higher plants. *In* "Thermobiology" (Ed. A. H. Rose), pp. 231–292. Academic Press, London and New York.

Larcher, W. (1975). "Physiological Plant Ecology". Springer-Verlag, Berlin.

Larcher, W., Heber, U. and Santarius, K. A. (1973). Limiting temperatures for life functions. *In* "Temperature and Life" (Eds. H. Precht, J. Christopherson, H. Hensel and W. Larcher), pp. 195–263. Springer-Verlag, Berlin.

Larkum, A. W. D. (1968). Ionic relations of chloroplasts *in vivo*. *Nature* **218**, 447–449.

Larsen, S. (1964). On the relationship between labile and non-labile phosphate in soils. *Acta agric. Scand.* **14**, 249–253.

Larsen, S. (1967). Soil phosphorus. *Adv. Agron.* **19**, 151–210.

Last, F. T., Cape, J. N. and Fowler, D. (1986). Acid rain—or pollution climate? *Span* **29**, 1–4.

Lawton, J. H. and McNeill, A. S. (1979). Between the devil and the deep blue sea: on the problem of being a herbivore. *In* "Population Dynamics". *Brit. Ecol. Soc. Symp.* **20**, 223–244.

Lazenby, A. (1955). Germination and establishment of *Juncus effusus* L. II. The interaction effects of moisture and competition. *J. Ecol.* **43**, 595–605.

Lee, D. W. and Graham, R. (1986). Leaf optical properties of rainforest sun and extreme shade plants. *Amer. J. Bot.* **73**, 1100–1108.

Lee, J., Reeves, R. D., Brooks, R. R. and Jaffre, T. (1978). The relation between nickel and citric acid in some nickel-accumulating plants. *Phytochem.* **17**, 1033–1035.

Lee, R. B. (1982). Selectivity and kinetics of ion uptake by barley plants following nutrient deficiency. *Ann. Bot.* **50**, 429–449.

Lee, R. B. and Ratcliffe, R. G. (1983). Phosphorus nutrition and the intracellular distribution of inorganic phosphate in pea root tips: a quantitative study using ^{31}P NMR. *J. exp. Bot.* **34**, 1222–1244.

Leigh, R. A., Wyn Jones, R. G. and Williamson, F. A. (1973). The possible role of vesicles and ATPases in ion uptake. *In* "Ion Transport in Plants" (Ed. W. P. Anderson), pp. 407–418. Academic Press, London and New York.

Lenzian, K. J. and Unsworth, M. H. (1982). Ecophysiological effects of atmospheric pollutants. *Encyclopaedia of Plant Physiology* **12D**, 465–502.

Leopold, A. C. and Kriedemann, P. E. (1975). "Plant Growth and Development", 2nd Edn. McGraw-Hill, New York.

Levin, D. A. (1976). The chemical defences of plants to pathogens and herbivores. *Ann. Rev. Ecol. Syst.* **7**, 121–160.

LEVITT, J. (1978) An overview of freezing injury and survival, and its inter-relationships to other stresses. *In* "Plant Cold Hardiness and Freezing Stress" (Eds. P. H. Li and A. Sakai), pp. 3–15. Academic Press. New York and London.

LEVITT, J. (1980). "Responses of Plants to Environmental Stress". 2nd Edition, Vol. 2. Academic Press, New York.

LEWIS, M. C. (1972). The physiological significance of variation in leaf structure. *Sci. Progr., Oxf.* **60,** 25–51.

LI, P. H. and SAKAI, A. (1978). "Plant Cold Hardiness and Freezing Stress". Academic Press, New York.

LOACH, K. (1970). Shade tolerance in tree seedlings. 2. Growth analysis of plants raised under artificial shade. *New Phytol.* **67,** 273–286.

LOCHHEAD, A. G. and ROUATT, J. W. (1955) The "rhizosphere effect" on the nutritional groups of soil bacteria. *Soil Sci. Soc. Amer. Proc.* **19,** 48–49.

LOCK, J. M. and MILBURN, T. R. (1971). The seed biology of *Themeda triandra* Forsk. in relation to fire. *Symp. Brit. Ecol. Soc.* **11,** 337–349.

LONG, R. C. and WOLTZ, W. G. (1972) Depletion of nitrate reductase activity in response to soil leaching. *Agron. J.* **64,** 789–792.

LOSCH, R. and TENHUNEN, J. D. (1981). Stomatal responses to humidity—phenomenon and mechanism. *In* "Stomatal Physiology" (Eds. P. G. Jarvis and T. A. Mansfield), pp. 137–161. Cambridge University Press, Cambridge.

LOUGHMAN, B. C. (1969). The uptake of phosphate and its movement within the plant. *In* "Ecological Aspects of the Mineral Nutrition of Plants". *Brit. Ecol. Soc. Symp.* **9,** 309–322.

LUDLOW, M. M. (1980). Adaptive significance of stomatal responses to water stress. *In* "Adaptations of Plants to Water and High Temperature Stress" (Eds. N. C. Turner and P. J. Kramer), pp. 123–138. Wiley-Interscience, New York.

LÜTTGE, U. and SMITH, J. A. C. (1982). Membrane transport, osmoregulation and the control of CAM. *In* "Crassulacean Acid Metabolism" (Eds. I. P. Ting and M. Gibbs), pp. 69–91. American Society of Plant Physiology, Rockville.

LYNCH, J. M. (1975a). The formation of ethylene by soil micro-organisms. *A.R.C. Letcombe Laboratory Ann. Rep.* 1974, 88–95.

LYNCH, J. M. (1975b). Ethylene in soil. *Nature* **256,** 576–577.

LYNCH, J. M. and ELLIOT, L. P. (1983). Minimizing the potential phytotoxicity of wheat straw by microbial degradation. *Soil Biol. Biochem.* **15,** 221–222.

LYONS, J. M. (1973). Chilling injury in plants. *Ann. Rev. Pl. Physiol.* **24,** 445–466.

MABRY, J. J., HUNZIKER, J. H. and DiFEO, D. R. (1977). "Creosote Bush. Biology and Chemistry of *Larrea* in New World Deserts". US/1BP synthesis series 6. Dowden, Hutchinson and Ross Inc., Stroudsburg, Pennsylvania.

McCORMICK, L. H. and BOWDEN, F. Y. (1972). Phosphate fixation by aluminium in plant roots. *Soil Sci. Soc. Amer. Proc.* **36,** 799–802.

McCREE, K. J. (1972). Test of current definitions of photosynthetically active radiation against leaf photosynthesis data. *Agric. Meteorol.* **10**, 443–453.

McCREE. K. J. and TROUGHTON, J. H. (1966). Prediction of growth rate at different light levels from measured photosynthesis and respiration rates. *Plant Physiol.* **41**, 559–566.

McLAREN, J. S. and SMITH, H. (1978). Phytochrome control of the growth and development of *Rumex obtusifolius* under simulated canopy light environments. *Plant Cell Env.* **1**, 61–68.

McMANMON and CRAWFORD, R. M. M. (1971). A metabolic theory of flooding tolerance: the significance of enzyme distribution and behaviour. *New Phytol.* **70**, 299–306.

McNAUGHTON, S. J., FOLSOM, T. C., LEE, T., PARK, F., PRICE, C., ROEDER, O., SCHMITZ, J. and STOCKWELL, C. (1974). Heavy metal tolerance in *Typha latifolia* without the evolution of tolerant races. *Ecology* **55**, 1163–1165.

McNEILLY, T. and BRADSHAW, A. D. (1968). Evolutionary processes in populations of copper-tolerant *Agrostis tenuis* Sibth. *Evolution.* **22**, 108–118.

McPHERSON, J. K. and MÜLLER, C. H. (1969). Allelopathic effects of *Adenostoma fasciculatum*, "chamise", in the California Chaparral. *Ecol. Monogr.* **39**, 117–198.

McWILLIAM, J. R., KRAMER, P. J. and MUSSER, R. L. (1982). Temperature-induced water stress in chilling-sensitive plants. *Aust. J. Plant Physiol.* **9**, 343–352.

MADHOK, P. O. and WALKER, R. B. (1969). Magnesium nutrition of two species of sunflower. *Plant Physiol.* **44**, 1016–1022.

MAHMOUD, A. and GRIME, J. P. (1974). A comparison of negative relative growth rates in shaded seedlings. *New Phytol.* **73**, 1215–1220.

MALLOCH, D. W., PIROZYNSKI, K. A. and RAVEN, P. H. (1981). Ecological and evolutionary significance of mycorrhizal symbioses in vascular plants. *Proc. Natl Acad. Sci. USA* **77**, 2113–2118.

MALLOTT, P. G., DAVY, A. J., JEFFERIES, R. L. and HUTTON, M. J. (1975). Carbon dioxide exchange in leaves of *Spartina anglica* Hubbard. *Oecologia* **20**, 351–358.

MALONE, C., KOEPPE, D. E. and MILLER, R. J. (1974). Localisation of lead accumulated by corn plants. *Plant Physiol.* **53**, 388–394.

MANCINELLI, A. L. and RABINO, I. (1978). The "high irradiance responses" of plant photomorphogenesis. *Bot. Rev.* **44**, 129–180.

MANSFIELD, J. W. and DEVERALL, B. J. (1974a). The rates of fungal development and lesion formation in leaves of *Vicia faba* during infection by *Botrytis cinerea* and *B. fabae*. *Ann. appl. Biol.* **76**, 77–89.

MANSFIELD, J. W. and DEVERALL, B. J. (1974b). Changes in wyerone acid concentration in leaves of *Vicia faba* after infection by *Botrytis cinerea* or *B. fabae*. *Ann. appl. Biol.* **77**, 227–235.

MANSFIELD, T. A. (1976). The role of stomata in determining the responses of plants to air pollutants. *In* "Commentaries in Plant Science" (Ed. H. Smith), pp. 13–22. Pergamon Press, Oxford.

MANSFIELD, T. A. (1983). Movements of stomata. *Sci. Prog. Oxf.* **68,** 519–542.
MANSFIELD, T. A. and DAVIES, W. J. (1981). Stomata and stomatal mechanisms. *In* "The Physiology and Biochemistry of Drought Resistance" (Eds. L. G. Paleg and D. Aspinall), pp. 315–347. Academic Press, London.
MANSFIELD, T. A. and FREER-SMITH, P. H. (1981). Effects of urban air pollution on plant growth. *Biol. Rev.* **56,** 343–368.
MARION, G. M. and MILLER, P. C. (1982). Nitrogen mineralization in a tussock tundra soil. *Arc. Alp. Res.* **14,** 287–293.
MARRS, R. H., ROBERTS, R. D., SKEFFINGTON, R. A. and BRADSHAW, A. D. (1983). Nitrogen and the development of ecosystems. *In* "Nitrogen as an Ecological Factor" (Eds. J. A. Lee, S. McNeill and I. H. Rorison), pp. 113–136. Blackwell Scientific Publications, Oxford.
MARSHALL, D. R., BROUÉ, P. and PRYOR, A. J. (1973). Adaptive significance of alcohol dehydrogenase isoenzymes in maize. *Nature New Biology* **244,** 16–17.
MARTIN, J. K. (1977). Factors influencing the loss of organic carbon from wheat roots. *Soil Biol. Biochem.* **9,** 1–7.
MARTIN, J. T. and JUNIPER, B. E. (1970). "The Cuticles of Plants". Arnold, London.
MARTIN, M. H. (1968) Conditions affecting the distribution of *Mercurialis perennis* in certain Cambridgeshire woodlands. *J. Ecol.* **56,** 777–793.
MATHYS, W. (1973). Vergleichende Untersuchungen der Zinkaufnahme von resistenten und sensitiven Populationen von *Agrostis tenuis* Sibth. *Flora* **162,** 492–499.
MATHYS, W. (1975). Enzymes of heavy-metal-resistant and non-resistant populations of *Silene cucubalus* and their interaction with some heavy metals *in vitro* and *in vivo*. *Physiol. Plant* **33,** 161–165.
MATHYS, W. (1977). The role of malate, oxalate, and mustard oil glucosides in the evolution of zinc resistance in herbage plants. *Physiol. Plant* **40,** 130–136.
MATSUMOTO, H., HIRASAWA, P., MORIMURA, S. and TAKAHASHI, E. (1976). Localisation of aluminium in tea leaves. *Plant Cell Physiol.* **18,** 890–885.
MAY, D. S. and VILLAREAL, H. M. (1974). Altitudinal differentiation of the Hill Reaction in populations of *Taraxacum officinale* in Colorado. *Photosynthetica* **8,** 73–77.
MAYER, A. M. and POLJAKOFF-MAYBER, A. (1975). "The Germination of Seeds", 2nd Edn. Pergamon, Oxford.
MAZELIS, M. and VENNESLAND, B. (1957). Carbon dioxide fixation into oxalacetate in higher plants. *Plant Physiol.* **32,** 591–600.
MEDINA, E. and TROUGHTON, J. H. (1974). Dark CO_2 fixation and the carbon isotope ratio in Bromeliaceae. *Plant Sci. Letters* **2,** 357–362.
MEIDNER, H. (1954). Measurements of water intake from the atmosphere by leaves. *New Phytol.* **53,** 423–426.
MEIDNER, H. and MANSFIELD, T. A. (1968). "The Physiology of Stomata". McGraw-Hill, London.

MEIDNER, H. and SHERIFF, D. W. (1976). "Water and Plants". Blackie, Glasgow.

MERZ, E. (1959). Pflanzen und Raupen. Über einigen Prinzipien der Futterwahl bei Grossschmetterlingsraupen. *Biol. Zentrbl.* **78**, 152–158.

MEWISSEN, D. J., DAMBLON, J. and BACQ, Z. M. (1959). Comparative sensitivity to radiation of seeds from a wild plant grown on uraniferous and non-uraniferous soil. *Nature* **183**, 1449.

MILLER, P. C. and 11 OTHERS (1984). Plant-soil processes in *Eriophorum vaginatum* tussock tundra in Alaska: a systems modelling approach. *Ecological Monographs* **54**, 361–405.

MILTHORPE, F. L. and NEWTON, P. (1963). Studies on the expansion of the leaf surface. III. The influence of radiation on cell division and expansion. *J. exp. Bot.* **14**, 483–495.

MINCHIN, F. R. and PATE, J. S. (1974). Diurnal functioning of the legume root nodule. *J. exp. Bot.* **25**, 295–308.

MOHR, H., DRUMM, H. and KASEMIR, H. (1974). Licht und Farbstoffe. *Ber. deutsch. Bot. Ges.* **87**, 49–70.

MONTEITH, J. L. (1977). Climate. In "Ecophysiology of Tropical Crops" (Eds. P. de T. Alvim and T. T. Kozlowski), pp. 1–27. Academic Press, New York.

MOONEY, H. A. (1980). Seasonality and gradients in the study of stress adaptation. In "Adaptation of Plants to Water and High Temperature Stress" (Eds. N. C. Turner and P. J. Kramer), pp. 279–294, Wiley-Interscience, New York.

MOONEY, H. A. and BILLINGS, W. D. (1961). Comparative physiological ecology of arctic and alpine populations of *Oxyria digyna*. *Ecol. Monog.* **31**, 1–29.

MOONEY, H. A. and JOHNSON, A. W. (1965). Comparative physiological ecology of arctic and alpine population of *Thalictrum alpinum*. *Ecology* **46**, 721–727.

MORGAN, D. C. and SMITH, H. (1979). A systematic relationship between phytochrome-controlled development and species habitat for plants grown in simulated natural conditions. *Planta* **145**, 253–258.

MORGAN, D. C. AND SMITH, H. (1981). Non-photosynthetic responses to light quality. *Encyclopaedia of Plant Physiology* **12A**, 109–134.

MORGAN, J. M. (1984). Osmoregulation and water stress in higher plants. *Ann. Rev. Plant Physiol.* **35**, 299–319.

MOSSE, B. (1972). The influence of soil type and *Endogone* strain on the growth of mycorrhizal plants in phosphate deficient soils. *Rev. Ecol. Biol. Sol* **9**, 529–537.

MOTT, J. J. (1972). Germination studies on some annual species from an arid region of Western Australia. *J. Ecol.* **60**, 293–304.

MUDD, J. B. and KOZLOWSKI, T. T. (EDS.) (1975). "Responses of Plants to Air Pollution". Academic Press, New York and London.

MUKERJI, S. K. (1936). Contribution to the autecology of *Mercurialis perennis* L. *J. Ecol.* **24**, 38–81.

MÜLLER, C. H. (1966). The role of chemical inhibition (allelopathy) in vegetational composition. *Bull. Torrey Bot. Club* **93**, 332–351.

MÜLLER, C. H. (1970). Phytotoxins as plant habitat variables. *Recent Advances in Phytochemistry* **3**, 106–121.

MÜLLER, W. H. (1965). Volatile materials produced by *Salvia leucophylla*: effect on seedling growth and soil bacteria. *Bot. Gaz.* **126**, 195–200.

MULROY, T. W. (1979). Spectral properties of heavily glaucous and non-glaucous leaves of a succulent rosette plant. *Oecologia* **29**, 301–310.

MUNNS, R., GREENWAY, H. and KIRST, G. O. (1983). Halotolerant eukaryotes. *Encyclopaedia of Plant Physiology* **12C**, 59–134.

MUNZ, P. A. (1974). "A Flora of Southern California". University of California Press, Berkeley.

MUSGRAVE, A., JACKSON, M. B. and LING, E. (1972). *Callitriche* stem elongation is controlled by ethylene and gibberellin. *Nature New Biology* **238**, 93–96.

MUSGRAVE, A. and WALTERS, J. (1973). Ethylene-stimulated growth and auxin transport in *Ranunculus sceleratus* petioles. *New Phytol.* **72**, 783–789.

MYERSCOUGH, P. J. and WHITEHEAD, F. H. (1966). Comparative biology of *Tussilago farfara*, *Chamaenerion angustifolium*, *Epilobium montanum* and *Epilobium adenocaulon*. 1. General biology and germination. *New Phytol.* **65**, 192–210.

NATIONAL ACADEMY OF SCIENCES (1979). Protection against depletion of stratospheric ozone by chlorofluorocarbons, Washington.

NASSERY, H. (1971). Phosphate absorption by plants from habitats of different phosphate status. III. Phosphate fractions in the roots of intact plants. *New Phytol.* **70**, 949–951.

NASYROV, Y S. (1978). Genetic control of photosynthesis and improving productivity. *Ann. Rev. Pl. Physiol.* **29**, 215–237.

NEILSON, R. E., LUDLOW, M. M. and JARVIS, P. G. (1972). Photosynthesis in Sitka spruce (*Picea sitchensis* (Bong.) Carr). II Response to temperature. *J. appl. Ecol.* **9**, 721–745.

NEWMAN, E. I. (1973). Competition and diversity in herbaceous vegetation. *Nature* **244**, 310.

NEWMAN, E. I. (1985). The rhizosphere: carbon sources and microbial populations. *In* "Ecological Interactions in Soil" (Eds. A. H. Fitter, D. Atkinson, D. J. Read and M. B. Usher), pp. 107-121. Blackwell Scientific Publications, Oxford.

NEWMAN, E. I. and ANDREWS, R. E. (1973). Uptake of P and K in relation to root growth and root density. *Plant and Soil* **38**, 49–69.

NEWMAN, E. I. and MILLER, M. H. (1977). Allelopathy among some British grassland species. II. Influence of root exudates on phosphate uptake. *J. Ecol.* **65**, 399–412.

NEWMAN, E. I. and ROVIRA, A. D (1975). Allelopathy among some British grassland species. *J. Ecol.* **63**, 727–738.

NEWMAN, E. I., CAMPBELL, R and ROVIRA, A D. (1977). Experimental alteration of soil microbial populations for studying effects on higher plant interactions. *New Phytol.* **79,** 107–118.

NEWTON, J. E. and BLACKMAN, G. E. (1970). The penetration of solar radiation through canopies of different structure. *Ann. Bot.* (N.S.) **34,** 329–348.

NEWTON, P. (1963). Studies on the expansion of the leaf surface. II. The influence of light intensity and photoperiod. *J. exp. Bot.* **14,** 458–482.

NISSEN, P. (1974). Uptake mechanisms: organic and inorganic. *Ann. Rev. Pl. Physiol.* **25,** 38–80.

NOBEL, P. S. (1974). "Introduction to Biophysical Plant Physiology". W. H. Freeman, San Francisco.

NOBEL, P. S. (1984). Extreme temperatures and thermal tolerances for seedlings of desert succulents. *Oecologia* **62,** 310–317.

NOBEL, P. S. and JORDAN, P. W. (1983). Transpiration stream of desert species: resistances and capacitances for a C_3, a C_4 and a CAM plant. *J. exp. Bot.* **34,** 1379–1391.

NOWAK, R. S. and CALDWELL, M. M. (1984). A test of compensatory photosynthesis in the field: implications for herbivory tolerance. *Oecologia* **61,** 311–318.

NYE, P. H. (1966). The effect of nutrient intensity and buffering power of a soil, and the absorbing power, size and root-hairs of a root, on nutrient absorption by diffusion. *Plant and Soil* **25,** 81–105.

NYE, P. H. (1969). The soil model and its application to plant nutrition. *In* "Ecological Aspects of the Mineral Nutrition of Plants". *Brit. Ecol. Soc. Symp.* **9,** 105–114.

NYE, P. H. (1973). The relation between the radius of a root and its nutrient absorbing power (α). *J. exp. Bot.* **24,** 783–786.

NYE, P. H. and MARRIOTT, F. H. C. (1969). A theoretical study of the distribution of substances around roots resulting from simultaneous diffusion and mass flow. *Plant and Soil* **30,** 459–472.

NYE, P. H. and TINKER, P. B. H. (1969). The concept of a root demand coefficient. *J. appl. Ecol.* **6,** 293–300.

NYE, P. H. and TINKER, P. B. H. (1977). "Solute Movement in the Soil-Root System". Blackwell, Oxford.

OAKS, A. and HIREL, B. (1985). Nitrogen metabolism in roots. *Ann. Rev. Plant Physiol.* **36,** 345–365.

O'BRIEN, T. A., MOORBY, J. and WHITTINGTON, W. J. (1967). The effect of management and competition on the uptake of ^{32}phosphorus by ryegrass, meadow fescue, and their natural hybrid. *J. appl. Ecol.* **4,** 513–520.

OERTLI, J. J. (1971). The stability of water under tension in the xylem. *Z. Pflanzenphysiol.* **65,** 195–209.

OLMSTED, C. E. (1944). Growth and development in range grasses. IV. Photoperiodic responses in twelve geographic strains of side-oats grama. *Bot. Gaz.* **100,** 46–74.

OLSEN, C. (1971). Selective ion absorption in various plant species and its ecological significance. *C. R. Trav. Lab. Carlsberg* **38,** 399–422.

OLSEN, S. R. and KEMPER, W. D. (1968). Movement of nutrients to plant roots. *Adv. Agron.* **20,** 91–151.

ORMROD, D. P. (1982). Air pollutant interactions in mixtures. *In* "Effects of Gaseous Air Pollution in Agriculture and Horticulture" (Eds. M. H. Unsworth and D. P. Ormrod), pp. 307–331. Butterworth, London.

ORSHAN, G. (1963). Seasonal dimorphism of desert and Mediterranean chamaephytes and its significance as a factor in their water economy. *Symp. Brit. Ecol. Soc.* **3,** 206–222.

OSBORNE, D. (1984). Ethylene and plants of aquatic and semi-aquatic environments: a review. *Plant Growth Regulation* **2,** 167–185.

OSMOND, C. B., WINTER, K. and ZIEGLER, H. (1982). Functional significance of different pathways of CO_2 fixation in photosynthesis. *Encyclopaedia of Plant Physiology* **12B,** 480–547.

OSMOND, C. B., ZIEGLER, H., STICHLER, W. and TRIMBORN, P. (1975). Carbon isotope discrimination in alpine succulent plants supposed to be capable of crassulacean acid metabolism (CAM). *Oecologia* **18,** 209–218.

OWEN, D. F. and WIEGERT, R. G. (1981). Mutualism between grasses and grazers: an evolutionary hypothesis. *Oikos* **36,** 376–378.

OWEN, P. C. (1952). The relation of germination of wheat to water potential. *J. exp. Bot.* **3,** 188–203.

OZANNE, P. G. AND SHAW, T. C. (1967). Phosphate sorption by soils as a measure of the phosphate requirement for pasture growth. *Aust. J. agric. Sci.* **18,** 601–612.

PACKHAM, J. R. and WILLIS, A. J. (1977). The effects of shading on *Oxalis acetosella*. *J. Ecol.* **65,** 619–642.

PAPAROZZI, T. and TUKEY, H. B. (1984). Characterization of injury to birch and bean leaves by simulated acid precipitation. *In* "Direct and Indirect Effects of Acidic Precipitation on Vegetation" (Ed. R. A. Linthurst), pp. 13–18. Butterworth, Boston.

PARKHURST, D. F. and LOUCKS, O. L. (1972). Optimal leaf size in relation to environment. *J. Ecol.* **60,** 505–537.

PAREEK, R. P. and GAUR, A. C. (1973). Organic acids in the rhizosphere of *Zea mays* and *Phaseolus aureus* plants. *Plant and Soil* **39,** 441–444.

PARKER, J. (1968). Drought-resistance mechanisms. *In* "Water Deficits and Plant Growth" (Ed. T. T. Kozlowski), vol. 1, pp. 195–234. Academic Press, New York and London.

PARKER, J. (1972). Protoplasmic resistance to water deficits. *In* "Water Deficits and Plant Growth" (Ed. T. T. Kozlowski), vol. 3, pp. 125–176. Academic Press, New York and London.

PASSIOURA, J. B. (1963). A mathematical model for the uptake of ions from the soil solution. *Plant and Soil* **18,** 225–238.

PASSIOURA, J. B. (1972). The effect of root geometry on the yield of wheat growing on stored water. *Aust. J. agric. Res.* **23,** 745–752.

PASSIOURA, J. B. (1976). The control of water movement through plants. *In* "Transport and Transfer Processes in Plants", pp. 373–380. Academic Press, London and New York.

Passioura, J. B. (1984). Hydraulic resistances of plants. 1. Constant or variable? *Aust. J. Plant Physiol.* **11**, 333–339.

Peacock, J. M. (1975). Temperature and leaf growth in *Lolium perenne*. II. The site of temperature perception. *J. appl. Ecol.* **12**, 115–123.

Pearce, F. (1986). The strange death of Europe's trees. *New Scientist* **1537**, 41–45.

Pearcy, R. W. (1983). The light environment and growth of C3 and C4 tree species in the understorey of a Hawaiian forest. *Oecologia* **58**, 19–25.

Pearcy, R. W. and Ehleringer, J. (1984). Comparative ecophysiology of C3 and C4 plants. *Plant. Cell Env.* **7**, 1–13.

Pearsall, W. H. (1938). The soil complex in relation to plant communities. *J. Ecol.* **26**, 180–193.

Pearsall, W. H. (1968) "Mountains and Moorlands", 2nd Edn. Collins, Glasgow.

Penman, H. L. (1963). "Vegetation and Hydrology". Technical Bulletin no. 53, Commonwealth Bureau of Soils.

Peterson, H. B. and Nielsen, R. F. (1978). Heavy metals in relation to plant growth on mine and mill wastes. *In* "Environmental Management of Mineral Wastes" (Eds. G. T. Goodman and M. J. Chadwick), pp. 297–310. Sijthoff and Noordhoff, Alphen.

Peterson, P. J. (1969). The distribution of zinc-65 in *Agrostis tenuis* Sibth. and *A. stolonifera* L. tissues. *J. exp. Bot.* **20**, 863–887.

Philipson, J. J. and Coutts, M. P. (1978). The tolerance of tree roots to waterlogging. III. Oxygen transport in lodgepole pine and Sitka spruce of primary structure. *New Phytol.* **80**, 341–349.

Pickett, S. T. A. and White, P. S. (1985). "The Ecology of Natural Disturbance and Patch Dynamics". Academic Press, New York.

Pigott, C. D. (1975). Experimental studies on the influence of climate on the geographical distribution of plants. *Weather for 1975*, 82–90.

Pisek, A., Larcher, W., Moser, W. and Pack, I. (1969). Kardinale Temperaturbereiche der Photosynthese und Grenztemperaturen des Lebens der Blätter verschiedener Spermatophyten. III. Temperaturabhängigkeit und optimaler Temperaturbereich der Netto-Photosynthese. *Flora* **B158**, 608–630.

Pisek, A., Larcher, W., Vegis, A. and Napp-Zin, K. (1973). The normal temperature range. *In* "Temperature and Life" (Eds. H. Precht, J. Christopherson, H. Hensel and W. Larcher), pp. 102–194. Springer-Verlag, Berlin.

Pitman, M. G. and Cram, W. J. (1973). Regulation of inorganic ion transport in plants. *In* "Ion Transport in Plants" (Ed. W. P. Anderson), pp. 465–482. Academic Press, London and New York.

Pollak, G. and Waisel, Y. (1970). Salt secretion in *Aeluropus litoralis* (Willd.) Parl. *Ann. Bot.* **34**, 879–888.

Polonenko, D. R. and Mayfield, C. I. (1979). A direct observation technique for studies of rhizoplane and rhizosphere colonisation. *Plant and Soil* **51**, 405–420.

PONNAMPERUMA, F. N. (1972). The chemistry of submerged soils. *Adv. Agron.* **24,** 29–96.

PONNAMPERUMA, F. N. (1984). Effects of flooding in soils. *In* "Flooding and Plant Growth" (Ed. T. T. Kozlowski), pp. 9–45. Academic Press, New York.

PORATH, E. and POLJAKOFF-MAYBER, A. (1964). Effect of salinity on metabolic pathways in pea root tips. *Israel J. Bot.* **13,** 115–121.

POSTHUMUS, A. C. (1982). Biological indicators of air pollution. *In* "Effects of Gaseous Air Pollution in Agriculture and Horticulture" (Eds. M. H. Unsworth and D. P. Ormrod), pp. 27–42. Butterworth, London.

POWELL, C. L. (1975). Rushes and sedges are non-mycotrophic. *Plant and Soil* **42,** 481–484.

POWELL, C. L. (1976). Mycorrhizal fungi stimulate clover growth in New Zealand hill country soils. *Nature* **264,** 436–438.

POWELL, C. L. and BAGYARAJ, D. J. (1984). "VA Mycorrhiza". CRC Press, Boca Raton.

PRECHT, H., CHRISTOPHERSON, J. HENSEL, H. and LARCHER, W. (1973). "Temperature and Life". Springer-Verlag, Berlin.

PRISCO, J. T. and O'LEARY, J. W. (1972). Enhancement of intact bean leaf senescence by NaCl salinity. *Physiol. Plant* **27,** 95–100.

PROCTOR, J. (1971a). The plant ecology of serpentine. II. Plant response to serpentine soils. *J. Ecol.* **59,** 397–410.

PROCTOR, J. (1971b). The plant ecology of serpentine. III. The influence of high magnesium/calcium ratio and high nickel and chromium levels in British and Swedish serpentine soils. *J. Ecol.* **59,** 827–842.

QUEIROZ, O. (1977). CAM: rhythms of enzyme capacity and activity in adaptive control mechanisms. *In* "Encyclopaedia of Plant Physiology: Photosynthesis Vol. 2" (Ed. M Gibbs and E. Latzko), pp. 126–137. Springer-Verlag, Berlin.

RADFORD, P. J. (1967). Growth analysis formulae—their use and abuse. *Crop Sci.* **7,** 171–175.

RAISON, J. K., BERRY, J. A., ARMOND, P. A. and PIKE, C. S. (1980). Membrane properties in relation to the adaptation of plants to high and low temperature stress. *In* "Adaptation of Plants to Water and High Temperature Stress" (Eds. N. C. Turner and P. J. Kramer), pp. 261–273. Wiley Interscience, New York.

RASCHKE, K. (1975). Stomatal action. *Ann. Rev. Pl. Physiol.* **26,** 309–340.

RASCHKE, K. (1976). How stomata resolve the dilemma of opposing priorities. *Phil. Trans. R. Soc. Lond.* **B273,** 551–560.

RAUNKIAER, C. (1907). "Planterigets Livsformer og deres Betydning for Geografien". Copenhagen. (Translated as "The Life-Forms of Plants and Statistical Plant Geography" (1934), Clarendon Press, Oxford.)

RAVEN, J. A. (1983). Phytophages of xylem and phloem: a comparison of animal and plant sap-feeders. *Adv. Ecol. Res.* **13,** 135–234.

RAYDER, L. and TING, I. P. (1983). CAM idling in *Hoya carnosa* (Asclepiadaceae). *Photosynthesis Res.* **4,** 203–211.

READ, D. J., KOUCHEKI, H. K. and HODGSON, J. (1976). Vesicular-arbuscular mycorrhiza in natural vegetation. I. The occurrence of infection. *New Phytol.* **77,** 641–653.

REID, D. M. and BRADFORD, K. J. (1984). Effects of flooding on hormone relations. *In* "Flooding and Plant Growth" (Ed. T. T. Kozlowski), pp. 195–219. Academic Press, New York.

REID, D. M. and WATSON, K. (1985). Ethylene as an air pollutant. *In* "Ethylene and Plant Development" (Eds. J. A. Roberts and G. A. Tucker), pp. 277–286. Butterworth, London.

REILLY, A. and REILLY, C. (1973). Copper-induced chlorosis in *Becium homblei.* *Plant and Soil* **38,** 671–674.

RHODES, I. (1969). Yield, canopy structure and light interception in two ryegrass varieties in mixed culture and monoculture. *J. Br. Grassld Soc.* **24,** 123–127.

RHODES, L. H. and GERDEMANN, J. W. (1975). Phosphate uptake zones of mycorrhizal and non-mycorrhizal onions. *New Phytol.* **75,** 555–562.

RICE, E. L. (1984). "Allelopathy". 2nd Edition. Academic Press, New York.

RICHARDS, A. J. (1972). The *Taraxacum* Flora of the British Isles. *Watsonia.* **9,** Suppl.

RICHTER, H. (1976). The water status in the plant. Experimental evidence. *In* "Water and Plant Life" (Eds. O. L. Lange, L. Kappen and E.-D. Schulze). *Ecological Studies* **19,** 42–58. Springer-Verlag, Berlin.

RIDGE, I. and AMARASINGH, I. (1984). Ethylene and growth in the fringed waterlily (*Nymphoides peltata*): stimulation of cell division and interaction with buoyant tension in petioles. *Plant Growth Regulation* **2,** 235–249.

RILEY, D. and BARBER, S. A. (1971). Effect of ammonium and nitrate fertilisation on phosphorus uptake as related to root-induced pH changes at the root-soil interface. *Soil Sci. Soc. Amer. Proc.* **35,** 301–306.

RITTER-STUDNIČKA, J. (1971). Zellensaft-Analysen zum Problem der Serpentinvegetation. *Österr. Bot. Ž.* **119,** 410–431.

ROBARDS, A. W., CLARKSON, D. T. and SANDERSON, J. (1979). Structure and permeability of the epidermal/hypodermal layers of the sand sedge (*Carex arenaria* L.). *Protoplasma* **101,** 331–347.

ROBERTS, A. M. and WALTERS, D. R. (1986). Stimulation of photosynthesis in uninfected leaves of rust-infected leeks. *Ann. Bot.* **56,** 893–896.

ROBERTS, J. A. and TUCKER, G. A. (EDS.) (1985). "Ethylene and Plant Development". Butterworth, London.

ROBERTSON, K. P. and WOOLHOUSE, H. W. (1984). Studies of the seasonal course of carbon uptake of *Eriophorum vaginatum* in a moorland habitat. *J. Ecol.* **72,** 423–435, 685–700.

ROBINSON, D. and RORISON, I. H. (1983). Relationships between root morphology and nitrogen availability in a recent theoretical model describing nitrogen uptake from soil. *Plant, Cell Env.* **6,** 641–648.

ROOSE, M. L., BRADSHAW, A. D. and ROBERTS, T. M. (1982). Evolution of resistance to gaseous air pollutants. *In* "Effects of Gaseous Air Pollution in Agriculture and Horticulture" (Eds. M. H. Unsworth and D. P. Ormrod), pp. 379–409. Butterworth, London.

RORISON, I. H. (1958). The effect of aluminium on legume nutrition. *In* "Nutrition of the Legumes" (Ed. E G. Hallsworth), pp. 43–61. Butterworth, London.

RORISON, I. H. (1965). The effect of aluminium on the uptake and incorporation of phosphate by excised sainfoin roots. *New Phytol.* **64,** 23–27.

RORISON, I. H. (1968). The response to phosphorus of some ecologically distinct plant species. I. Growth rates and phosphorus absorption. *New Phytol.* **67,** 913–923.

RORISON, I. H. (1971). The use of nutrients in the control of the floristic composition of grassland. *In* "The Scientific Management of Plant and Animal Communities for Conservation". *Brit. Ecol. Soc. Symp.* **11,** 65–77.

RORISON, I. H. (1975). Nitrogen source and metal toxicity. *J. Sci. Fd. Agric.* **26,** 1426.

RORISON, I. H., SUTTON, C. D. and HALLSWORTH, E. G. (1958). The effects of climatic conditions on aluminium and manganese toxicities. *In* "Nutrition of the Legumes" (Ed. E. G. Hallsworth), pp. 62–68. Butterworth, London.

ROSS, J. (1970). Mathematical models of photosynthesis in a plant stand. *In* "Prediction and Measurement of Photosynthetic Productivity" (Ed I. Setlik), pp. 29–45. Centre for Agricultural Publishing and Documentation, Wageningen.

ROSS, J. (1981). "The Radiation Regime and Architecture of Plant Stands". Junk, The Hague, Netherlands.

ROSSWALL, T. and HEAL, O. W. (EDS.) (1975) "Structure and Function of Tundra Ecosystems". *Ecol. Bull.* **20.** Swedish Natural Science Research Council, Stockholm.

ROSSWALL, T., FLOWER-ELLIS, J. G. K., JOHANSSON, L. G., JONSSON, S., RYDEN, B. E. and SONESSON, M. (1975). Stordalen (Abisko). Sweden. *In* "Structure and Function of Tundra Ecosystems". (Eds. T. Rosswall and O. W. Heal). *Ecol. Bull.* **20,** 265–294. Swedish Natural Science Research Council, Stockholm.

ROUGHGARDEN, J. and DIAMOND, J. (1986). Overview: the role of species interactions in community ecology. *In* "Community Ecology" (Eds. J. Diamond and T. J. Case), pp. 333–343. Harper & Row, New York.

ROVIRA, A. D. (1969). Plant root exudates. *Bot. Rev.* **39,** 35–37.

ROVIRA, A. D. & DAVEY, C. B. (1974). Biology of the rhizosphere. *In* "The Plant Root and its Environment" (Ed. E. W. Carson), pp. 153–240. University Press of Virginia.

ROVIRA, A. D., NEWMAN, E. I., BOWEN, H. J. and CAMPBELL, R. (1974). Quantitative assessment of the rhizoplane microflora by direct microscopy. *Soil Biol. Biochem.* **6,** 211–216.

ROZEMA, J. (1976). An ecophysiological study of the response to salt of four halophytic and glycophytic *Juncus* species. *Flora* **165,** 197–209.

RUSSELL, E. W. (1973). "Soil Conditions and Plant Growth", 10th Edn. Longman, London.

RUSSELL, R. S. and CLARKSON, D. T. (1976). Ion transport in root systems. *In* "Perspectives in Experimental Biology" (Ed. N. Sunderland), Vol. 2, pp. 401–411. Pergamon, Oxford.

RUSSELL, R. S. and SANDERSON, J. (1967). Nutrient uptake by different parts of the intact roots of plants. *J. exp. Bot.* **18**, 491–508.

RUTTER, A. J. (1968). Water consumption by forests. In "Water Deficits and Plant Growth" (Ed. T. T. Kozlowski), vol. 2, pp. 23–84. Academic Press, New York and London.

SAKAI, A. (1970). Freezing resistance in willows from different climates. *Ecology* **51**, 485–491.

SAKAI, A. and WEISER, C. J. (1973). Freezing resistance of trees in North America with reference to tree regions. *Ecology* **54**, 118–126.

SALISBURY, E. J. (1916). The oak-hornbeam woods of Hertfordshire. *J. Ecol.* **4**, 83–117.

SALISBURY, E. J. (1952). "Downs and Dunes: their Plant Life and its Environment". Bell, London.

SALISBURY, F. B. (1963). "The Flowering Process". Pergamon, Oxford.

SALISBURY, F. B. and SPOMER, G. G. (1964). Leaf temperatures of alpine plants in the field. *Planta* **60**, 497–505.

SANDERS, F. E. and TINKER, P. B. (1973). Phosphate flow into mycorrhizal roots. *Pestic. Sci.* **4**, 385–395.

SAVILE, D. B. O. (1972). "Arctic Adaptations in Plants". Monograph 6, Canada Department of Agriculture, Ottawa.

SCHILLER, W. (1974). Versuche zur Kupferresistenz bei Schwermetall-ökotypen von *Silene cucubalus* Wib. *Flora* **163**, 327–341.

SCHOLANDER, P. F., VAN DAM, L. and SCHOLANDER, S. I. (1955). Gas exchange in the roots of mangroves. *Am. J. Bot.* **42**, 92–98.

SCHULZE, E.-D. (1972). Die Wirkung von Licht und Temperatur auf den CO_2-Gaswechsel verschiedener Lebensformen aus der Krautschicht eines montanen Buchenwaldes. *Oecologia* **9**, 235–238.

SCHULZE, E.-D. (1982). Plant life forms and their carbon, water and nutrient relations. *Encyclopaedia of Plant Physiology* **12B**, 615–676.

SCHULZE, E.-D., FUCHS, M. and FUCHS, M. I. (1977). Spatial distribution of photosynthetic capacity and performance in a mountain spruce forest of northern Germany. 3. The significance of the evergreen habit. *Oecologia* **30**, 239–248.

SCHULZE, E.-D. and HALL, A. E. (1982). Stomatal responses, water loss and CO_2 assimilation rates of plants in contrasting environments. *Encyclopaedia of Plant Physiology* **12B**, 181–230.

SCHULZE, E.-D., KUPPERS, M. and MATYSSEK, R. (1986). The roles of carbon balance and branching pattern in the growth of woody species. In "On the Economy of Plant Form and Function" (Ed. T. J. Givnish), pp. 585–602. Cambridge University Press, Cambridge.

SHANTZ, H. L. and PIEMEISAL, L. N. (1927). The water requirements at Akron, Colorado. *J. agric. Res.* **34**, 1093–1190.

SHEEHY, J. E. and COOPER, J. P. (1973). Light interception, photosynthetic activity and crop growth rate in canopies of six temperate forage grasses. *J. appl. Ecol.* **10**, 235–250.

SHEEHY, J. E. and PEACOCK, J. M. (1975). Canopy photosynthesis and crop growth rate of eight temperate forage grasses. *J. exp. Bot.* **26**, 679–691.

SHELDON, J. C. (1974). The behaviour of seeds in soil. III. The influence of seed morphology and the behaviour of seedlings on the establishment of plants from surface-lying seeds. *J. Ecol.* **62,** 47–66.

SHERIFF, D. W. (1977). Evaporation sites and distillation in leaves. *Ann. Bot.* (N.S.) **41,** 1081–1082.

SHERWIN, T. and SIMON, E. W. (1969). The appearance of lactic acid in *Phaseolus* seeds germinating under wet conditions. *J. exp. Bot.* **20,** 776–785.

SHIRLEY, H. L. (1929). The influence of light intensity and light quality on growth of plants. *Am. J. Bot.* **16,** 354–390.

SHREVE, F. and WIGGINS, I. L. (1964). "Vegetation and Flora of the Sonoran Desert". Stanford University Press, Stanford.

SHROPSHIRE, W. (1971). Photoinduced parental control of seed germination and the spectral quality of solar radiation. *Solar Energy* **15,** 99–105.

SILANDER, J. A. and ANTONOVICS, J. (1982). Analysis of interspecific interactions in a coastal plant community—a perturbation approach. *Nature* **298,** 557–560.

SINGH, B. B., HADLEY, H. H. and BERNARD, R. L. (1971). Morphology of pubescence in soybeans and its relation to plant vigour. *Crop Sci.* **11,** 13–16.

SKENE, MACGREGOR (1924). "The Biology of Flowering Plants". Sidgwick and Jackson, London.

SLATYER, R. O. (1963). Climatic control of plant water relations. *In* "Environmental Control of Plant Growth" (Ed. L. T. Evans), pp. 33–52. Academic Press, New York.

SLATYER, R. O. (1967). "Plant-Water Relationships". Academic Press, London and New York.

SLAVIK, B. (1974). Methods of studying plant water relations. *Ecological Studies* **9**. Chapman and Hall, London.

SMALL, E. (1972). Photosynthetic rates in relation to nitrogen recycling as an adaptation to nutrient deficiency. *Can. J. Bot.* **50,** 2227–2233.

SMIRNOFF, N. and CRAWFORD, R. M. M. (1983). Variation in the structure and response to flooding of root aerenchyma in some wetland plants. *Ann. Bot.* **51,** 237–249.

SMITH, A. M. (1976). Ethylene in soil biology. *Ann. Rev. Phytopathology* **14,** 53–73.

SMITH, A. M. and AP REES, T. (1979). Pathways of carbohydrate fermentation in the roots of marsh plants. *Planta* **146,** 327–334.

SMITH, A. M., KALSI, G. and WOOLHOUSE, H. W. (1984). Products of fermentation in the roots of alders (*Alnus* Mill.). *Planta* **160,** 272–275.

SMITH, F. A. (1973). The internal control of nitrate uptake into excised barley roots with differing salt contents. *New Phytol.* **72,** 769–782.

SMITH, K. A. (1977). Soil aeration. *Soil Sci.* **123,** 284–291.

SMITH, K. A. (1978). Ineffectiveness of ethylene as a regulator of soil microbial activity. *Soil Biol. Biochem.* **10,** 269–272.

SMITH, K. A. and DOWDELL, R. J. (1974). Field studies of the soil atmosphere. I. Relationship between ethylene, oxygen, soil moisture content and temperature. *J. Soil Sci.* **25,** 219–230.

SMITH, K. A. and JACKSON, M. B. (1974). Ethylene, waterlogging and plant growth. *A.R.C. Letcombe Lab. Ann. Rep.* 1973, 60–75.

SMITH, K. A. and ROBERTSON, P. D. (1971). Effect of ethylene on root extension of cereals. *Nature* **234**, 148–149.

SMITH, W. K. (1978). Temperatures of desert plants: another perspective on the adaptability of leaf size. *Science* **201**, 614–616.

SNAYDON, R. W. (1970). Rapid population differentiation in a mosaic environment. I. The response of *Anthoxanthum odoratum* populations to soils. *Evolution* **24**, 257–269.

SNAYDON, R. W. (1971). An analysis of competition between plants of *Trifolium repens* populations collected from contrasting soils. *J. appl. Ecol.* **8**, 687–698.

SNAYDON, R. W. and BRADSHAW, A. D. (1961). Differential responses to calcium within the species *Festuca ovina* L. *New Phytol.* **60**, 219–234.

SNAYDON, R. W. and BRADSHAW, A. D. (1969). Differences between natural populations of *Trifolium repens* L. in response to mineral nutrients. II. Calcium, magnesium and potassium. *J. appl. Ecol.* **6**, 185–202.

SNELLGROVE, R. C., SPLITSTOESSER, W. E., STRIBLEY, D. P. and TINKER, P. B. (1982). The distribution of carbon and the demand of the fungal symbiont in leek plants with vesicular-arbuscular mycorrhizas. *New Phytol.* **92**, 75–88.

SOUTHWOOD, T. R. E. (1973). The insect–plant relationship—an evolutionary perspective. *In* "Insect–plant Relationships" (Ed. H. F. van Emden). *Roy. Ent. Soc. Symp.* **6**, 3–30.

SPANSWICK, R. M. (1976). Symplastic transport in tissues. *In* "Transport in Plants IIB". *Encyclopaedia of Plant Physiology* **2**, 35–53. Springer-Verlag, Berlin.

SPARLING, G. P. and TINKER, P. B. (1975). Mycorrhizas in Pennine grassland. *In* "Endomycorrhizas" (Eds. F. E. Sanders, B. Mosse and P. B. Tinker), pp. 545–560. Academic Press, London and New York.

SPEDDING, D. J. (1969). Uptake of sulphur dioxide by barley leaves at low sulphur dioxide concentrations. *Nature* **224**, 1229–1231.

SPENCE, D. H. N. (1981). The zonation of plants in freshwater lakes. *Adv. Ecol. Res.* **12**, 37–125.

SPRENT, J. I. (1983). Adaptive variation in legume nodule physiology resulting from host rhizobial interactions. *In* "Nitrogen as an Ecological Factor" (Eds. J. A. Lee, S. McNeill and I. H. Rorison), pp. 29–42. Blackwell Scientific Publications, Oxford.

STARKEY, R. L. (1929). Some influences of the development of higher plants upon the micro-organisms in the soil. *Soil Sci.* **27**, 319-334.

STEBBINS, G. L. (1981). Co-evolution of grasses and herbivores. *Ann. Missouri Bot. Gdn.* **68**, 75–86.

STEINMANN, F. and BRÄNDLE, R. (1981). Die Überflutungstoleranz der Teichbinse (*Schoenoplectus lacustris* (L.) PALLA): 3. Beziehungen zwischen der Sauerstoffversorgung und der "Adenylate Energy Charge" der Rhizome in Abhängigkeit von der Sauerstoffkonzentration in der Umgebung. *Flora* **171**, 307–314.

STEWART, G. R. and LEE, J. A. (1974). The role of proline accumulation in halophytes. *Planta* **120**, 279–289.

STEWART, G. R., LEE, J. A. and OREBAMJO, T. (1972). Nitrogen metabolism of halophytes. I. Nitrate reductase activity in *Suaeda maritima*. *New Phytol.* **71**, 263–267.

STEWART, G. R., LEE, J. A. and OREBAMJO, T. (1973). Nitrogen metabolism of halophytes. II. Nitrate availability and utilisation. *New Phytol.* 539–546.

STOCKER, O. (1976). The water-photosynthesis syndrome and the geographical plant distribution in the Sahara deserts. *In* "Water and Plant Life" (Eds. O. L. Lange, L. Kappen and E.-D. Schulze). *Ecological Studies* **19**, 506–521. Springer-Verlag, Berlin.

STOLZY, L. H., FOCHT, D. D. and FLUHER, H. (1981). Indications of soil aeration status. *Flora* **171**, 236–265.

STONE, E. C. (1957a). Dew as an ecological factor. I. A review of the literature. *Ecology* **38**, 407–413.

STONE, E. C. (1957b). Dew as an ecological factor. II. The effect of artificial dew on the survival of *Pinus ponderosa* and associated species. *Ecology* **38**, 414–422.

STOREY, R., AHMAD, N. and WYN JONES, R. G. (1977) Taxonomic and ecological aspects of the distribution of glycine-betaine and related compounds in plants. *Oecologia* **27**, 319–332.

STOUTJESDIJK, P. H. (1972). Spectral transmission curves of some types of leaf canopies with a note on seed germination. *Acta. Bot. Neerl.* **21**, 185–191.

STOUTJESDIJK, P. H. (1974). The open shade, an interesting micro-climate. *Acta Bot. Neerl.* **23**, 125–130.

SUTCLIFFE, J. (1977). "Plants and Temperature". Arnold, London.

SUTCLIFFE, J. (1981). Sap in the treetops. *New Scientist* **90**, 682–684.

SZEICZ, G. (1974). Solar radiation in crop canopies. *J. appl. Ecol.* **11**, 1117–1156.

TADANO, T. (1975). Devices of rice roots to tolerate high iron concentrations in the growth media. *Jpn. Agric. Res. Quart.* **9**, 34–39.

TAERUM, R. (1970). Comparative shoot and root growth studies of six grasses in Kenya. *East Afr. Agr. and For. J.* **36**, 94–113.

TAL, M. (1971). Salt tolerance in the wild relatives of the cultivated tomato: response of *Lycopersicon esculentum*, *L. peruvianum* and *L. esculentum minor* to sodium chloride solutions. *Aust. J. agric. Res.* **22**, 631–638.

TANSLEY, A. G. (1917). On competition between *Galium saxatile* L. (*G. hercynicum* Weig.) and *Galium sylvestre* Poll. (*G. asperum* Schreib.) on different types of soil. *J. Ecol.* **5**, 173–179.

TAYLOR, G. E. and MURDY, W. H. (1975). Population differentiation of an annual plant species, *Geranium carolianum*, in response to sulphur dioxide. *Bot. Gaz.* **136**, 212–215.

TAYLOR, G. E. and TINGLEY, D. T. (1983). Sulphur dioxide flux into leaves of *Geranium carolinianum*. *Plant Physiol.* **72**, 237–244.

TAYLOR, G. J. and FOY, C. D. (1985). Mechanisms of aluminium tolerance in

Triticum aestivum L. (wheat). 1. Differential pH induced by winter cultivars in nutrient solutions. *Am. J. Bot.* **72**, 695–701.

TAYLOR, O. C., THOMPSON, C. R., TINGLEY, D. T. and REINERT, R. A. (1975). Oxides of nitrogen. *In* "Responses of Plants to Air Pollution" (Eds. J. B. Mudd and T. T. Kozlowski), pp. 121-139. Academic Press, New York and London.

TEAL, J. M. and KANWISHER, J. W. (1966). Gas transport in the marsh grass *Spartina alterniflora*. *J. exp. Bot.* **17**, 355–361.

THOMPSON, K. and GRIME, J. P. (1983). A comparative study of germination responses to diurnally-fluctuating temperatures. *J. appl. Ecol.* **20**, 141–156.

THOMSON, W. W., DUGGER, W. M. and PALMER, R. L (1966). Effects of ozone on the fine structure of the palisade parenchyma cells of bean leaves. *Can. J. Bot.* **44**, 1677–1682.

THURSTON, J. M. (1960). Dormancy in weed seeds. *Symp. Brit. Ecol. Soc.* **1**, 69–82.

TIESZEN, L. L. and WIELAND, N. K. (1975). Physiological ecology of arctic and alpine photosynthesis and respiration. *In* "Physiological Adaptation to the Environment" (Ed. F. J. Vernberg), pp. 157–200. Intext Educational Publishers, New York.

TIFFIN, L. D. (1972). Translocation of micronutrients in plants. *In* "Micronutrients in Agriculture" (Eds. J. J. Mortrest, P. M. Giodano and W. L. Lindsay), pp. 199–229. Soil Science Soc. Amer., Madison.

TIKU, B. L. and SNAYDON, R. W. (1971). Salinity tolerance within the grass species *Agrostis stolonifera*. *Plant and Soil* **35**, 421–431.

TILMAN, D. (1982). "Resource Competition and Community Structure". Princeton University Press, Princeton.

TILMAN, D. (1985). Nitrogen-limited growth in plants from different successional stages. *Ecology* **67**, 555–563.

TINKER, P. B. (1969). The transport of ions in the soil around plant roots. *In* "Ecological Aspects of the Mineral Nutrition of Plants". *Brit. Ecol. Soc. Symp.* **9**, 135–148.

TING, I. P. (1985). Crassulacean acid metabolism. *Ann. Rev. Plant Physiol.* **36**, 595–622.

TOMOS, A. D. (1985). The physical limitations of leaf cell expansion. *In* "Control of Leaf Growth" (Eds. N. R. Baker, W. J. Davies and C. K. Ong), pp. 1–33. Cambridge University Press, Cambridge.

TRANQUILLINI, W. (1964). The physiology of plants at high altitudes. *Ann. Rev. Pl. Physiol.* **15**, 345–362.

TRANQUILLINI, W. (1979). "Physiological Ecology of the Alpine Timberline". Springer-Verlag, Berlin.

TRAPPE, J. M. (1962). Fungus associates of ectotrophic mycorrhizae. *Bot. Rev.* **28**, 538–606.

TREICHEL, S. (1975). Der Einfluss von NaCl auf die Prolinkonzentration verschiedener Halophyten. *Z. Pflanzenphysiol.* **76**, 56–68.

TRENBATH, B. R. and HARPER, J. L. (1973). Neighbour effects in the genus *Avena*. I. Comparison of crop species. *J. appl. Ecol.* **10**, 379–400.

TRINICK, M. J. (1973). Symbiosis between *Rhizobium* and the non-legume *Trema aspera*. *Nature* **244**, 459–460.

TRIPEPI, R. R. and MITCHELL, C. A. (1984). Metabolic response of river birch and European birch roots to hypoxia. *Plant Physiol.* **76**, 31–35.

TROUGHTON, J. H., MOONEY, H. A., BERRY, J. A. and VERITY, D. (1977). Variable carbon isotope ratios of *Dudleya* species growing in natural environments. *Oecologia* **30**, 307–311.

TUCKER, D. J. (1975). Far-red light as a suppressor of side-shoot growth in the tomato. *Plant Sci. Letters* **5**, 127–130.

TURESSON, G. (1922). The genotypical response of the plant species to the habitat. *Hereditas* **3**, 211–350.

TURNER, R. G. and MARSHALL, C. (1972). The accumulation of Zn by subcellular fractions of roots of *Agrostis tenuis* Sibth. in relation to zinc tolerance. *New Phytol.* **71**, 671–676.

TYLER, G. (1976). Soil factors controlling metal ion absorption in the wood anemone *Anemone nemorosa*. *Oikos* **27**, 71–80.

TYREE, M. and JARVIS, P. G. (1982). Water in tissues and cells. *Encyclopaedia of Plant Physiology* **12B**, 35–77.

UNSWORTH, M. H. (1981). The exchange of carbon dioxide and air pollutants between vegetation and the atmosphere. *In* "Plants and their Atmospheric Environment" (Eds. J. Grace, E. D. Ford and P. G. Jarvis), pp. 111–138. Blackwell Scientific Publications, Oxford.

UNSWORTH, M. H. (1982). Exposure to gaseous pollutants and uptake by plants. *In* "Effects of Gaseous Air Pollution in Agriculture and Horticulture" (Eds. M. H. Unsworth and D. P. Ormrod), pp. 43–63. Butterworth, London.

UNSWORTH, M. H., BISCOE, P. V. and BLACK, V. (1976). Analysis of gas exchange between plants and polluted atmospheres. *In* "Effects of Air Pollutants on Plants" (Ed. T. A. Mansfield), pp. 5–16. Cambridge University Press, Cambridge.

VAARTAJA, O. (1954). Photoperiodic ecotypes of trees. *Can. J. Bot.* **32**, 392–399.

VAN BEERS, W. F. J. (1962). Acid sulphate soils. *Int. Inst. Land Reclam. Imp. Bull.* **3**.

VAN DEN BERGH, J. P. (1969). Distribution of pasture plants in relation to the chemical properties of the soil. *In* "Ecological Aspects of the Mineral Nutrition of Plants". *Brit. Ecol. Soc. Symp.* **9**, 11–23.

VAN DEN HONERT, T. H. (1948). Water transport as a catenary process. *Disc. Faraday Soc.* **3**, 146–153.

VAN EMDEN, H. F., EASTOP, V. F., HUGHES, R. D. and WAY, M. J. (1969). The ecology of *Myzus persicae*. *Ann. Rev. Ent.* **14**, 197–270.

VAN STEVENINCK, R. F. M. (1965). The significance of calcium in the apparent permeability of cell membranes and the effects of substitution with other divalent cations. *Physiol. Plant* **18**, 54–59.

VEIHMEYER, F. J. and HENDRICKSON, A. H. (1948). Soil density and root penetration. *Soil Sci.* **65**, 487–493.

VIETS, F. G. (1965). The plant's need for and use of nitrogen. *In* "Soil Nitrogen" (Eds. W. V. Bartholomew and F. E. Clark), pp. 543–554. Agronomy Series 10. Academic Press, London and New York.

VINCE-PRUE, D. (1983). The perception of light–dark transitions. *Phil. Trans. R. Soc. London.* **303**, 523–536.

VOGEL, J. C., FULS, A. and DANIN, A. (1986). Geographical and environmental distribution of C3 and C4 grasses in the Sinai, Negev and Judean deserts. *Oecologia* **70**, 258–265.

VON WILLERT, D. J. (1974). Der Einfluss von NaCl auf die Atmung und Aktivität der Malatdehydrogenase bei einigen Halophyten und Glykophyten. *Oecologia* **14**, 127–137.

VON WILLERT, D. J. (1985). *Welwitschia mirabilis*—new aspects in the biology of an old plant. *Adv. Bot. Res.* **11**, 157–191.

VON WILLERT, D. J., TREICHEL, J., KIRST, G. O. and CURDTS, E. (1976). Environmentally controlled changes of phosphoenolpyruvate carboxylases in *Mesembryanthemum*. *Phytochem.* **15**, 1435–1436.

WAINWRIGHT, S. J. and WOOLHOUSE, H. W. (1975). Physiological mechanisms of heavy metal tolerance in plants. *In* "The Ecology of Resource Degradation and Renewal". *Brit. Ecol. Soc. Symp.* **15**, 231–258.

WALLEY, K. A., KHAN, M. S. I. AND BRADSHAW, A. D. (1974). The potential for evolution of heavy metal tolerance in plants. I. Copper and zinc tolerance in *Agrostis tenuis*. *Heredity* **32**, 309–319.

WALTER, H. (1963). The water supply of desert plants. *Symp. Brit. Ecol. Soc.* **3**, 199–205.

WARDLAW, I. F. (1979). The physiological effects of temperature on plant growth. *Proc. Agron. Soc. NZ* **9**, 39–48.

WARDLE, P. (1974). Alpine timberlines. *In* "Arctic and Alpine Environments" (Eds. J. D. Ives and R. G. Barry), pp. 371–402. Methuen, London.

WAREING, P. F. (1969). The control of bud dormancy in seed plants. *Symp. Soc. Exp. Biol.* **23**, 241–262.

WARMING, E. (1909). "Oecology of Plants". Oxford University Press, Oxford.

WASSINK, E. C. (1972). Some notes on temperature relations in plant physiological processes. *Meded. Landbouw. Wageningen* **72** (25), 1–15.

WEAVER, J. E. (1958). Classification of root systems of forbs of grassland and a consideration of their significance. *Ecology* **39**, 393–401.

WEBSTER, G. L., BROWN, W. V. and SMITH, B. N. (1975). Systematics of photosynthetic carbon fixation pathways in Euphorbia. *Taxon* **24**, 27–34.

WEBSTER, R. and BECKETT, P. H. T. (1972). Matric suctions to which soils in South Central England drain. *J. agric. Sci.* **78**, 379–387.

WELBANK, P. J. (1961). A study of the nitrogen and water factors in competition with *Agropyron repens* (L.) Beau. *Ann. Bot.* **25**, 116–137.

WELLBURN, A. R., MAJERNIK, O. and WELLBURN, A. M. (1972). Effects of SO_2 and NO_2 polluted air upon the ultrastructure of chloroplasts. *Environ. Pollut.* **3**, 37–49.

WENT, F. W. (1953). The effect of temperature on plant growth. *Ann. Rev. Pl. Physiol.* **4**, 347–362.

WERK, K. S. and EHLERINGER, J. (1984). Non-random leaf orientation in *Lactuca serriola*. *Plant, Cell Env.* **7**, 81–88.

WHEELER, B. D., AL-FARRAJ, H. M. and COOK, R. E. D. (1985). Iron toxicity to plants in base-rich wetlands: comparative effects on the distribution and growth of *Epilobium hirsutum* L. and *Juncus subnodulus* Schrank. *New Phytol.* **100**, 653–669.

WHIPPS, J. M. (1984). Environmental factors affecting the loss of carbon from the roots of wheat and barley seedlings. *J. exp. Bot.* **35**, 767–773.

WHIPPS, J. M. and LYNCH, J. (1983). Substrate flow and utilisation in the rhizosphere of cereals. *New Phytol.* **96**, 605–623.

WHITE, J. and STREHL, C. E. (1978). Xylem feeding by periodical cicada nymphs on tree tops. *Ecol. Entomol.* **3**, 323–328.

WHITE, R. E. (1973). Studies on mineral ion absorption by plants. II. The interaction between metabolic activity and the rate of phosphorus uptake. *Plant and Soil* **38**, 509–523.

WHITE, R. E. (1979). "Introduction to the Principles and Practice of Soil Science". Blackwell Scientific Publications, Oxford.

WHITEHEAD, F. H. (1971). Comparative autecology as a guide to plant distribution. *In* "The Scientific Management of Animal and Plant Communities for Conservation". *Symp. Brit. Ecol. Soc.* **11**, 167–176.

WHITEHEAD, F. H. (1973). The relationship between light intensity and reproductive capacity. *In* "Plant Response to Climatic Factors" (Ed. R. O. Slatyer), pp. 73–75. UNESCO, Paris.

WHITMORE, M. E. (1985). Relationships between dose of SO_2 and NO_2 mixtures and growth of *Poa pratensis*. *New Phytol.* **99**, 545–553.

WIGNARAJAH, K. and GREENWAY, H. (1976). Effect of anaerobiosis on activities of alcohol dehydrogenase and pyruvate decarboxylase in roots of *Zea mays*. *New Phytol.* **77**, 575–584.

WILD, A., SKARLOU, V., CLEMENT, C. R. and SNAYDON, R. W. (1974). Comparison of potassium uptake by four plant species grown in sand and in flowing solution culture. *J. appl. Ecol.* **11**, 801–812.

WILLIAMS, D. E., VLAMIS, J., HALL, H. and GOVANO, K. D. (1971). Urbanisation, fertilisation, and manganese toxicity in Alameda mustard. *Calif. Agric.* **25**, 8–14.

WILLIAMSON, R. E. and ASHLEY, C. C. (1982). Free Ca^{2+} and cytoplasmic streaming in the alga *Chara*. *Nature* **296**, 647–651.

WILLIS, A. J. (1963). Braunton Burrows: the effects on the vegetation of the addition of mineral nutrients to the dune soils. *J. Ecol.* **51**, 353–374.

WILLMER, C. (1983). "Stomata". Longman, London.

WILSON, G. B. and BELL, J. N. B. (1985). Studies on the tolerance to SO_2 of grass populations in polluted areas. 3. Investigations on the rate of development of tolerance. *New Phytol.* **100**, 63–77.

WILSON, J. W. (1966). An analysis of plant growth and its control in arctic environments. *Ann. Bot.* **30**, 383–402.

WINTER, K. (1974). Einfluss von Wassertress auf die Aktivität der Phosphoe-nolypyruvat—Carboxylase bei *Mesembryanthemum crystallinum*. *Planta* **121**, 147–153.

WINTER, K., LÜTTGE, U., WINTER, E. and TROUGHTON, J. H. (1978). Seasonal shift from C3 photosynthesis to crassulacean acid metabolism in *Mesembryanthemum crystallinum* growing in its natural environment. *Oecologia* **34**, 225–237.

WOLEDGE, J. (1971). The effect of light intensity during growth on the subsequent rate of photosynthesis of leaves of tall fescue. *Ann. Bot* (N.S.) **35**, 311–322.

WOLDENDORP, J. W. (1983). The relationship between the nitrogen metabo-lism of *Plantago* species and the characteristics of the environment. *In* "Nitrogen as an Ecological Factor" (Eds. J. A. Lee, S. McNeill and I. H. Rorison), pp. 137–166. Blackwell Scientific Publications, Oxford.

WOOLHOUSE, H. W. (1966). The effect of bicarbonate on the uptake of iron in four different species. *New Phytol.* **65**, 372–375.

WOODWARD, F. I. (1983). The significance of interspecific difference in specific leaf area to the growth of selected herbaceous species from different altitudes. *New Phytol.* **95**, 313–323.

WRATTEN, S. D., EDWARDS, P. J. and DUNN, I. (1984). Wound-induced changes in the palatability of *Betula pubescens* and *B. pendula*. *Oecologia* **61**, 372–375.

WRIGHT, K. E. (1943) Internal precipitation of phosphorus in relation to aluminium toxicity. *Plant Physiol.* **18**, 708.

WU, L. and ANTONOVICS, J. (1975). Zinc and copper uptake by *Agrostis stolonifera*, tolerant to both zinc and copper. *New Phytol.* **75**, 231–237.

WU, L. and BRADSHAW, A. D. (1972). Aerial pollution and the rapid evolution of copper tolerance. *Nature* **238**, 167.

WYN-JONES, R. G. and STOREY, R. (1981). Betaines. *In* "Physiology and Biochemistry of Drought Resistance in Plants" (Eds. L. G. Paleg and D. Aspinall), pp. 172–204. Academic Press, London.

YEO, A. R., LAUCHLI, A., KRAMER, D. and GULLASCH, J. (1977). Ion measurements by X-ray microanalysis in unfixed, frozen, hydrated plant cells of species differing in salt tolerance. *Planta* **134**, 35–38.

ZIEGLER, I. (1972). The effect of SO_2 on the activity of ribulose-1, 5-diphosphate carboxylase in isolated spinach chloroplasts. *Planta* **103**, 155–163.

ZIMMERMANN, M. H. (1978). Hydraulic architecture of some diffuse-porous trees. *Can. J. Bot.* **56**, 2286–2295.

ZIMMERMANN, M. H. and BROWN, C. L. (1971). "Trees: Structure and Function". Springer-Verlag, New York.

ZIMMERMANN, M. H. and McDONOUGH, J. (1978). Dysfunction in the flow of food. *In* "Plant Disease" (Eds. J. G. Horsfall and E. B. Cowling), vol. 3, pp. 117-140. Academic Press, New York and London.

ZUCKER, W. V. (1983). Tannins—does structure determine function? *Am. Nat.* **121**, 735–765.

AUTHOR INDEX

SYSTEMATIC INDEX

SUBJECT INDEX